D0365513

Agricultural Policy
for the 21st Century

Agricultural Policy
for the 21st Century

Edited by
Luther Tweeten and Stanley R. Thompson

Foreword by
D. Gale Johnson

Iowa State Press
A Blackwell Publishing Company

Luther Tweeten is Professor Emeritus of Agricultural Policy and Trade at Ohio State University. He is editor of four books and author or co-author of seven books and over 500 journal articles and published papers. He is a former President and a current Fellow of the American Agricultural Economics Association. Recent awards include the Charles Black Award from CAST, the Henry A. Wallace Distinguished Alumni Award from the Iowa State University College of Agriculture, the Distinguished Scholar Award from Ohio State University, and the Lifetime Achievement Award from the Southern Agricultural Economics Association.

Stanley R. Thompson is a professor at Ohio State University. He received his Ph.D. from Cornell University and was a professor at Michigan State University for 15 years prior to joining the Ohio State faculty. He teaches and conducts research in applied econometrics, international trade, agricultural marketing and policy, commodity futures markets, generic promotion and transportation economics. He was department chair from 1990 to 1998, a visiting scholar at ABARE, Canberra, Australia (1985), a guest professor at the University of Giessen, Germany (1999), and a Fulbright scholar to Vienna, Austria (2002). He has served as associate editor of the *American Journal of Agricultural Economics*, edited and contributed to several books, and published more than 30 articles in professional journals.

© 2002 Iowa State Press
A Blackwell Publishing Company
All rights reserved

This text was produced from camera-ready copy provided by the author.

Iowa State Press, 2121 State Avenue, Ames, Iowa 50014

Orders:	1-800-862-6657
Office:	1-515-292-0140
Fax:	1-515-292-3348
Web site:	www.iowastatepress.com

Authorization to photocopy items for internal or personal use, or the internal or personal use of specific clients, is granted by Iowa State Press, provided that the base fee of $.10 per copy is paid directly to the Copyright Clearance Center, 222 Rosewood Drive, Danvers, MA 01923. For those organizations that have been granted a photocopy license by CCC, a separate system of payments has been arranged. The fee code for users of the Transactional Reporting Service are 0-8138-0899-5/2002 $.10.

Printed on acid-free paper in the United States of America

First edition, 2002

International Standard Book Number: 0-8138-0899-5

CIP data available on request

The last digit is the print number: 10 9 8 7 6 5 4 3 2 1

Contents

Foreword   ix
     D. Gale Johnson, Professor Emeritus, University of Chicago

Preface   xi

1. Farm Commodity Programs:
   Essential Safety Net or Corporate Welfare?   1
   *Luther Tweeten, Professor Emeritus of Agricultural Policy and
   Trade, Ohio State University, Columbus*

2. Agricultural Policy: Pre- and Post-FAIR Act Comparisons   35
   *Bruce L. Gardner, Professor, Department of Agricultural and
   Resource Economics, University of Maryland*

3. The Content of Farm Policy in the Twenty-first Century   55
   *David Orden, Professor, Department of Agricultural and
   Applied Economics, Virginia Polytechnic Institute and State
   University, Blacksburg*

4. An Empirical Analysis of the Farm Problem:
   Comparability in Rates of Return   70
   *Jeffrey W. Hopkins, Economist, Economic Research Service,
   U.S. Department of Agriculture, Washington, DC
   Mitchell Morehart, Economist, Economic Research Service,
   U.S. Department of Agriculture, Washington, DC*

5. Income Variability of U.S. Crop Farms and Public Policy   91
   *Carl Zulauf, McCormick Professor of Agricultural Marketing
   and Policy, OhioState University*

6. Crop Insurance: Inherent Problems and Innovative Solutions 109
   Shiva S. Makki, Ohio State University and Economic Research
   Service, U.S. Department of Agriculture

7. Impact of Agribusiness Market Power on Farmers 127
   Suresh Persaud, Agricultural Economist for the U.S.
   Department of Agriculture, Washington, DC
   Luther Tweeten, Professor Emeritus of Agricultural Policy and
   Trade, Ohio State University, Columbus

8. Do Farmers Receive Huge Rents for Small Lobbying Efforts? 146
   David Bullock, Associate Professor in the Department of
   Agricultural and Consumer Economics at the University of
   Illinois at Urbana-Champaign
   Jay S. Coggins, Associate Professor, Department of Applied
   Economics, University of Minnesota

9. Coalitions and Competitiveness:
   Why Has the Sugar Program Been Resilient? 160
   Charles B. Moss, Professor of Food and Resource Economics,
   University of Florida, Gainesville
   Andrew Schmitz, Eminent Scholar: Ben Hill Griffin Jr.
   Endowed Chair, Food and Resource Economics, University of
   Florida, Gainesville

10. Farmland Is Not Just for Farming Any More:
    The Policy Trends 184
    Lawrence W. Libby, C. William Swank Professor of Rural-
    Urban Policy, Department of Agricultural, Environmental,
    and Development Economics, Ohio State University

11. Kuznets Curves For Environmental Degradation
    and Resource Depletion 204
    Aref A. Hervani, Visiting Assistant Professor of Economics,
    Dalton State University, Dalton, Georgia
    Luther Tweeten, Professor Emeritus of Agricultural Policy and
    Trade, Ohio State University, Columbus

12. Food Security, Trade, and Agricultural Commodity Policy 231
*Daniel A. Sumner, Frank H. Buck Jr. Chair Professor,
Department of Agricultural and Resource Economics,
University of California, Davis, and Director, University of
California, Agricultural Issues Center*

13. Competing Paradigms in the OECD and
Their Impact on the WTO Agricultural Talks 245
*Tim Josling, Professor and Senior Fellow, Institute for
International Studies, Stanford University*

14. The Changing Economics of Agriculture and the Environment 265
*David E. Ervin, Professor, Portland State University and
Senior Policy Analyst, Wallace Center for Agricultural and
Environmental Policy, Winrock International
Frank Casey, Natural Resources Economist, Defenders of
Wildlife*

15. Rational Policy Processes for a Pluralistic World 286
*Alan Randall, Chair, Department of Agricultural,
Environmental,* and Development Economics, Ohio State
University

Index 302

Foreword

D. Gale Johnson

Perhaps a more appropriate title for this collection of essays might be "Why so few can get so much without really trying." For a long time our farm programs were guided by the Puritan ethic—in order to get a payment you had to do something such as idle land or restrict production. The FAIR Act of 1996 rejected the Puritan ethic and indicated that the government would make payments to you so long as you could prove that you had control over land that had produced certain crops in the past—you were no longer required to produce anything. Will this be the agricultural policy of the twenty-first century? Bullock and Coggins argue that such may be the case. Since the subsidy payments have been capitalized into the value of the land, eliminating the subsidies will result in a sharp fall in land values. Thus, the subsidy programs may be more or less permanent.

I thought that the FAIR Act of 1996 would provide the transition from the government farm programs as we had known them for the previous six decades to something approaching reliance on the market for decisions concerning production and consumption of farm products. But these hopes were dashed by the decision that the Agricultural Act of 1949 was the permanent farm legislation that would come into effect if there was no new legislation at the expiration of the 1996 act. I must admit that in spite of my pessimism, I did not imagine that within a very short time the subsidies paid to agriculture would exceed anything seen before.

Luther Tweeten argues that there has been a new paradigm guiding agricultural policy while David Orden argues that the policy decisions have not really been in response to that paradigm. Orden recognizes that the labor market has been able to reach equilibrium but he makes an argument about the land market that I do not find convincing. He seems to argue that since land has not left agricultural production, except in response to government programs such as the Conservation Reserve Program, it does not fit with the new paradigm. What equilibrium theory tells us with respect to land is that changes in demand for land will be reflected in its price and not in its supply because the elasticity of supply is so low, approaching zero. Barring governmental intervention there is no significant alternative use for agricultural land—in spite of all the talk, urban and related land use constitute a very small percentage of national land use. Thus, it is not the return to a unit of farm land that is equalized in the market, since there is no alternative use for the vast majority of the land, but the rate of return per dollar value of the land that would be equalized to the return on similar investments in the rest of the economy. The returns to labor and capital in agriculture are equalized with returns to comparable factors elsewhere in the economy through changes in their supply, but in the case of land the equalization occurs through changes in its price and not in its quantity.

But whether there has been a new paradigm or not, the very substantial increases in both the absolute and relative returns to farm resources, especially labor, over the past half century, have been due to the functioning of markets and not to the operation of governmental price policies. Agricultural policy formulation has never recognized the extent to which agriculture had to adjust to economic growth, neither in this country nor in any other. It has not been much more than a century ago that half of the labor in the U.S. economy was engaged in farming; today it is about 2 percent.

Over most of that century the return to labor engaged in farming was less than in nonagriculture. Why? Because the slow growth of demand for farm output due to its low income elasticity of demand and the rapid growth of labor productivity in agriculture continuously reduced the demand for farm labor at a given wage. There had to be a continuous transfer of workers from agriculture to nonagriculture and a significant earnings difference was required to induce as much as 3 or 4 percent of the farm workforce to transfer out of farming year after year. It was not until the end of the 1970s that the annual labor transfer became very small and the incomes of farm and nonfarm people were approximately equalized. Our government had no significant role in assisting farm people to make the transfer to nonfarm jobs, but on this score it has been no different than governments throughout the world. Farmers everywhere have suffered because governments have been unwilling to recognize that farming must decline as economic growth occurs— nothing has been or is done to assist farm people to adjust.

The essays in this volume represent significant contributions to our understanding of recent developments in farm policy. While you may not agree with everything that has been written here, I am confident that you will find much from which you can learn and some things that will challenge your firmly held beliefs.

Preface

The chapters in this book were originally presented at a symposium, "Challenging the Agricultural Economics Paradigm," held at Ohio State University in September 2000. The symposium honored the career of Luther Tweeten, Anderson Professor of Agricultural Marketing, Policy, and Trade at Ohio State University for thirteen years. Before that, he was on the faculty of agricultural economics at Oklahoma State University for twenty-six years. Some papers were invited and others were selected by a review committee from competitive submissions.

Authors were encouraged to address two issues raised by Tweeten in his four decades as a professional agricultural economist: Do agricultural markets work and what is the appropriate role for government in these markets? Authors, many of the leading lights in agricultural economics, would not have consented to and were not asked to adhere to a thesis—they were asked to be as critical or supportive as their own individual special insights led them. Thus, topics range from farm policy, resource economics, international trade, and welfare economics to food security—topics Tweeten frequently addressed in his career. The insights and in-depth analysis offered by the authors are intended for all students of farm policy, including professors, classroom students, informed laypersons, and others grappling with the important economic issues of contemporary agriculture.

Although no central thesis pervades this book, the chapters cannot be read without awareness that U.S. farm commodity programs have lost their way. They cannot be defended on economic equity grounds—income per farm household set successively higher all-time records in each of the five years from 1996 through 2000 and is well above income per nonfarm household. Wealth per farm household averages about double that of nonfarm households. Income and wealth of commercial farmers and landlords—the ultimate beneficiaries of commodity programs—are even more favorable.

Commodity programs also fail economic efficiency tests. By distorting savings and investment decisions of taxpayers and encouraging excessive resource use and commodity production by farmers, commodity programs reduce not only farm crop and livestock receipts but also national income. This book makes the case that commodity programs do not effectively address very real problems of family farm loss, risk, and environmental degradation.

What accounts for the disarray in current public policy and difficulties in formulating agricultural policy for the twenty-first century? Authors of this book blame a partisan political bidding war, strategic alliances, and, most important, a disengaged public. A better-informed public will be essential in pressuring members of Congress to address effectively serious problems at far less cost to taxpayers than incurred in the 1990s. These essays are intended to boost that effort.

We are deeply grateful to the many individuals, firms, and organizations that made this book possible. We thank Wen Chern and Carl Zulauf for their service on the program committee with Stanley R. Thompson, chair. Shawn Carpenter provided all-around assistance in processing the manuscript. The manuscript would not have been possible without financial assistance from the Andersons, Farm Foundation, Ohio Farm Bureau Federation, OSU Swank Program in Rural-Urban Policy, and the Department of Agricultural, Environmental, and Development Economics at Ohio State University. Alan Randall, chair of the department, was especially supportive. Views presented in the chapters of this book are solely the responsibility of the authors, however.

1

Farm Commodity Programs:
Essential Safety Net or Corporate Welfare?

Luther Tweeten

Contemporary commodity programs poorly serve publicly stated objectives for agricultural policy. These programs have become an exercise in politics rather than in economics. In a bidding war using taxpayers' money to win votes, Congress has ignored the new agricultural paradigm that farm commodity markets are efficient, that farm households have higher income and wealth than nonfarm households, and that farm people and natural resources are more appropriate targets than commodities for public policy. This chapter contends that the most effective anecdote for government failure lies not in political science but in economic education—a better informed public pressuring Congress for policy reform that recognizes the new agricultural policy paradigm.

INTRODUCTION

From 1933 to 2000, taxpayers spent $561 billion (year 2000 dollars) to support farm prices and incomes (Luttrell 1983, 17; Spitze 1996; U.S. Department of Agriculture March 2001, and earlier editions). Spending since 1950 alone totaled $451 billion or nearly $9 billion per year. Whether federal funds for farm price and income support have been well spent depends on the public objectives for those funds and whether these objectives were served. Farm numbers dropped from 6.5 million in 1933 to 2.0 million in year 2000 or by 69 percent (U.S. Department of Agriculture July 1960 40; February 2000, 39), but preserving farms is only one objective of commodity programs.

In the best tradition of public policy economics, it is customary for an economist to list positivistic options (means) to meet the needs of people. The media make clear that society wants policies that improve well-being of people through greater economic efficiency (more real income), economic equity (if transfers are made they best go from the wealthy to the poor), and freedom to make decisions. Somewhat more objective sociopsychological scales of well-being constructed by social scientists also indicate that these objectives contribute to well-being of society (Blue and Tweeten 1997).[1] So the first objective of this chapter is to explore whether farm commodity programs contribute to well-being of people by serving economic equity, efficiency, and freedom. The second objective is to examine whether commodity programs cost-effectively alleviate economic instability, environment degradation, cash-flow squeeze, family farm loss, and other farm problems.

The conclusion of this chapter is that commodity programs, whatever virtue they once may have had, no longer serve their intended economic objectives. Contemporary commodity programs appear to be much about politics and little about economics. How are the relatively few farmers who receive most of the benefit from farm programs (0.2 percent of the nation's population in 1999) able to extract billion of dollars annually from taxpayers and voters?[2] Are commodity programs merely corporate welfare or are they entitlements essential to preserve the nation's heritage and family farmers for survival and maintaining income parity with nonfarmers? This chapter addresses these and other policy issues.

THE NEW AGRICULTURAL PARADIGM
American agriculture is operating under a new economic and policy paradigm (Tweeten 1989; Tweeten and Zulauf 1997). By paradigm, I refer to the presumptions, problem-puzzles, and problem-solving prescriptions of mainstream agricultural policy. The paradigm shift is certainly not a Kuhnian revolution in economic theory. (See chapter 3 by David Orden for further discussion of this issue and a somewhat dissenting assessment of my interpretation of paradigms in farm policy.) To be sure, economic theory has been in flux since Adam Smith's *Wealth of Nations* published in 1776, but the economic theory of Smith was not fundamentally different from the neoclassical economic theory of today. That neoclassical economic paradigm is the *normative proposition* that well-being of society is increased by pursuing policies providing social benefits in excess of social costs and the *positive* (predictive) *proposition* that individuals and firms will act to raise their well-being by pursuing actions providing them perceived benefits in excess of costs (for a more nuanced treatment, see chapter 15 by Alan Randall).

The new paradigm emphasizes that agricultural commodity markets work. To be sure, the government needs to play a role in provision of public goods (e.g., grades, standards, basic research, information systems, infrastructure, competition) so the market can function well. But compelling historical experience demonstrates that when these public goods are provided, the market rarely can be improved upon for economically efficient provision of food and fiber and for economic growth, international competitiveness, and food security. Farmers respond to prices set by supply and demand to clear markets, using all publicly available information.

Farms like other firms are always in very short-run equilibrium but never fully achieve long-term equilibrium because markets are dynamic. However, equilibrium is close enough so that able commercial farmers with or without subsidies on average earn returns comparable to what their capital and labor resources would earn if employed in other sectors of the economy.

This new paradigm contrasts with the traditional wisdom that held that government must perennially intervene in agricultural commodity markets to avoid surplus production and to raise prices, resource returns, and incomes that inevitably would be low without interventions. (For a superb review of historic development of the old paradigm, see Bonnen and Schweikhardt [1998].) The old paradigm held that, because farm markets do not work, chronic income transfers from taxpayers and consumers (the latter from commodity prices inflated by supply controls) are essential for farm resources to earn a "parity" return. The old paradigm contended that farmers could not adjust rapidly enough to avoid surplus output and chronic low returns as agribusinesses released a continuing torrent of new laborsaving, output-increasing technologies. The old paradigm held that farmers could not survive in a farming economy at the mercy of shocks from nature and man—the latter from misguided macroeconomic policy, use of food as a weapon, and the like. The new paradigm recognizes that market failure is much more frequent with natural resources than with commodities and that government failure in providing public goods is more common than price-system failure in supplying market goods.

The old paradigm dominated agricultural economists' thinking for decades. The presumption that the farming economy was in chronic disequilibrium and unable to earn returns comparable to resource returns in other sectors was supported by an elaborate conceptual superstructure. Most notable was Willard Cochrane's (1965) *treadmill theory* that technology continually increases food and fiber supply relative to demand, forcing commodity prices and revenues down, the latter because of inelastic product demand. Glenn Johnson's (see Johnson and Quance 1972) *fixed asset theory* held that the process of moving low-paid labor and other redundant resources out of farming in response to lower revenue was thwarted because resources were fixed to farming—tractors and combines had no use, no salvage value, and hence no mobility outside agriculture even if returns on them in agriculture were low.

These theories are indeed useful to explain annual and cyclical instability of farm prices, resource returns, and income, but do not explain chronic low returns for one simple reason: data unequivocally establish that able operators of commercial farms earn as favorable returns as their resources would earn elsewhere when averaged over almost any five-year period since the 1930s.[3] I have explained in detail elsewhere the shortcomings of the treadmill and fixed asset theories (Tweeten 1970, 1989).

The "new" agricultural paradigm may be said to be old in that a few perceptive agricultural economists such as Don Paarlberg in the 1960s noted the tendency for markets to properly reward resources on commercial farms. The term "new" refers to accumulation of enough years of data to definitively validate the paradigm.

Many, perhaps most, agricultural economists now subscribe to the new paradigm. Some agricultural economists remain agnostic, contending that presence of interventions in the past does not provide a real test that (free) markets work. But we have had abundant tests: only about two-fifths of farmers receive payments from government and only about half of all farm commodity output is covered by government commodity programs. Markets for these nonprogram commodities are characterized by considerable instability, but competent commercial operators earn favorable returns over time if they do not pay excessive prices for land as in the asset-inflation "bubble" of 1975-1985. Not surprisingly, individual farmers and farm organizations benefiting from commodity programs tend to reject the new policy paradigm.

The general public seems to be unaware of the new paradigm. Orden et al. (1999) state, "We found that agrarian mythology [the old paradigm?] played little or no role in the 1996 FAIR Act outcome" (p. 227). That may be what they observed but the power of agrarian mythology is truly manifested in its enabling of costly transfers to agriculture without raising a high profile. The 0.2 percent of America's population (commercial farmers) who receive most of the benefits of commodity programs is in no position to dictate farm policy to the remaining 99.8 percent of the population. The public must be submissive. The myths of farm fundamentalism and the old paradigm immobilize the nonfarm public so that commercial farm interests can prevail in farm policy. Thus farm policy reform requires a more informed public operating in the political arena.

The propensity for equilibrium emphasized in the new paradigm means that farm income is now set by income of nonfarmers. The farming economy, a vital but small part of the nation's economy, is much influenced by but does not much influence the nonfarm economy. It follows that the farming industry and the nation have a stake in sound public policies that raise national income, and not in commodity programs that lower national income. But before returning to the role of the public in the farm economy and politics, I review current farm policy in light of policy objectives and farm problems.

ECONOMIC EFFICIENCY

Economic efficiency is apparent when resources are allocated to their highest and best use. Such allocation is apparent in rates of return on resources—markets will move resources from low return uses to higher return uses. This movement draws down high returns and lifts low returns until returns to a resource are equal among uses—adjusted for risk and pleasure of working in an occupation. Thus economic efficiency is evident if resources on commercial farms with competent managers earn rates of returns on average (not necessarily each year) comparable to what those resources would earn in nonfarm uses. (Of course, part-time and poorly managed resources receive less reward.)

Economic efficiency means that over time the public is getting the most out of resources used to meet its food and fiber needs. Too many resources in agriculture would mean that society is being denied resources for education, health care, recreation, or other favored purposes. Too few resources in agriculture could cause food shortage. Several means are available to judge economic efficiency. One is to apply principles from welfare economics:

1. *Market goods* are rival, excludable, and transparent (see box 1.1). A huge body of historic evidence indicates that markets almost never fail to provide efficient allocations of such market goods—helped of course by basic research, infrastructure, information, antitrust, and sound macroeconomic policies provided by the public sector. *Public goods* are characterized by externalities and are candidates for government interventions to correct these externalities. Externalities mean that private returns (costs) differ from social returns (costs) so that private firms do not act in the public interest. Consequently, presence of externalities is the first welfare economics condition for a public good justifying government intervention.

2. The second condition for government involvement is that interventions to correct market failure must cost less than the market failures being corrected. An example is crop yield risk in agriculture. Asymmetry of information (farmers know more than insurers), adverse selection (only high-risk farmers tend to sign up for insurance), moral hazard (insured farmers can plant a high-risk rather than a low-risk crop knowing they will be covered if the high-risk crop fails), and unequal discount rates (the public can borrow at a lower interest rate than can private firms to support insurance, storage, etc.) all point to a possible role for the public to cushion farmers against instability. On the other hand, the record of public mismanagement of crop insurance is legendary. For that reason the social cost (waste, etc.) of publicly supplied insurance, storage, and forward pricing is quite likely greater than the social cost to the public of relying on private tools of risk management.

The "bottom line," however, is that farm commodities are market goods that are allocated most efficiently by prices set through market supply and demand. Thus, efficiency is served by public intervention in some natural resource markets but not in commodity markets.

Box 1.1

Markets work well where goods are rival, excludable, and transparent. The latter terms cannot be separated from externalities: where goods are nonexcludable, freeloaders cannot be kept away, thus marginal private returns to firms fall short of marginal social returns to the public at large. Consequently, the private sector producers too little to maximize social benefits.

Where goods are nonrival so that consumption by one consumer does not reduce consumption available to another, marginal cost of good A to a consumer is zero because no goods and services are forgone to expand output of A. Hence, a private firm that sets a positive price to cover overhead costs will charge private marginal costs in excess of social marginal costs. The result is too little output to maximize net benefits for society.

Lack of market transparency including lack of knowledge also constrains market choices and the role of public policy. For example, many farmers do not know and probably underestimate their soil erosion or water pollution rate. Hence, they are likely to underestimate payoffs from conservation and underinvest in "best management practices" for sustainable agriculture. The public sector supplying information can help producers to make better decisions.

Competition also contributes to well-functioning markets. Experience indicates that only a few firms are necessary to provide a competitive and efficient market as noted in chapter 7 by Persaud and Tweeten. The important conclusion, however, is that farm commodities are rival, excludable, and transparent, and that markets are competitive. Such markets have been found to operate efficiently the world over as well as in American agriculture.

This conclusion contrasts sharply with the situation for natural resource markets. Environmental issues of soil conservation and of air, water, and food quality or safety are closely tied to externalities driving a wedge between private and social costs so that the public interest is not served. Where externalities abound and markets are scarce, public intervention may be essential to avoid excessive soil, water, and air degradation (see Tweeten and Amponsah 1998).

Rate of Return on Farm Resources

I have elsewhere summarized a large number of farm management and farming industry studies indicating favorable rates of return on resources of competent commercial farm operators (Tweeten 1989, ch. 4). At least since 1970, rates of return on resources have averaged 3-4 percent over all farm assets but 15-20 percent on large farms (Tweeten 1989, 118-23). Compilations from data used by Hopkins and Morehart in   chapter 4 of this volume indicate that three-fifths of farm households had negative rates of return on assets in the late 1990s, but three-fifths of farm output had positive rates of return. Returns are somewhat comparable among types of farms of similar size, although returns vary by year and among farms of a given type.

Lost National Income

Rates of return on assets in a well-functioning market will be equal with or without market interventions because benefits to farmers from market distortions are bid into rents and land prices until returns are equalized among resources. This subsection measures economic inefficiency by the loss in national income (deadweight loss or cost) caused by commodity program market distortions. Before turning to that topic, however, I briefly review why aggregate farm commodity price or terms of trade (parity ratio defined as the ratio of the index of price received by farmers for crops and livestock to the index of prices paid by farmers for inputs) is a poor measure of market efficiency and fairness.

Farm commodity terms of trade or parity in 1999 was only 40 percent of the 1910-14 average (U.S. Department of Agriculture March 2001, 27). This number has been interpreted by some to mean that farmers are only receiving 40 percent of a "fair" price. In fact, as explained below, real farm prices have risen!

Technology is not the villain it is often portrayed to be. It has raised real economic terms of trade for farmers and nonfarmers alike—including for farmers who left their operations in midcareer (see Perry et al. 1991; Bentley et al. 1989). Aggregate farm output and aggregate production input are measured over time by weighting physical quantities by constant dollar prices. These constant-dollar values are added over all inputs to measure aggregate production input and are added over all crops and livestock to measure aggregate farm output. The multifactor productivity index (ratio of aggregate output of crops and livestock to aggregate farm production input) in 1999 was 3.94 times the 1910-14 level.[4] Thus, the factor terms of trade index, defined as the real price (purchasing power) of farm output per unit of farm production input, was 158 percent (40 x 3.94) of the 1910-14 level. In other words, real buying power (factor terms of trade) of the average production input was 58 percent greater in 1999 than in the 1910-14 base period!

Alternatively, a productivity index of 3.94 in 1999 meant that farmers were "growing 4 blades of grass where one grew" in 1910-14 with the same input volume. Hence, farmers needed only 25 percent as high a real price in 1999 to achieve the same real income per unit of production input as in 1910-14. In fact, however, real farm commodity price in 1999 was 40 percent of that in 1910-14, hence real price received per factor input in 1999 was 40/25 or 158 percent of that in 1910-14 after the parity ratio is adjusted for productivity growth. It follows that, comparing 1999 with 1910-14, the real parity ratio (factor terms of trade) was *up* by 58 percent rather than down 60 percent as implied by the conventional parity ratio (commodity terms of trade) of 40 percent.

Technology not only allowed real farm prices to improve but also helped real farm income to improve. Real personal income per capita of the U.S. population increased to 3.5 times its 1930 level by 1999, an average annual gain of 1.8 percent per year (Council of Economic Advisors 2000, 335, and earlier issues). Meanwhile, per capita income of farmers increased from 40 percent that of the average American to 117 percent of the average American (U.S. Department of Agriculture 1960, 38; March 2001, 46). It follows that real income per person on farms was approximately ten times (3.5 x 117/40) higher in 1999 than in 1930.

In an earlier study (Tweeten 1994, 7), I apportioned the 3.4 percent annual growth rate in real per capita income of farm residents to five principal sources for the 1930-1990 period as noted in table 1.1.

Table 1.1. Proportion of Farm Household Income Growth by Source, 1930–1990

	%
Parity price ratio (commodity terms of trade)	-5
Multifactor productivity growth	19
Farm size growth	28
Government payments growth	2
Off-farm income growth	56
	100

Source: Tweeten 1994, 7.

Technology played a role in several of these sources of income growth. Partly because of productivity gains in excess of the increase in demand for farm output, a falling parity ratio would have reduced farm household income from 1930 to 1990, ceteris paribus. Multifactor productivity gains more than compensated. Other things equal, multifactor terms of trade accounted for approximately 14 percent (19 percent gain from productivity less a 5 percent loss from falling parity price ratio) of the increase in farm income per capita from 1930 to 1990. Government payments and farm commodity programs in general contributed little to growth in productivity and income of farm people (Makki et al. 1999).

Technology was central to the two major sources of farm income growth—farm size and off-farm income. Mechanization technology accounted for much of the 28 percent increase in farm income attributed to greater farm size. Mechanization along with improved roads and vehicles also facilitated off-farm employment and earnings, which accounted for 56 percent of the income gain of farmers per capita.

Market distortions caused by commodity programs reduce national income in several ways. Federal income taxes used to support government programs lose on average at least $16 of real national income (deadweight loss) per $100 collected (Ballard et al. 1985, 13). Thus, government payments to farmers averaging $18.6 billion per year from 1998 to 2000 lost $3.0 billion of national income annually because the public made different savings, investment, and labor use decisions than they would have in the absence of taxes that finance farm programs.

Commodity programs reduced national income by distorting use of farmers' as well as taxpayers' resources. Prior to the 1996 farm bill, commodity programs emphasized idling cropland, causing output to fall short of output of a well-functioning economy by an average of 5-6 percent in the 1960s and 1980s—periods of considerable excess capacity (Tweeten 1989). Annual loss in national income averaged $4-6 billion (about 4 percent of farm receipts according to estimates by Tweeten (1989, 351) and the Council of Economic Advisors (1987, 159).

In contrast to earlier farm bills, the 1996 farm bill induced too much rather than too little commodity production for economic efficiency. Estimates of excessive farm production induced by direct payments, marketing loans, and insurance subsidies are shown in table 1.2. Only grain, oilseed, and cotton supports are considered.

Direct Government Payments
Production flexibility contract payments and disaster loss payments presumably don't affect producers' current and future production decisions. Payments are said to be decoupled from future production. However, some portion even of payments not specifically tied to current or future production decisions find their way into production inputs. One reason is because farmers are short of capital. Payments lessen that restraint by providing security for production loans or by directly purchasing production inputs. Direct payments have raised farm output over competitive market levels by 0.15 percent (Westcott and Young 2000) to 0.25 percent (interpreted from Burfisher, Robinson, and Thierfelder 2000) as noted in table 1.2.

Table 1.2. Annual Production of Farm Output above Competitive Market Levels Induced by the 1996 Farm Bill and Crop Insurance Programs, United States, 1998–2000

Program feature	Contribution to farm output (%)		Source
	Low	High	
Direct payments			
	0.15		Midrange estimate (Westcott and Young)
		0.25	Assumes doubling the 50% reduction specified by Burfisher et al.
Marketing loans and loan deficiency payments			
	0.68		Assumes crop output 51% of total farm output. Westcott and Young, p. 12
		1.38	Assumes crop output determines livestock output. Westcott and Young, p. 12
Insurance subsidies			
	0.28		Westcott and Young, p. 12
		4.10	Assumes 25 million acre increase, and Conservation Reserve Program yields; Skees, p. 2
Total, all sources			
	1.11		Sum of low estimates
		5.73	Sum of high estimates
Loss in farm receipts ($ billion)			
Short run ($E=-0.3$)	4.93	25.46	Percentage for total additional output
Intermediate run ($E=-0.6$)	1.42	7.38	multiplied by gross receipts elasticity $1+1/E$ times 1998–2000
Long run	0.00	0.00	farm receipts averaging $193.2 billion. E is aggregate adjustment elasticity defined as the sum of demand and supply elasticities (see later text).

Sources: Westcott and Young (2000), Burfisher et al. (2000), Skees (2000).

Marketing Loans and Loan Deficiency Payments

Even loan support prices set below the average market price distort production because the anticipated or expected (mean) commodity price (on which producers make production decisions) is raised when the lower part of the price distribution is removed. Loan rates for crops under the 1996 farm bill are well above operating costs of production, hence supports induce farmers to produce because operating costs are sure to be covered. In some cases, as in soybeans, loan rates are above total unit cost (including land and overhead expenses) of production on efficient commercial farms. Hence loan supports distort production and trade compared to production and trade in well-functioning competitive markets.

Westcott and Young (2000, 12) estimate that marketing loan and loan deficiency payments have added four to five million crop acres. The high estimate in table 1.2 assumes that acres added are as productive as average cropland and that all crop and livestock production depends ultimately on crop production, hence production is increased by 4.5 million acres on 325 million acres of cropland or by (4.5/325) or 1.38 percent. The lower estimate of 0.68 percent added output adjusts for possible lower productivity of added acres and for resources beyond crops required to produce livestock. This lower estimate recognizes that livestock can be produced from grass and imported feeds as well as from domestic crops.

In the 1990s, the government shifted from nonrecourse loan rate supports (which tended to hold market prices at the loan support rate) to marketing loans that provided payments to farmers on the shortfall of market prices below the loan support rate. The former nonrecourse loan program was faulted for holding prices at high levels that provided an umbrella under which our export competitors produced to take over our world markets. In contrast, the marketing loan program has been faulted for dumping our commodities in world markets at subsidized prices below production costs. Either type of loan support distorts markets by generating excess production. An alternative would be a *recourse loan* that farmers could obtain from government for the loan rate value of a commodity at harvest but would have to be repaid at full loan value plus interest before the next harvest. A recourse loan would enable operators to avoid selling on an oversold harvest market and relieve cash-flow pressures while not distorting market incentives.

Insurance Subsidies

In year 2000, crop and revenue insurance subsidies totaled approximately $2.5 billion and accounted for 60 percent of crop revenue insurance cost, encouraging output. The nation gets more of what it pays for with insurance subsidies—risk. Crop risk is especially great in the Plains states, thus it is no surprise that Jerry Skees (2000) finds the contribution of insurance subsidies to crop acreage harvested is especially large in the Great Plains states such as Texas and North Dakota. (See Shiva Makki's detailed analysis in chapter 6.) Crop and revenue insurance loss ratios (program costs for indemnity payments and administration relative to premiums) averaged 1.88 for the United States from 1981 to 1999 and averaged over 2.0 in several states including Arkansas (2.97), Texas (2.72), Georgia (2.68), North Carolina (2.40), and North Dakota (2.16).

Crop and revenue insurance causes more land to be in crops and causes land to be cropped more intensively. For example, risky corn may be planted in place of less risky grain sorghum in the semiarid plains because a high corn yield will earn more than grain sorghum and a low corn yield will "earn" an insurance payment. Insurance subsidies hold land in crops that otherwise would be unprofitable to farm and would revert to grassland or forest. The land not only is marginal for farming but also may be environmentally fragile, prone to wind and water erosion. Westcott and Young (2000, 12) estimated that crop and revenue insurance premium subsidies have added approximately 900,000 acres to aggregate plantings of eight major crops. This translates into a 0.28 percent increase in farm output (table 1.2).

Jerry Skees (2000) estimated that crop and revenue insurance subsidies have added 25 million to 30 million acres of crops in the United States—an area nearly as large as that enrolled in the Conservation Reserve Program (CRP) and about one-tenth of total cropland harvested in the nation. The high estimate in table 1.2 assumes that twenty-five million crop acres are added by insurance subsidies and that they are only as productive as CRP acres—about half the productivity of an average acre cropped (see Tweeten 1989, 350). Thus, insurance subsidies add as much as 4.1 percent to farm output.

Total Impact

Excessive output and resources committed to farming cost the nation $0.93 billion in lost income based on the average, 3.4 percent, between the low (1.11 percent) and high (5.73 percent) estimate of excessive production in table 1.2. The national income (deadweight) loss is $2.64 billion with the high estimate.[5] This average for 1998-2000 is not much less in real terms than the cost of commodity program distortions in the 1960s and 1980s. Adding previously uncounted annual loss from tax distortion ($3.0 billion) and of administrative and lobbying resources (about $2 billion that could have been better used elsewhere), the total loss in national income is $5.93 billion based on 3.4 percent excess output in grains, oilseeds, and cotton only. Income loss of $600 million from peanut, tobacco, sugar, and dairy programs (Tweeten et al. 1997) brings the total to $6.53 billion, or 3 percent of farm receipts (see Council of Economic Advisors 1987, 159 for cost in 1985). The above estimates and alternative calculations by Bruce Gardner in chapter 2 indicate that losses in national income from farm programs could be greater from distortions in the way taxpayers use their income than from distortions in the way farmers use their resources.[6]

Several other observations follow from estimates in table 1.2:

- Based on adjustment elasticities shown in table 1.2, additions to output reduced farm receipts from $4.93 billion to $25.46 billion in the short run of one to two years, and by $1.42 billion to $7.38 billion in an intermediate run of three to four years. The impact on farm receipts is zero in the long run because farm aggregate demand is near unitary elasticity.
- Termination of grain, soybean, and cotton commodity programs might not reduce aggregate farm income. Less production attending termination of programs could raise farm commodity prices and gross farm receipts to compensate for the loss of government payments, which totaled $23.3 billion in year 2000. This number is less than the "high" estimate of $25.5 billion added to farm receipts in the short run from termination of 1996 farm bill production incentives.
- A notable conclusion from table 1.2 is that most of the excess output under the 1996 farm bill comes from excessive loan rates and (especially) from insurance subsidies. These production incentives lower farm prices and receipts while raising costs, creating pressure to return to supply management programs (Schnittker 2001, 93-98). Because sizable direct payments induce relatively little excess output, it follows that direct payments could be set as deemed necessary to maintain desired farm income after terminating loan rates (or setting them to cover only variable production costs or to provide the basis for recourse loans) and insurance subsidies that distort farm resource allocation. Thus liberalization of domestic and export markets potentially could make no group worse off but make the public better off while being supportive of the World Trade Organization "green box" rules for nondistorted trade. An extended period of direct payments to farmers, however, would lose substantial national income.

Because U.S. agricultural import tariffs average only one-fourth the 45 percent average for other industrialized countries, the United States stands to gain $20 billion of farm receipts annually from the 12 percent gain in word agricultural prices expected with world trade and commodity program liberalization (see Burfisher et al., 5, 10). Domestic farm program and trade liberalization could add $5-7 billion annually to U.S. national income and $31 billion annually to global income in the short run. Gains would be double that each year in the long run as dynamic gains from additional savings, investments, and productivity work their way through the system (see Burfisher et al. 2000, 5).

The 1996-type direct payment program makes freer world trade less onerous to U.S. farmers. One reason is because pre-1996 supply management programs caused farm output to be too low while post-1996 programs caused output to be too high relative to a well-functioning market. It follows that with an inelastic short- and intermediate-run output demand, liberalization with a pre-1996 program would lower farm receipts and with a post-1996-type program would raise farm receipts. While price support and insurance reform along with trade liberalization would add receipts to compensate for an end to direct payments, thereby benefiting American farmers and the nation, continuation of some decoupled payments could help farmers through the transition.

ECONOMIC EQUITY

Economists have avoided economic equity issues about as assiduously as other social scientists have avoided economic efficiency issues. Perhaps the most controversial issue of my professional career (at least as measured by attacks from economists whom I respect) is my effort to analyze as objectively as possible a presumably taboo topic: the marginal utility of income. Neal Blue and I (1997)estimated that the value of another dollar is 50 percent higher for low-income families than for families with median income. In turn, the latter families derive about five times as much utility from another dollar as a family with five times the median income. Of course, before concluding that wholesale income redistribution is warranted, issues of property rights and investment incentives must be confronted. The conclusion of importance here, however, is that government transfers of income/wealth from households with low income/wealth to those with high income/wealth reduce well-being of society.

If the first principle of economic equity is to make transfers only from higher to lower income/wealth households, then something is amiss when massive transfers were made to farmers whose income per household set successively higher all-time records in each of the years 1996 through 2000 (U.S. Department of Agriculture March 2001, 46).[7] Income per farm household averaged 6 percent to 17 percent above that of nonfarm households in each of the years 1996 through 2000. Government payments alone did not account for greater income of farm than nonfarm households. In fact, from 1998 to 2000, *nonfarm income alone* per farm household exceeded income from all sources per nonfarm household! If the definition of what is a farm remains unchanged, that relationship may continue for the foreseeable future, further eroding arguments for income transfers to farmers.

Table 1.3. Income and Net Worth of U.S. Farm Operator and All Households Compared in 1998

Classification of farm	Number of operator households	Total household income				Total net worth	
		From off-farm sources		From all sources		Average amount	% of US average household
		$ per household	% total	$ per household	% of US average household %	$ per household	%
All operator households	2,022,413	52,628	88.1	59,734	115.2	492,195	174.2
Small family farms							
Limited-resource [a]	150,268	13,153	132.5	9,924	19.1	78,718	27.9
Retirement (retired)	290,938	47,158	103.3	45,659	88.1	535,943	189.7
Residential/lifestyle (nonfarmer)	834,321	76,390	106.0	72,081	139.0	347,909	123.2
Farming occupation:							
Lower sales (under $100,000)	422,205	37,186	106.9	34,773	67.1	576,402	204.0
Higher-sales ($100,000–250,000)	171,469	28,717	57.2	50,180	96.8	669,458	237.0
Large family farms ($250,000–499,000)	91,939	47,252	44.4	106,541	205.5	944,533	334.3
Very large family farms (Over $500,000)	61,273	33,240	15.9	209,105	403.2	1,508,151	533.9

Source: U.S. Department of Agriculture, September 1999, pp. 19, 24; February 2000, p. 39.

[a] Household income under $20,000, farm assets under $150,000, and gross sales under $100,000.

Farm operator household income averaged 115 percent of U.S. average household income, and wealth averaged 174 percent of U.S. average household wealth in 1998 (table 1.3). Landowners are the ultimate beneficiaries of farm programs. Data are unavailable on their income and wealth, but it is almost certainly well above that of farm operators households shown in table 1.3.

The four classes of farms (lower four rows in table 1.3) whose operators classify themselves as farmers each have wealth averaging at least double that of the average U.S. household, hence do not warrant receiving farm commodity program payments if the equity criterion is to transfer wealth only from higher to lower wealth households.[8] The remaining three farm categories are limited-resource farms (household income under $20,000, farm assets under $150,000, and gross farm sales under $100,000), retirement farms (operators retired), and residential/lifestyle farms (operators list their occupation as other than farming). These three categories of farms account for 63 percent of all farms, but for only 10 percent of the value of farm production. Because of high income of residential/lifestyle farms, the high wealth of retirement farms, and the limited production of each of these two classes of farm, neither farm class would seem to be helped much by commodity programs distributing benefits nearly in proportion to farm sales.

That leaves limited-resource farms as the only category passing the equity test in table 1.3. These farms account for 7 percent of all farms but for only 1 percent of farm output and government payments. Only 18 percent of them received payments compared with 70 percent on midsized farms in 1998. Payments for limited-resource households averaged only $722 in 1998. It follows that commodity programs dispensing benefits according to farm output are of little or no benefit to limited-resource farms. Because over $100 of payments to all farmers were required to direct $1 to limited-resource households, commodity programs are a highly cost-ineffective means to help poor farmers. Rather, households on limited-resource farms can benefit from counseling, education, training, job search and relocation assistance, and public assistance programs.

Especially with the 1996 farm bill, commodity programs have become more market oriented. Providing more freedom for farmers to make production and marketing decisions is not a pressing current issue. Nonetheless, an end to market-distorting program features discussed earlier could enhance producers' freedom to make efficiency-enhancing production and marketing decisions.

Standard measures of poverty are so flawed they have little meaning in agriculture. A commercial farmer in poverty will not long be in business. Hobby farmers account for most small farm numbers but on average they are not poor. Even when nonmoney income, income averaging, and net worth (not currently included in poverty measurement) are considered along with conventional sources of income, many of the limited-resource farm households in table 1.3 are in poverty.

As noted earlier, however, their needs are not met by commodity programs. Bruce Gardner (2000, 1072) states that "commodity policy just cannot be justified as a remedy for any identifiable set of low income people." Instead, human resource development and public assistance efforts are called for.

FAMILY FARM LOSS

Some 80 percent of American adults agree that "the family farm must be preserved because it is an essential part of our heritage" (Jordan and Tweeten 1987). Surveys also indicate that families on farms more than families in the towns and cities display virtues prized by society such as self-reliance, independence, honesty, churchgoing, marital cohesion, and obedience to laws (Drury and Tweeten 1996). The 0.2 percent of the nation's families benefiting most from the billions of dollars spent each year on commodity programs could not sustain such outlays politically without these farm fundamentalist beliefs and values pervading society. The irony is that commodity programs erode the very virtues of self-reliance, independence, and honesty they presume to preserve. Shrill demands by some of today's farmers that government programs are an entitlement and that taxpayers owe them a good living belie Thomas Jefferson's claim that "dependence begets subservience."

Empirical evidence indicates that government commodity programs help preserve family farms in the short run such as during the financial crisis in the early 1980s (Tweeten 1993). On the other hand, programs provide capital and security, allowing farms to leverage equity to purchase machinery and to buy out their neighbors and consolidate holdings in the long run—thereby reducing farm numbers and increasing farm size.

Some clues regarding the contribution of farm price supports to the goal of preserving family farms are afforded by the experiences of New Zealand, the European Union, and Japan. The United States lost farms at a rate of 1.1 percent per year on average from 1980 to 1995. Meanwhile, Japan lost farms at the rate of 2.2 percent annually (private communication from Mitsuhiro Nakagawa, Ibaraki University, Japan) despite economic support four times the U.S. level. The European Union with support rates nearly double the U.S. rate lost farms at a 1.8 percent annual rate in the 1980-95 period (private communication from Mark Krum, University of Giessen, Germany). New Zealand experienced an increase in number of farms and decrease in size of farm after it ended its commodity programs supporting agriculture (Wright 1995, 20).

Although these international comparisons suggest that higher supports speed consolidation of farms, I conclude from empirical analysis that commodity supports have little net influence on farm numbers in the long run (Tweeten 1993). Forces of technology and markets rather than government programs eventually dominate to establish farm sizes and numbers. Government commodity programs can temporarily slow but not avoid adjustments in farm size and numbers to economic pressures from the marketplace.

Direct payments emphasized by the 1996 farm bill offer unprecedented opportunities to target small family farms for preservation, unlike earlier supply control programs, which could not easily target small farms (see Gundersen et al. 2000). For pre-1996 supply control programs, larger farms producing most output needed to be included for production controls to be effective. Current farm programs would need to be restructured nearly beyond recognition, however, to target the financially vulnerable operations in saving family farms. Such restructuring seems politically impossible based on historic evidence. Compared to commodity programs, credit programs could more easily target vulnerable farms.

FARMERS AND FARMING INDUSTRY ADJUSTMENT CAPABILITY
Farmers' economic salvation comes not from expansion of exports, value-added enterprises, nonfood use of farm output, or even from rapid introduction of improved technology. Instead, farmers' economic salvation rests with their ability to adjust to an inevitably changing environmental, technological, social, political, and economic milieu.

The new economic paradigm recognizes farmers' ability to adjust. To be sure, some farmers would face financial shocks and the need to adjust resources in response to termination of commodity programs. Agriculture will continue to be subject to major annual and cyclical shocks from humans and nature that buffet farm prices and incomes. Weather and business and commodity cycles will continue to plague agriculture. Farmers are well aware of the risks endemic to agriculture.

A critical issue is how rapidly farmers adjust to market signals. From an industry perspective, the issue is not how rapidly one enterprise adjusts in response to a change in the price of the commodity produced by that enterprise. Rather, the issue is how fast the *industry* responds—recognizing that resources specialized to farming cannot readily and profitably be moved to other sectors of the economy. Fortunately, econometric estimates provide insight into adjustment rates.

The short-run (one to two years) aggregate farm supply (output) elasticity is at least 0.1 and the aggregate farm output demand elasticity is approximately -0.2 (see box 1.2 and Arnade and Gopinath 1998; Tweeten 1989, 118; Henneberry and Tweeten 1991). The aggregate adjustment elasticity is the sum (absolute value) of these two elasticities, or 0.3. The intermediate-run (four years) supply elasticity is approximately 0.3 and the demand elasticity is of similar (absolute) magnitude, hence the intermediate-run aggregate adjustment elasticity is 0.6. Thus, a 10 percent drop in aggregate prices of farm commodities reduces excess supply of farm output by approximately 3 percent in one to two years and by 6 percent in four years.

I noted earlier that government-induced disequilibrium averaged about 6 percent in the 1960s and the early 1980s—periods of unusual shocks to the farming economy. Farm operating and financial inputs are much more easily adjusted and farmland is less easily adjusted than the capital Arnade and Gopinath (1998) calculate to adjust at 2 percent per year. But if overall agricultural resources adjust at only 2 percent per year, one of the very lowest numbers found among many empirical estimates, then excess capacity of 6 percent could be eliminated in three years if no additional inputs are added! This does not indicate severe resource fixity.

Not only is the farming industry dynamic, it has the good fortune of not being called on to make quantum adjustments characteristic of many small nonfarm businesses. Based on annual reports from agricultural banks, on average 3.0 percent of U.S. farmers went out of business each year from 1982 to 1999 and 2.4 percent for the 1992 to 1997 period (U.S. Department of Agriculture February 2000, 48). Thus, approximately 15 percent go out of business in five years.[9]

These numbers contrast with the dissolution rate of 40 percent in five years and 79 percent in ten years for all U.S. businesses (Goodwin 2000, 74). For businesses hiring no workers, and thus most like family farms, the dissolution rate is 65 percent in four years and over 90 percent in ten years. Approximately 15 percent of U.S. business dissolutions represent actual financial failures (Goodwin 2000, 74). Only approximately 10 percent of the farms going out of business experience foreclosure; most leave for other reasons, such as death and retirement, which annually average about 2 percent of all farmers (Tweeten and Zulauf 1995).

Box 1.2

In a comparatively recent study, Arnade and Gopinath (1998, 94) lament that "agriculture's capital behaves almost like a fixed factor. Our estimated rate of adjustment of agricultural capital is 2 percent per year." They cite three previous studies with higher adjustment rates. Earlier, I (Tweeten 1989, 115-18) noted that farm machinery capital may indeed adjust slowly, but highly productive financial and operating capital inputs can adjust rapidly. Thus, it is notable that Arnade and Gopinath calculated the aggregate agricultural output supply elasticity to be 0.2 in the short run and 1.4 in the long run—surprisingly high numbers closely in line with my estimates made in 1972 (see Henneberry and Tweeten 1991).

Table 1.4. Percent Change Per Year, 1992–97

Farm gross entrants		2.3
Farm gross exit due to:	foreclosures	-0.2
	retirements	-2.0
	other exits	-0.2
Net exit		-0.1

Source: U.S. Department of Agriculture, March 1999. Also, see text.

Box 1.3

Future adjustments in farm numbers are likely to occur at a higher rate than occurred in the 1992–1997 period. Farm technology moves inexorably, like tectonic plates. Continuing tremors can dissipate the pent-up energy without suffering, but in the absence of tremors a major "earthquake" strikes, causing massive trauma. Agricultural labor productivity is likely to expand at least 3 percent annually on average (it increased on average 4 percent per year from 1976 to 1996 according to the Council of Economic Advisors 2000, 418) and output is likely to increase 2 percent annually, hence annual labor and farm numbers need to decline (tremor) by at least 1 percent per year to avoid a later painful farm operator and family "earthquake" exodus.

Generous farm programs of recent years avoided tremors but are unsustainable. Eventually an "earthquake" to adjust labor will be unavoidable. Adjustment pressure will mount even with continuing payments because of rising rents and land prices. Given that laborsaving technology will not be stopped and that large farms don't need help, the only affordable means for taxpayers (and consumers) to maintain a stable number of farms in the face of relentless technological change may be to subsidize retention of small farms with sales of under (say) $50,000 per year. But, as indicated earlier, targeting of small farms appears to be politically unacceptable.

Using mandatory farm-size restrictions rather than payments to save family farms also will not work as laborsaving technology advances. Advancing laborsaving technology with a regulated fixed farm size will forgo some benefits to producers and society of size economies, but still will free labor for off-farm work. Thus, U.S. farmers would be driven even more to be "weekend warriors," working off-farm all week while doing their farm work on weekends with laborsaving technology—a phenomenon especially prominent in Japan, which does not have a functioning farmland market.

Farms were lost at an unusual average net rate of only 0.1 percent from 1992 to 1997 according to the Census of Agriculture (U.S. Department of Agriculture March 1999, 10). The dynamics of farm numbers adjustments for the 1992-97 period are summarized in table 1.4.[10]

Thus, farming was a positively dynamic industry in the 1992-97 period, attracting far more entrants than exits because of financial failure. That time period was not turbulent. At issue is how these numbers would change with phasing out of farm commodity programs. If government payments (totaling $20.6 billion or 11 percent of cash farm receipts in 1999) were gradually rather than precipitously phased out and if other farm income did not change, farm numbers are predicted to fall 1.0 percent in the short run and 5.8 percent in the long run based on the elasticity with respect to farm-nonfarm income ratios (Heady and Tweeten 1963, 423). Such results based on dated parameters need to be viewed cautiously, but even casual observation cannot escape the reality that American farmers adjust to change.

What farm households would be most likely to fail financially and exit from agriculture when exposed to the market? All farm categories with sales under $100,000 and accounting for 85 percent of farms had, on average, negative returns on assets and farm economic costs in excess of gross farm income in 1998. These farms, accounting for only one-sixth of farm output and receiving very little on average from government payments, are mostly hobby operators who seek tax and rural amenity advantages. Their high off-farm income sustains their hobby with or without taxpayers' generosity. Limited-resource small farms will not be well served by any policy designed as in past programs to provide payments proportional to farm receipts, especially crop receipts, because many produce beef cattle and calves.

Midsized farms with sales of $100,000 to $250,000, accounting for 8 percent of farms and 16 percent of crop and livestock sales in 1998, are especially vulnerable financially to termination of commodity programs. One reason is because households on such farms in 1998 received over half of their net farm income from government payments. The farms are only marginally profitable and efficient. They were helped by off-farm income of $28,717 per household in 1998. Despite wealth averaging twice that of U.S. households in 1998, household income from all sources averaged just below that of all U.S. households. Although only 4.3 percent were rated as financially vulnerable in 1998, many of these households would suffer financially with termination of farm commodity programs. With or without commodity programs, large numbers of households in this "disappearing middle" category will seek more off-farm income, will grow larger and more efficient, or will exit farming. They constitute perhaps the strongest case for help from federal/ state extension services and from other public agricultural as opposed to welfare agencies to assist with adjustments. These farms can be helped for a fraction of current program costs by targeting government credit and payments to them.

What happens to farmers who fail? Conventional wisdom holds that farm households are so wed to the farm by culture and specialized skills that they cannot adjust to alternative employment without trauma. That conjecture, if it ever were true, is no longer a valid generalization. Most farm operators have had experience in off-farm employment. By a 3:1 margin, Oklahoma farmers who left farming in midcareer said they were better off (Perry, Schreiner, and Tweeten 1991). Similar results were found by Bentley et al. (1989) for other states.

CASH FLOW

Family farms refinanced each generation face a cash-flow squeeze as they struggle to service debt and pay living expenses. Farmers, it is said, "live poor and die rich." To be sure, commodity programs help farmers to meet cash-flow needs in the short run. But the expected future benefits of commodity programs are capitalized into land prices, raising debt-service costs. Thus, commodity programs can intensify cash-flow problems of new farm operators. The appropriate response to cash-flow problems is not more generous commodity programs, but rather is leasing of farm real estate and equipment, contract farming, off-farm work, raising livestock, and the like.

Countercyclical payments to farmers also could reduce cash-flow problems, but it is difficult to design programs that do not create incentives for excessive production (see Commission on 21st Century Production Agriculture 2001). The bill passed in October 2001 by the U.S. House of Representatives would return agriculture to the pre-1996 target price, deficiency payment system. If implemented, the legislation could induce overproduction and a return to supply management. By paying producers the shortfall of the market price below a high target price, deficiency payments would offset some or all of the market-price signal to produce less. By basing deficiency payments on historic *program* yield and acreage, farmers have reason to raise their current yield and acreage in preparation for the time when historic program parameters are updated.

Farm real estate values climbed 15 percent despite the lowering of crop receipts by farm programs from 1996 to 2001. Cash-flow pressures from higher debt-service costs come gradually to landowners because 3 percent or less of farmland changes ownership annually. Pressures come almost immediately for renters, however. Ryan, Barnard, and Collender (2001, 23) note that nearly all the value of production flexibility contract payments was passed almost immediately by renters to landowners as higher rents. Because 43 percent of farmland is rented, it follows that nearly half of program payment benefits was lost quickly by renters and new landowners.

The proportion of farmland value attributed to government payments increased from 13 percent in 1990-1997 to 25 percent in 1998-2001 (Ryan et al. 2001). The latter is one estimate of the loss in farmland value if payments were terminated without reform in price support, insurance subsidy, and trade policies. Because withdrawal of commodity programs would cause major wealth losses, a phased rather than immediate withdrawal of programs seems warranted (see Orden, Paarlberg, and Roe 1999). Congress was unable to adhere to the phased withdrawal called for in the 1996 farm bill. But it is difficult to justify maintaining farm commodity programs indefinitely just because the government has led farmers to believe they will continue to receive transfers from taxpayers that have been capitalized into land prices.

INSTABILITY

I regard instability as the major problem of commercial agriculture (see chapter 6 by Makki and chapter 5 by Zulauf for an excellent overview). Risk is not in itself a case for subsidies from taxpayers, however. We don't provide government payments to lottery players, Las Vegas gamblers, Wall Street plungers, futures market speculators, day traders, and small businesses. Still, many consumers are nervous about relying strictly on the private sector to supply risk-management tools to farmers and to store food reserves for consumers.

The best economic argument for subsidizing storage to reduce variation in farm and food prices is that the public discount rate is less than the private firm discount rate. The contention is that, because of high cost, private firms hold too little stock for socially optimal price and food supply stabilization. Similarly, it may be argued that systemic (industrywide) risk is so high for private firms that only the government has sufficient scale to underwrite (reinsure) crop and farm revenue insurance.

The global record of government mismanagement of stocks, price supports, and insurance is not reassuring, however (Reinsel 1993). The U.S. House Agriculture Committee's proposal to provide countercyclical deficiency payments based on the shortfall of market prices (or loan rates) below target prices could increase farm-income instability. As Carl Zulauf notes in chapter 5, national price and yield of major farm commodities are negatively correlated and hence tend to self-insure—helping to stabilize farm receipts. Stabilizing on price may increase variation in receipts.

Government-financed stocks and insurance to manage risk appear to fail the second efficiency criterion noted earlier for public intervention. That is, the social cost of a private sector supplying less than "optimal" amounts of reserve stocks, insurance, and forward pricing is less than the administrative and mismanagement costs of public stabilization policies. Langemeier and Patrick (1990) show that farmers are remarkably good at stabilizing consumption despite unstable income from year to year. Farmers are extremely adept at self-insurance, are not very risk averse on average, and nowhere in the world are they willing to pay for unsubsidized all-risk crop insurance (see Reinsel 1993). Operators voluntarily enter farming and assume risks they well know characterize the industry.

Current commodity programs would have to be restructured massively to cost-effectively address problems of instability. Such restructuring would recognize the observation made earlier that most small farms have adjusted to risk relying on off-farm income to stabilize their finances, and commercial farms on average have sufficient wealth to pay for the many private risk management tools available to them.

The midsized family farms that frequently are least able to cope with risk can be provided with a risk safety net most cost-effectively by focusing stability on the "bottom line," net income, rather than on price, yield, gross revenue, or cost components of income that can vary to offset each other and hence stabilize net income. An investment retirement account (IRA) type program with the government partially matching a farmer's contribution and giving tax-exempt status to interest revenue is an option to address farming instability at minimal cost, including the low transaction cost of administration by the Internal Revenue Service. That type of program could easily be extended to all farmers and regions and would not need to be restricted to crops.

The program, recognizing that farming is not a low-income/wealth sector, can focus on facilitating the shift of earnings of farm people from high- to low-income years. By targeting net income of farmers, an insurance program patterned after Canada's *Net Income Stabilization Account* (NISA) can be cost-effective in stabilizing farm income. Such a program could be very cost-effective if payouts could be triggered only by very low or variable income of farm people from all sources—off-farm as well as farm.

U.S. representatives Kenny Hulshof (R-MO) and Karen Thurman (D-FL) originated legislation for a related nontaxed, super-savings, income-averaging farm and ranch risk management account (FARRM), allowing individual farmers and ranchers to put up to 20 percent of their income into a nontaxed savings account and draw on that account in bad years. Congress could afford to support this type of program more generously if it ended commodity price support and insurance program subsidies. Indeed, other farm-income supports would need to be terminated with introduction of a net income (from all sources) stabilization program to avoid compounding problems of administering an already unmanageable farm policy.

ENVIRONMENT

Instability may be the number-one problem of commercial farmers but environmental degradation is the number-one problem of farmers affecting nonfarmers. Farm operating and durable capital input markets for the most part work well, but externalities preclude reliance solely on markets to address problems of soil erosion and water and air quality in agriculture. That is, costs to society exceed the costs to the farms on which soil erosion and water and air problems originate. Commodity programs are not a cost-effective means to address environmental problems, however. Use of environmental programs such as U.S. Senator Tom Harkin's Conservation Security Act of 2001 as a "Trojan horse" to channel direct payments to farmers serves neither the environment nor farm economic welfare well.

Economic growth fostered by letting markets work can benefit the environment (see chapter 11 by Hervani and Tweeten). And markets can be used more widely to address environmental and cultural/lifestyle concerns, including preservation of farmland, as pointed out by Lawrence Libby in chapter 10. Labeling and certification, now widely employed for organic foods, can be extended so that consumers can vote with dollars in the marketplace to support food produced nonintensively and to protect wildlife and biodiversity. Thus, all farmers and consumers are not forced into one lockstep method of food production. A disadvantage is the free-rider problem and the reality that low-yield farming, by requiring more total cropland, diminishes land for wildlife and raises aggregate erosion.

Helpful public efforts include the Environmental Quality Improvement Program (EQIP) to address externalities of livestock production, the wetlands and conservation reserve programs to address environmental problems in land and water management, and *conservation compliance* programs to improve water quality and reduce soil erosion. The conservation compliance program is in disarray and needs rethinking. It may be argued that the public deserves much more in environmental protection than it is now getting from the tens of billions of dollars taxpayers are providing to farmers. Proposals for programs to more carefully target problems of soil conservation and air and water quality are addressed in more detail elsewhere (Tweeten and Zulauf 1997). In chapter 14, David Ervin and Frank Casey present additional options for improving delivery of environmental protection services to agriculture.

MISCELLANEOUS CONCERNS

Because of their prominence, the foregoing issues and problems were treated in some detail. Other concerns potentially addressed by commodity programs (rural community loss, food and farm operator supply, food security, and international and agribusiness competitiveness) are examined only briefly in the following pages because of space limitations or because they warrant only terse treatment.

Rural Community Loss

Some persons have proposed that farm commodity programs are necessary to preserve rural communities—small towns and cities. Rural areas, defined here as nonmetropolitan counties (no cities of over fifty thousand residents) have been growing in population. Farming-dependent counties, defined as those in which at least 20 percent of income is derived from farm labor and proprietor income, numbered 555 or 24 percent of nonmetro counties and 18 percent of all U.S. counties in 1990 (personal communication from Calvin Beale). Many are losing population. Less than one-tenth of the rural (nonmetropolitan) labor force works in farming, and 93 percent of the rural population resides in non-farming-dependent counties (Wright 1995, 17). Many farming-dependent counties are located in the Great Plains, a region suited by climate and sparse population to deal with environmental problems associated with livestock feeding-processing clusters to which the nation is headed (Tweeten and Flora 2001). Many such counties will attract livestock feeding and processing operations to raise income and employment.

Farm safety-net programs are not cost-effective means to support rural towns and cities. Many farming-dependent communities are best helped with extension programs that assist rural communities in deciding how they wish to use their resources. In many cases, more federal and state financing of schools can be justified to better prepare local rural youth for employment at home or elsewhere. Increased government aid to rural schools reduces the burden on rural communities of paying for human- resource-development programs that benefit communities elsewhere—often growing urban areas—where former rural residents live and work.

Food and Farm Operator Supply

It has been said that in the absence of commodity programs the nation will run out of food and farm operators. I am not aware of recent studies, but older studies indicate that overall farm productivity is raised with consolidation because operators taking over vacated farms are more efficient than the operators who exit. The 3 percent average decrease in farm output attending a phase out of farm commodity programs as noted in table 1.2 might raise farm commodity prices up to 10 percent. Given that American consumers spend only 2 percent of their income on farm food ingredients, they would hardly notice—especially if the change were spread over (say) five years. The overall cost of food and fiber plus public outlays for commodity programs would fall substantially with a phase out of commodity programs because government costs would fall more than food costs would rise.

Relatively few commercial farm operator replacements are needed annually and the supply of potential operators from smaller farms and from off-farm sources is large (Tweeten and Zulauf 1995). The nation is in no jeopardy from running out of food or farmers with or without commodity programs although some marginal farmers would exit farming in the absence of government subsidies.

In 1999, some eighty-five thousand farms (4.2 percent) were classified as vulnerable to financial failure, having negative farm cash flow and a debt-asset ratio over 40 percent (U.S. Department of Agriculture September 2000, 20,21). However, off-farm income would sustain many if not most of these farms in the absence of a public safety net. Many financially fragile farms will fail eventually with or without farm safety-net programs.

Food Security

Food insecurity issues and ways to address them are treated with insight by Daniel Sumner in chapter 12. At issue here is whether American farm commodity safety-net programs are essential to ensure future food security at home and abroad . The answer is no. The world has been blessed with food availability, even abundance, since World War II. The food insecurity problem traces to lack of productivity and buying power in poor countries rather than to U.S. commodity programs.

America has provided helpful humanitarian assistance in the form of food aid from surpluses generated by farm safety-net programs. On the other hand, subsidized food exports and international commodity prices depressed by agricultural support programs in developed countries have lowered prices and discouraged local food production in poor countries. The U.S. government could continue to maintain its four-million-ton grain reserve as a ready source of humanitarian food assistance, but need not maintain current commodity programs for food security at home and abroad. As the world's largest exporter of food, the United States will remain food secure without a farm safety net.

International Competitiveness and Agribusiness Concentration

It is said that a farmer can compete with other farmers at home or abroad, but cannot compete with foreign governments subsidizing farm exports. Similarly, many farmers view a safety net as essential to countervail the market power of agribusinesses that are growing larger and more concentrated.

Several observations are warranted. First, neither economic theory nor empirical evidence indicates that American farmers are systematically exploited by foreign governments or domestic agribusiness firms (see chapter 7). To be sure, imperfect competition characterizes many agribusinesses. If they do indeed exercise market power, fewer resources will be used in farming than if agribusiness industry were competitive. However, the oligopolistic (few firms) market structure that characterizes much agribusiness is recognized for massive advertising and innovation to expand food and fiber sales. This expansion of markets along with the prominence of cooperatives in agribusiness points to a contemporary farming sector as large and paid as high commodity prices as would prevail if the agribusiness sector were comprised of many more firms.

In chapter 13, Tim Josling finds evidence that some countries are embracing the new agricultural economic paradigm even as other entities such as the European Union seem to be moving in the opposite direction. Considerable progress has been made in reducing trade barriers with major competitors such as Australia, Canada, and New Zealand. More open trade encourages global competition among agribusinesses to price farm inputs and commodities more competitively for farmers.

CONCLUSIONS: POLITICS TRIUMPHS OVER SOUND ECONOMICS

Conclusions regarding the overall contribution of commodity programs to society could be thwarted because the economic and other farm program objectives considered in this study are incommensurate—they cannot be added together or simultaneously achieved. However, this potentially troublesome shortcoming need not be confronted because public commodity-program interventions fail essentially all economic tests.

Commodity programs have lost their economic justification. They are neither economically equitable nor efficient. American agriculture is not a welfare case.[11] Agricultural problems of family farm loss, cash flow, instability, poverty, and environment are real, but current commodity programs do not address these problems cost-effectively.

Commodity programs also have lost their social justification. The irony is that commodity programs erode the very virtues such programs were originated to preserve. Government commodity programs have created an insidious culture of dependency, with farmers depending on supports from taxpayers because programs have reduced farm receipts. Farmers have been placed by Congress on an ever higher income- and- asset-value pedestal that eventually will collapse at great loss in farm wealth and personal trauma.

The 1996 FAIR Act and subsequent legislation in the 1990s brought fundamental, useful reform: an end to set-asides and coupled deficiency payments, and to massive government stock accumulation and subsidized exports. These worthy economic reforms and planting flexibility can be retained while flaws in the 1996 act are corrected.

The 1996 farm bill works at cross-purposes. Direct payments helping to maintain income of farmers are partly or wholly offset by high commodity-price loan rates and crop-insurance subsidies generating output to reduce commodity market prices and farm receipts. If direct payments are retained, loan rates and insurance subsidies at the least need to be ended or sharply lowered to avoid a return to supply management set-asides and government stock accumulation. Long-term restructuring would retain resource programs and environmental protection but not current commodity programs.

If commodity programs have lost their justification based in the new economic paradigm and the traditional welfare criteria outlined above, why is "good" economics such "bad" politics? Why is "bad" economics such "good" politics? Answers to these questions are important because commodity programs are now an exercise in politics rather than economics.

The billions of dollars spent each year on commodity programs could not be sustained politically without an uninformed and inattentive public. The new paradigm of agricultural policy may be accepted by most economists and may be known by many farmers, farm organizations, and members of Congress. However, the general public appears to be unaware of the economic position of farm households or of the legislative measures being used to address farm problems.

David Bullock and Jay Coggins document in chapter 8 how farmers have been unusually successful in the political arena, parlaying a relatively small lobbying investment into large payouts from government. One reason for the success, as highlighted in chapter 9 by Charles Moss and Andrew Schmitz, is that farmers and agribusinesses often have common interests and work together for favorable legislation.

A more important factor in recent years is that Republicans and Democrats in a hotly contested race for presidential and congressional leadership have entered a bidding war of "I'll see you and raise you one ($billion)" for farm programs and votes. That's an attractive strategy indeed for politicians when the race is tight and other people's (taxpayers') money is being spent. Farmers are thought to be switch voters, willing to vote for politicians who respond to their "pocketbook" needs. Many states with small populations are farm states with more seats in Congress and electoral votes per capita than large urban states. Thus, farm votes are viewed as cheaper to gain than most. The strategy worked for normally conservative Republicans bidding against populist Democrats, giving Republicans the Congress and presidency in the 2000 election. The bidding war could continue because it is a "prisoners' dilemma"—the nation would be better off without it but neither party can afford to stop bidding.

The basis for farm-policy reform has been weakened by shifts in the agricultural political landscape. Voices for commodity-policy reform have been lowered. One is agribusiness, which thrives on farm output volume and got what it wanted with an end to set asides in 1996. Another voice is the American Farm Bureau Federation, which became an active proponent of commodity programs after its membership base shifted from the cornbelt to the Sun Belt—the latter's agriculture is more committed than most regions to commodity programs (see Guither et al. 1989, 15).

The conclusion is that farm policy appears to be an example of government failure. In our democratic-capitalist system, government fiat trumps market solutions. Given that farm policy is an exercise in politics rather than in economics, the time appears to have come for economists to turn over farm policy to political scientists. The time appears to have come to give Congress a twelve-step program for obsessive-compulsive behavior rather than give another workshop of economic education.

That view is cynical and I reject it. None the less, at its core, failed farm policy could trace to an uninformed and inattentive general public. Would a public more literate on economic policy allow 0.2 percent of the nation's population to command a large chunk of the treasury? More time can be spent usefully on educating the general public regarding the new agricultural policy paradigm before that question can be answered.

NOTES

1. Satisfaction derived from another dollar declines on average (diminishing marginal utility of income) as income rises, but additional income is essential to draw competent people to demanding jobs that meet people's needs. Hence, diminishing marginal utility is not a case for egalitarianism (equal incomes among people).

2. In 1999, there were two million farm households but 263,537 farm households with farm crop and livestock sales over $100,000 comprising only 0. 2 percent of the nation's population accounted for 73 percent of the $15.2 billion total payments to farmers. Payments averaged $42,020 on these farms. Some 42 percent of all farmers received payments but other farmers participated in support programs not providing payments—in all, over half of all farmers participated in government support programs.

3. Most small operations are unable to achieve economies of size. Less-efficient operators who produce at high cost per unit (or market for low commodity prices per unit) receive low returns whether they are in farm or nonfarm occupations and whether they receive government payments or not. Government payment support is of transitory benefit because land prices are bid up so that higher land costs or rents offset benefits of programs to operators. An exception is targeted payments to small farmers (see Gunderson et al. 2000). Other sectors of the economy also have some small and inefficient firms earning low returns but few sectors indeed have the unique situation where a large share of firms in the sector are small and earn negative returns due to tax and hobby benefits of an occupation as a way of life. The nation ordinarily does not subsidize consumption goods such as yachts or pleasure cruises.

4. Data were available only through 1996 (Council of Economic Advisors 2000, 418, and earlier issues), but were extrapolated to 1999.

5.   The formula is $DW = .5R\left(\dfrac{1}{\alpha} - \dfrac{1}{\beta}\right)\left(\dfrac{\Delta q}{q}\right)^2$

where DW is deadweight cost in $ billion, R = $193.2 billion farm receipts, α = intermediate-run supply elasticity of 0.2, ß is intermediate-run demand elasticity of –0.3, and $\Delta q/q$ = 0.034 is the addition to output as measured by an average of the high and low estimate in table 1.1 of excess farm output induced by programs.

6.   Government outlays, which are mostly transfers between taxpayers and producers, should not be confused with deadweight costs, which are a loss in national income.

7.   Farm crop and livestock receipts averaged $198 billion annually over the five years from 1996 to 2000, also a record. Prior to 1995, receipts had never exceeded $181 billion in nominal dollars. As apparent from data in table 1.1, farming expenses are inflated by and farm receipts are reduced by the 1996 farm bill programs, therefore pre- and post-1996 farm bill expenses and net farm income cannot be compared usefully. Farm subsidies *cause* the low farm gross and net receipts they ostensibly are designed to cure.

8.   Households listing farming as their occupation but with sales under $100,000 per year accounted for only 8 percent of crop and livestock sales in 1998. They are mostly marginal farming operations for which commodity programs provide too little assistance to relieve a basically untenable situation. Commodity programs are not a cost-effective solution to their problems. Instead, they need to seek supplemental off-farm income, add livestock or leased land, or move to situations offering a brighter future.

9.   The attrition for farmers and small nonfarm businesses are not strictly comparable—we do not but would like to know what percentage of new starts in farming survives for five years.

10. Retirements (including deaths) were estimated from the aged operator cohort in 1987 (Tweeten and Zulauf 1995, 216), hence are only an approximation for 1992-97.

11. Farmers are fortunate indeed that commodity programs are not welfare for the poor, because, under 1996 federal law, welfare recipients are only entitled to payments for five years.

ACKNOWLEDGMENT
Comments of Allan Lines, Jasper Womach, and Carl Zulauf are much appreciated.

REFERENCES
Arnade, C., and M. Gopinath. "Capital Adjustment in U.S. Agriculture and Food Processing." *Journal of Agricultural and Resource Economics* 23(July 1998): 85-98.
Ballard, C. L., J. B. Shoven, and J. Whalley. "General Equilibrium and Computations of the Marginal Welfare Costs of Taxes in the United States." *American Economic Review* 75(1985): 128-38.

Bentley, S., P. Barlett, F. Leistritz, S. Murdock, W. Saupe, D. Albrecht, B. Ekstrom, R. Hamm, A. Leholm, R. Rathge, and J. Wanzek. *Involuntary Exits from Farming: Evidence from Four Studies.* Agricultural Economic Report No. 625. Washington, DC: Economic Research Service, U.S. Department of Agriculture, November 1989.

Blue, E. N., and L. Tweeten. "The Estimation of Marginal Utility of Income for Application to Agricultural Policy Analysis." *Agricultural Economics* 16 (1997): 155-69.

Bonnen, J., and D. Schweikhardt. "The Future of U.S. Agricultural Policy." *Review of Agricultural Economics* 20 (Spring/Summer 1998): 2-37.

Burfisher, M., S. Robinson, and K. Thierfelder. "North American Farm Programs and the WTO." *American Journal of Agricultural Economics* 82 (August 2000): 768-74.

Cochrane, W. *The City Man's Guide to the Farm Problem.* Minneapolis: University of Minnesota Press, 1965.

Commission on 21st Century Production Agriculture. *Directions for Future Farm Policy.* Washington, DC: Whitten Federal Building, January 2001.

Council of Economic Advisors. *Economic Report of the President.* Washington, DC: U.S. Government Printing Office, 1987 and 2000 issues.

Drury, R., and L. Tweeten. "Have Farmers Lost Their Uniqueness?" *Review of Agricultural Economics* 19 (Spring/Summer 1996): 1217-18.

Gardner, B. "Economic Growth and Low Incomes in Agriculture." *American Journal of Agricultural Economics* 82 (December 2000): 1059-74.

Goodwin, B. "Instability and Risk in U.S. Agriculture." *Journal of Agribusiness* 18 (March 200): 71-89.

Guither, H., B. Jones, M. Martin, and R. Spitze. *U.S. Farmers Preference for Agricultural and Food Policies in the 1990s.* Bulletin 787. Champaign-Urbana: Agricultural Experiment Station, University of Illinois, 1989.

Gunderson, C., M. Morehart, L. Whitener, L. Ghelfi, J. Johnson, K. Kassel, B. Kuhn, A. Mishra, S. Offutt, and L. Tiegen. *A Safety Net for Farm Households.* Agricultural Economic Report 788. Washington, DC: Economic Research Service, USDA, October 2000.

Heady, E., and L. Tweeten. *Resource Demand and Structure of the Agricultural Industry.* Ames: Iowa State University Press, 1963.

Henneberry, S., and L. Tweeten. "A Review of International Agricultural Supply Response." *Journal of International Food and Agribusiness Marketing* 2 (1991): 49-68.

Johnson, G., and C. L. Quance. *The Overproduction Trap in U.S. Agriculture.* Baltimore: Johns Hopkins University Press, 1972.

Jordan, B., and L. Tweeten. *Public Perceptions of Farm Problems.* Research Report
    P-894. Stillwater: Agricultural Experiment Station, Oklahoma State University,
    1987.

Langemeier, M., and G. Patrick. "Farmers' Propensity to Consume: An Application
    to Illinois Grain Farmers." *American Journal of Agricultural Economics* 72
    (1990): 309-25 .

Luttrell, C. *Down on the Farm with Uncle Sam.* Original Paper 43. Minneapolis:
    Color Master Press, June 1983.

Makki, S., L. Tweeten, and C. Thraen. "Investing in Research and Education Versus
    Commodity Programs." *Journal of Productivity Analysis* 12 (1999): 77-94.

Orden, D., R. Paarlberg, and T. Roe. *Policy Reform in American Agriculture.*
    Chicago: University of Chicago Press, 1999.

Paarlberg, D. *American Farm Policy.* New York: Wiley, 1964.

Perry, J., D. Schreiner, and L. Tweeten. *Analysis of the Characteristics of Farmers
    Who Have Curtailed or Ceased Farming in Oklahoma.* Research Report P-
    919. Stillwater: Agricultural Experiment Station, Oklahoma State University,
    1991.

Reinsel, R., ed. *Managing Security in Unregulated Markets.* Boulder, CO: Westview
    Press, 1993.

Ryan, J., C. Barnard, and R. Collender. "Government Payments to Farmers
    Contribute to Rising Land Values." Pp. 22-26 in *Agricultural Outlook.* AGO-
    282. Washington, DC: Economic Research Service, USDA, June/July 2001.

Schnittker, J. "Fixing the FAIR Act to Reduce Budget Costs. In John Schnittker
    and Neil Harl, eds., *Fixing the Farm Bill, 93-98.* Ames: Department of
    Economics, Iowa State University, 2001.

Skees, J. *The Potential Influence of Risk Management Programs on Cropping
    Decisions.* Selected paper presented at American Agricultural Economics
    Association meeting in Nashville, Tennessee, August 2000. Lexington:
    Department of Agricultural Economics, University of Kentucky, September
    2000.

Spitze, R. G. F. *Evaluating Cost of Governmental Food and Agricultural Policies.*
    86 E-353. Champaign: Department of Agricultural Economics, University of
    Illinois, May 1996.

Tweeten, L. *Foundations of Farm Policy.* Lincoln: University of Nebraska Press,
    1970.

———. *Farm Policy Analysis.* Boulder, CO: Westview Press, 1989.

———. "Government Commodity Program Impacts on Farm Numbers."   In
    Arne Hallam, ed., *Size, Structure, and the Changing Face of American
    Agriculture,* 123-54. Proceedings of program sponsored by NC-181 Committee.
    Boulder, CO: Westview Press, 1993.

————. "Is It Time to Phase Out Commodity Programs?" In Luther Tweeten, ed., *Countdown to 1995: Perspectives for a New Farm Bill.* Anderson Publication ESO 2122. Columbus: Department of Agricultural Economics, Ohio State University, 1994.

————. "The Economics of Global Food Security." *Review of Agricultural Economics* 21 (1999): 473-88.

————. *Impacts of Unilateral Liberalization of Farm Programs.* Paper presented at annual meeting of the Southern Agricultural Economics Association in Forth Worth, Texas, January, 2001. Columbus: Department of Agricultural, Environmental, and Development Economics, Ohio State University, 2001.

Tweeten, L., and C. Flora. *Vertical Coordination in Agriculture.* Task Force Report No. 137. Ames, IA: Council for Agricultural Science and Technology, 2001.

Tweeten, L., and C. Zulauf. "Farm Succession: Who Will Farm in the Twenty-First Century?" In Ray Goldberg, ed., *Research in Domestic and International Agribusiness Management,* Vol. 11, 213-32. Greenwich, CT: JAI Press, Inc., 1995.

————. "Public Policy for Agriculture after Commodity Programs." *Review of Agricultural Economics* 19,2 (Fall/Winter 1997): 263-80.

Tweeten, L., and W. Amponsah. *"Sustainability: The Role of Markets Versus the Government."* In G. D'Souza and T. Gebremedhin, *Sustainability in Agricultural and Rural Development, ch. 3.* Aldershot, UK: Ashgate, 1998.

Tweeten, L., J. Sharples, and L. Evers-Smith. *Impact of CFTA/NAFTA on U.S. and Canadian Agriculture.* Working Paper 97-3. St. Paul, MN: International Agricultural Trade Research Consortium, March 1997.

U.S. Department of Agriculture. *The Farm Income Situation.* FIS-179. Washington, DC: Agricultural Marketing Service, USDA, July 1960.

————. *1997 Census of Agriculture.* "United States Summary and State Data." AC97-A-51. Washington, DC: National Agricultural Statistics Service, March 1999.

————. *Agricultural Income and Finance.* AIS-72 Sept. 1999, AIS-74 Feb. 2000 and AIS-75 Sept. 2000. Washington, DC: Economic Research Service, USDA.

————. *Agricultural Outlook.* AGO-279. Washington, DC: Economic Research Service, USDA, March 2001.

Westcott, P., and C. Young. "U.S. Farm Program Benefits: Links to Planting Decisions and Agricultural Markets." In *Agricultural Outlook,* 10-13. AGO-275. Washington, DC: Economic Research Service, USDA, October 2000.

Wright, B. "Goals and Realities for Farm Policy." In Daniel Sumner, ed., *Agricultural Policy Reform in the United States, ch. 2.* Washington, DC: AEI Press, 1995.

2

Agricultural Policy:
Pre- and Post-FAIR Act Comparisons

Bruce L. Gardner

The 1996 FAIR Act was a novel departure in farm policy, with fixed payments, absence of acreage set-asides, and avoidance of Commodity Credit Corporation (CCC) commodity stockpiles that provide a possible means of transition to a market-based agriculture. Some now argue that the FAIR Act has failed, on the grounds that the federal government has spent too much, while at the same time this spending is not effectively targeted at situations and people where help is most needed. This chapter analyzes the welfare economics of FAIR Act commodity programs in 1999–2000. While the FAIR has not been a sterling example of policy at its finest, in terms of net social benefit it is an improvement over the target price and acreage reduction approach that preceded it.

INTRODUCTION
The subject of agricultural policy is a broad one, covering topics from the production of agricultural raw materials to the consumption of final food products. Luther Tweeten has had something to say on almost all of them, but I will focus on some that have been central to his work. These topics concern appropriate policies for commercial agriculture and farm people, with an emphasis on commodity programs. But first I want to discuss briefly the broader issue of the "agricultural economics paradigm" to which this book calls attention, and the question of how that paradigm generates policy implications.

PARADIGM AND POLICY
In chapter 1, Tweeten contrasts an old and a new paradigm about the economics of agriculture. The old one highlights chronic disequilibrium as technology changes, low returns to resources in farming, market power of agribusiness, and unfair

practices by countries competing with the United States in international trade. The new theory highlights factor-market equilibrium in which nonfarm opportunities govern returns to resources in farming.[1] I agree that the new theory has more explanatory power than the old one in understanding incomes in U.S. agriculture in recent decades, although I don't fully agree that it is new. D. Gale Johnson (1958, 1973), for example, takes essentially the same view that factor-market equilibration between the farm and nonfarm sectors is key in explaining why farm as compared to nonfarm incomes differ, and what would have to be done to achieve income equilibration. What we see in the 1990s, as Tweeten was among the first to recognize, is evidence not available in the 1950s and 1960s that factor-return equilibration has actually worked in both labor and capital markets.

Tweeten refers to his approach to policy issues as "positivistic," an approach which, as I understand it, occupies a halfway house between the positive and the normative. He states that he looks for options that promote efficiency, equity, and freedom, and that cost-effectively address "the very real farm problems such as family farm loss, cash flow, economic instability, environmental degradation, and poverty." Selection of even the most cost-effective options, however, remains an exercise in normative economics. Even if, for example, you spend a billion dollars on poverty reduction quite cost-effectively, it takes a definite normative stance to recommend carrying out such a policy. And, suppose someone asks why the interests of migrant farmworkers, or more radically, animal welfare, are left off the list of things to be taken into account in policy choice. Isn't the decision not to count animal welfare a normative rather than positive one? We can estimate any of these gains or losses as positive economists, but there is no way to use any combination of them to order policies in a recommending vein without being normative. If we just specify criteria, and then estimate the ranking of policies according to those criteria, we have not departed from positive economics explicitly. But we can easily be read as making implicit recommendations, as Tweeten is easily read. I don't object to this, and have I no objection to economists making recommendations. In fact I believe we are dodging real responsibilities if we do not, and Tweeten has been exemplary in not dodging this responsibility.[2]

The principal analytical agenda of this chapter is the benefit-cost analysis of current commodity programs. This is a drastic narrowing down from the list of food and farm policies suggested in the introductory paragraph. This narrowing reflects a subtext in the way Luther Tweeten approaches policy issues, with respect to which I agree with him. It stems from his predilection to look first and foremost to efficiency—to give high marks only to policies that provide remedies for market failure. Tweeten is in general partial to market solutions. His recent writings on price instability tend to emphasize how difficult it is for government storage programs to achieve better results than private storage, and how government storage tends to interfere with the stabilization functions that private storage would otherwise perform (Tweeten 1996). On international food aid Tweeten also cautions quite strongly against the adverse incentive effects that such programs can have in the receiving countries. With respect to regulatory action to assist smaller family farms as against larger commercial farms and agribusiness, here too I would put Tweeten in the camp opposing the populist views of the need to rescue farmers from the oppression and exploitation. There are two good reasons for these positions. First is the lack of demonstrable market failure to remedy (so deadweight losses damaging to the economy would likely be increased by intervention). Second, with respect to incomes as returns to farm resources, labor earnings and returns to investment are largely determined by off-farm opportunity costs, exogenous to agriculture, so that market intervention is quite unlikely to generate sufficient improvement in farm incomes to compensate for the efficiency losses caused. If you want to address problems of low incomes in agriculture, it is preferable to do so by means of general welfare and income transfer programs, and policies that promote growth in the economy as a whole.

On all those matters I agree with Tweeten (if I have accurately stated his views). But they are normative recommendations and have been the subject of much dispute. I cannot consider them all, and will focus on recent commodity policy. On this subject I may have some disagreement with Luther. He berates the FAIR Act commodity policies of the late 1990s for their deadweight losses, which he places at over $3 billion annually (Tweeten 2000); but I will argue that this is precisely the area in which those programs have generated real gains.

FAIR ACT COMMODITY PROGRAMS
A call to rally on Capitol Hill stated: "The 'Freedom to Farm' Act has failed, causing economic disaster in much of rural America" (Farm Aid 1999). As stated more recently by someone with ostensibly more clout, "the 1996 Freedom to Farm bill fails to provide an effective safety net for American farmers. The President, Vice President, and I have implored Congress to avoid costly, ad hoc, emergency aid by addressing the fundamental flaws in the farm bill" (Glickman 2000). Economists have pointed to the high costs of payments under the FAIR Act as compared to payments that would have been made under the previous law. Payments to farmers have been several billion dollars higher than would have been the case under the 1990 Act's deficiency payments (which most likely would have been decreased to meet budgetary requirements if the FAIR Act approach had not been introduced). Nonetheless, there has been surprisingly little published benefit-cost analysis of the Agricultural Market Transition Act (AMTA), the main commodity title of the FAIR Act.

Gains and Losses from the FAIR Act's Commodity Programs

The basic differences between the pre-1996 and AMTA programs for grains and cotton are illustrated in figure 2.1. In the pre-1996 program, a target price P_T was guaranteed on a U.S. average basis. This was accomplished by deficiency payments equal to the difference between the target price and the U.S. average market price received by farmers during the marketing season. In order to discourage overproduction in response to the target price, and to save budgetary outlays (especially under the 1990 legislation that introduced a 15 percent nonpayment base), deficiency payments were paid on a limited quantity of output, Q^1. Moreover, in years when commodity stocks had become large (or threatened to become large),[3] the secretary of agriculture was empowered, and in some circumstances required, to impose an annual acreage reduction program, in which a specified percentage of each farm's acreage eligible for deficiency payments—varying from 5 to 35 percent depending on the crop and year—had to be held idle (and certain weed control and conservation practices followed) in order for that farm to qualify for payments. Moreover, after 1985 payments were made only on a "program yield" established for each farm, so that a farmer could not increase payments by, for example, applying additional fertilizer.

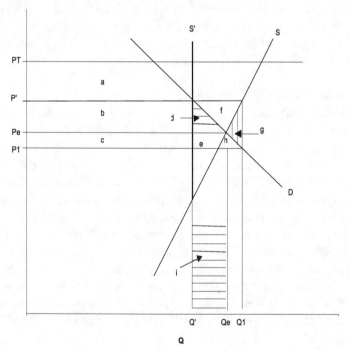

Fig. 2.1. Pre-1996 and AMTA comparison.

Those provisions resulted in a close approximation to a nondistorting payment system in which deficiency payments were received on a preestablished quantity, shown as Q^l in figure 2.1. A farm could increase its program crop acreage through base-building over a five-year period, so the market supply curve under the program was not perfectly inelastic at Q^l but this effect is not accounted for in the diagram. The market demand curve D intersects S^l to determine the market price P^l under the constraints of the program. If there had been no program, the market would have cleared at Pe where D and the unconstrained supply curve S intersect. The gains to producers are measured by the area a+b minus i, the costs of acreage idled under the Acreage Reduction Program (ARP). These costs are essentially the rental value of ARP acreage, the opportunity cost of not using the land, plus required weed control and conservation costs minus the subsequent productivity gain from fallowing). Costs to taxpayers are $(PT - P^l) * Q^l$ (area a) and costs to commodity buyers are measured by area $b + j$. The deadweight loss is the sum of those three components, which add up to area $i + j$ (shown as horizontally hatched areas in fig. 2.1).

Both the pre-1996 and AMTA programs maintain remnants of the Commodity Credit Corporation (CCC) loan programs that formerly supported market prices through governmental willingness to take delivery of commodities at the "loan rate" price. Ever since the costly CCC stock buildups of the early 1980s, these programs have been administered in such a way that CCC would not acquire stocks of commodities and would not support the season-average market price (although they might support the harvest-season price). The primary mechanism is "marketing loans," under which the farmer is paid the loan-level price and is then free to sell at the market price. This mechanism evolved in the 1990s into "loan deficiency payments (LDPs)," under which the farmer picks a day on which the crop would have been placed under loan, and instead of doing so accepts a payment equal to the difference between the loan rate and the "posted county price" intended to represent that locality's market conditions on that day.

LDPs are a lot like the former deficiency payment, with five important differences:

1. Loan levels are lower than the former target prices.
2. The loan level varies geographically, being higher where market prices are normally higher.
3. Because the LDP is maximized when market prices are lowest, farmers tend to take them early in the marketing year while selling the commodity later in the year when market prices are on average higher, thus providing an average farm price received under support that is higher than the loan-rate level.

4. LDPs are subject to a substantially higher payment limit per farm, and provide payment benefits to commodities that could not be used in earlier CCC loan programs, such as corn used for silage.
5. Most important, LDPs are paid on all that a producer grows, while deficiency payments were made on only a limited quantity.

The price received by farmers under the FAIR Act, when LDPs are being made as in 1997–2000, have been less than the target price guarantee of pre-1996 programs (albeit LDPs are made on a larger quantity). This situation is shown with farm price supported at P^1 in figure 2.1. The LDP program is essentially a classical production subsidy, with its price support generating output of Q_1, which clears the market at price P_1. The program generates producer gains of $b + j + f$, buyers' gains of $c + e + h$, and taxpayer costs that are larger than total gains, adding up to a deadweight loss of the vertically hatched area g.

In exchange for giving up deficiency payments, the FAIR Act provided "production flexibility contract payments." The payments were set for a seven-year declining schedule that gave each producer a pro rata share of the payments they received or would have received under the pre-1996 program if they had participated. Once a farmer has signed up, payments accrue each year until 2002 in a prescheduled amount regardless of the farmer's planting decisions. AMTA payments were set up to be fixed regardless of market prices, but in 1998, 1999, and 2000 AMTA payments were augmented by 50 percent, 100 percent, and 100 percent, respectively, in emergency market-loss assistance payments to compensate producers for low prices.

Although details are complicated, the basic economic analysis of the FAIR Act programs is simpler than was the case with the former programs. AMTA payments arguably have no effect on farmers' acreage or other production decisions, since the payment is the same no matter what the farmer does. Therefore the producer gain equals the taxpayer cost, and deadweight loss is zero.[4] It can be argued that in the absence of AMTA payments, some producers would have stopped producing and their acreage would have left crop production altogether. One reason might be that only AMTA payments forestall foreclosure. But for such farms, the absence of AMTA payments would then have meant transfer of the assets to another person, not the shift of land out of crop production (a shift that can occur without loss of AMTA payments anyway). Another possibility is that a farmer would like to grow program crops rather than use land in another way even though program crops lose money compared to the best alternative, and AMTA payments finance this quasi-consumption activity. But I will assume, as economic analysis generally does, that farmers allocate their land among uses in such a way as to maximize returns from it.

In any case, AMTA payments are no more distorting of farmers' production decisions than deficiency payments were under the pre-1996 programs. Since those programs were assumed in the figure 2.1 depiction to have no distortionary effects (beyond the costs of idled acreage), it is appropriate to maintain that assumption in FAIR Act assessment. Therefore, the schematic comparison of the horizontally hatched and vertically hatched areas of figure 2.1 as representing the deadweight losses of pre-1996 and FAIR Act programs, respectively, is my "final answer," in theory.

Empirical Analysis

Ten years ago I carried out a cost-benefit analysis of the main commodity programs of the late 1980s, with the results by commodity shown in table 2.3. The estimate is that the 1987 programs generated $17.5 billion in gains to producers, at a cost of $22.5 billion to taxpayers and buyers of commodities. (The situation was roughly similar in 1986 and 1988 also.) The gains are estimates of producers' surplus under the programs, at normal crop yields, compared to an estimate of producers' surplus under a situation of no programs, with commodity prices that would clear the markets after adjustment to the no-program situation (see Gardner 1990 for details). The difference between the gains and losses is the deadweight loss, $5 billion annually. This is the sum of losses due to buyers and sellers responding to different prices and, most importantly, the opportunity cost of idled cropland under annual acreage reduction programs. Although taxpayer costs and payments to producers varied considerably between different years of the late 1980s, the deadweight losses, our measure of economic inefficiency, were in the neighborhood of the $5 billion average magnitude throughout these years.

Now consider the comparable programs of the late 1990s. An analytically difficult aspect of this exercise is determining, on the *ex ante* basis that determines production decisions, what the expected producer price is, including the subsidy (the LDP) that corresponds to price P' in figure 2.1. We can't just compare the loan levels ($1.89 for corn, $2.58 for wheat, $5.26 for soybeans, and 51.92 cents per pound for cotton during 1998–2000) with average farm prices received. One reason is that, as mentioned earlier, loan deficiency payments and marketing loan gains have provided prices to producers that exceed the loan rate in each of the last three years, so farmers can be expected to count on this in making their planting decisions. See table 2.1 for data for the 1999 crops, with all data in dollars per bushel, except cotton, which is cents per pound.[5]

These calculations indicate corn growers can expect to receive 8 percent more than the loan rate, soybean growers 5 percent, wheat growers 13 percent, and cotton growers 23 percent. The price wedge between producer price received and buyer price paid is even larger, because farm-level market prices are less than the loan rates. The percentage price wedges, calculated as price differences divided by the average farm price paid as estimated by USDA, are shown in the right-hand column.

Table 2.1. Data for 1999 Crops[a]

Crop	Average farm price	Average marketing loan benefit	Average farmer revenue	Percentage price wedge
Corn	1.80	0.25	2.05	14
Soybeans	4.65	0.87	5.52	19
Wheat	2.50	0.41	2.91	16
Cotton	44.90	19.00	63.90	42

Source: U.S. Department of Agriculture (September 2001 and earlier issues)
[a] Dollars per bushel except cotton, which is cents per pound.

A second complication is that an estimated 8 percent of the 1999 corn crop, 2 percent of the soybean crop, and 12 percent of the wheat crop were neither placed under loan nor received a loan deficiency payment for the 1999 crop.[6] Those estimates are quite uncertain, but it does appear that some farm output does not get loan-program support. Some grain and oilseeds are of specialized or high-quality varieties that sell for more than the loan rate; nonetheless, a producer can get an LDP for them (so there is less reason for nonparticipation than under the old loan program where the farmer actually had to deliver grain to retain the loan-level price). Payment limits could possibly make a large farm ineligible for loan program benefits, but the farm would have to be quite large. The LDP payment limit for 1999–2000 was $150,000 ($75,000 doubled under provisions of the Agricultural Appropriations Act of 2000) and a "person" for payment purposes can benefit from up to a 50 percent interest in two additional farming entities, so the effective limit is typically $300,000, and a farm jointly owned by husband and wife could double this.[7] For the crop with the highest average LDP per bushel in the program, wheat, the average 1999-crop payment was $0.47 per bushel. To obtain the maximum payment, a producer would need a beneficial interest in 638,000 bushels, which would require an operation of fifteen thousand acres at the U.S. average 1999 wheat yield (42.7 bushels per acre). There are very few of these.[8] So payment limits must be a quite negligible factor.

Assuming the loan program was expected to work the same way in 2000, and assuming price projections for the 2000 crops as of August 2000 were producers' expectations when they made planting decisions, the price wedge relevant to supply distortion would be the "average farm revenue" prices for each crop as listed above minus $1.65 for corn, $4.35 for soybeans, and $2.50 for wheat (note that the corn and soybean market price projections are lower than the 1999 actual prices shown earlier). These data indicate a producer incentive price 22 percent above the market price for corn, 24 percent for soybeans, and 16 percent for wheat. To avoid misleading precision, let us consider a 20 percent subsidy rate for grains and soybeans as an aggregate.

Assuming aggregate elasticities of supply of 0.2 and of demand of –0.5, the output increase is $0.2/(1/0.2 + 1/0.5) = 2.9$ percent, and the deadweight loss triangle is $0.5 * 0.029 * 0.2 = 0.3$ percent of crop revenue or about $100 million for all corn, wheat, and soybeans.[9] Of the taxpayer cost of $5.2 billion for these three crops plus another $1.1 billion for cotton, rice, grain sorghum, barley, and oats (all of which get about the same rate of protection on average), producers get $5/(5+2)$ of the $6.3 billion or $4.5 billion and buyers save $1.8 billion.

Westcott and Price (2000) made a more detailed and refined estimate of the effects of the FAIR Act's marketing loan provisions. They remove price wedges attributable to LDPs and other marketing loan provisions as of 1998 and then simulate the effects for each commodity in the simulation model that USDA uses for its baseline commodity market projections to 2005. This model embodies a complete set of commodity supply and demand elasticities and cross-elasticities, with baseline projections of yields and export demand (so it is not a comparative statics exercise like my crude calculations but rather an exercise in comparative dynamics). Taking a snapshot of their results for 2000, that is, two years after the loan program is taken away, they estimate the percentage changes in prices and quantities attributable to the program (see table 2.2).

Table 2.2. Percentage Changes in Selected Variables in 2000 Attributed to 1996 Farm Bill

	Percent change		
Commodity	Producer price	Market price	Quantity
Wheat	12	-3	2.5
Corn-Soybeans	15	-3	1.5
Cotton[a]	25	-8	10.0

Source: Westcott and Price (2000)
[a] The cotton figures are Westcott-Price estimates only for quantity. I estimate producer price based on a supply elasticity of 0.5 and demand elasticity of –0.67.

Harberger (1964) derives a remarkably simple formula for multimarket deadweight losses resulting from a set of tax/subsidy rates in each market: it incorporates the effects of all own- and cross-price elasticities of supply and demand and requires data only for the subsidy rates and the resulting changes in quantities. Expressed in percentage change terms the formula is (equation 4' of Harberger, p. 70):

$$DW = 0.5 \, \Sigma_i \, S_i \, \%\Delta X_i \, V_i$$

where DW is the deadweight loss, the S_i are subsidy rates, the $\%\Delta X_i$ are the changes in output that result from the whole set of subsidies, and the V_i are the market values of the commodities in billion dollars. Applying the Westcott-Price quantity data to the 1999-crop price wedges above, the deadweight loss estimate for wheat, corn-soybeans, and cotton is:

$$DW = 0.5 \, (0.16 * 0.025 * 6 + 0.16 * 0.015 * 22 + 0.42 * 0.10 * 4) = \$0.12 \text{ billion}$$

Thus, the deadweight loss of the marketing loan program in 2000 is estimated at $120 million.

Producers also received about $11 billion in AMTA payments for all crops ($5.5 billion originally contracted, doubled in the emergency Market Loss Assistance program—a supplement that was repeated for the 2000 crops in legislation signed by President Clinton in July 2000). In the short run this is essentially a pure transfer from taxpayers, without buyer benefits or deadweight losses. That is the assumption in calculating that current programs result in grain output only 3 percent above the no-program, free-market level. However, it is likely that payments as large as these will keep some farmers in business who would otherwise leave farming. AMTA and LDP payments on corn in 2000 will total in the neighborhood of $8 billion, or 40 percent of the value of the corn crop. The effect is to further delay output adjustments that are the most promising way to return corn prices to profitable levels.

With respect to other commodities, I will be even less precise. GAO (2000) recently revisited the sugar program and estimated it generated $1.1 billion in gains to producers in 1998 at a cost to consumers and taxpayers of $1.7 billion. The larger deadweight loss of $0.6 billion arises because the price wedge (as of 1998) is about 100 percent of the no-program price rather than 20 percent as was used for the grains, and because of the value of import quotas, which transfer about $400 million from U.S. buyers of sugar to foreign sugar suppliers who obtain the quotas.

The dairy program was substantially scaled back in the 1990s. In 1998 and 1999 it spent an annual average of about $400 million on dairy products for price support. Assuming this is equivalent to a production subsidy and that supply and demand elasticities are about equal, it would have generated $200 million annually for producers at a cost of slightly more than $200 million to consumers/taxpayers, with deadweight losses of less than $10 million (and so lost to rounding in summing gains and losses).[10]

Table 2.3 sums up the preceding calculations, permitting a rough comparison between pre-1990 and post-1996 policies during relatively low-price periods. Despite talk of the disappearing safety net under the FAIR Act, producers received more support in 1999–2000 than they did in 1987. My 1999–2000 calculations omit crop insurance subsidies, which were substantially larger in 1999–2000 than in 1987. On the other hand, the calculations also omit export subsidies, which were more costly in 1987.

Economists' first order of normative business in the traditional view, which I

Table 2.3. Gains and Losses from Commodity Programs, 1987 and 1999

Commodity	Producer gains	Buyer + taxpayer gains	Deadweight loss total
(Billion Dollars)			
1987:			
Grains and oilseeds	11.8	-14.6	-2.8
Cotton	0.9	-1.5	-0.6
Sugar	2.7	-3.1	-0.4
Dairy	1.3	-2.6	-1.3
Tobacco, peanuts, wool	0.8	-0.7	0.1
Total	17.5	-22.5	**-5.0**
1999–2000:			
Grains and oilseeds	15.5	-15.6	-0.1
Cotton	1.8	-1.9	-0.1
Sugar[a]	1.1	-1.7	-0.6
Dairy	0.2	-0.2	-0.0
Other[b]	0.6	-0.6	-0.0
Total	19.2	-20.0	**-0.8**

[a]Estimates of GAO (2000).

[b]Wool program, costing about $100 million annually, eliminated in 1993.

share, is efficiency rather than redistribution. On this score, the 1999–2000 calculations indicate an almost $4 billion gain in 1999–2000 over 1987, a quite remarkable improvement. Is this believable? The reasons for reduction in deadweight loss are: (1) demise of the acreage reduction program, which was idling from 20 percent (corn) to 35 percent (rice) of participating producers' acreage bases in 1987—giving a deadweight loss rectangle of over $2 billion; (2) the dairy program has lower price support activity and is no longer losing value in stored commodities and dairy herd buyouts as was occurring in 1987; (3) the sugar program has greater deadweight losses now mainly because its effect in raising corn prices in 1987 cut the costs of the corn program, an effect we would still see today on loan deficiency payments except that GAO estimates the sugar program no longer has an appreciable price-supporting effect on corn sweeteners. These estimates are all shaky, but nonetheless a substantial reduction in deadweight losses remains my final answer because of the absence now of supply management through idled resources.[11] If we handled current market conditions the way it was done in the pre-1996 programs, we would be idling in the neighborhood of 30 million acres. At an average rental value of $70 per acre (the U.S. average in USDA 2000), this would cost our economy $2.1 billion annually.

Income Redistribution in Commodity Programs
The redistributional issue dominates discussion of the FAIR Act, even among economists. As Bruce Babcock's (2000) recent contribution puts it, "Are FAIR's Payments Fair?" He quotes Secretary Glickman's view, "We should not make payments to farmers who have not planted a crop and do not need the help. Instead, as the President proposes, we should target assistance to family farmers who really are struggling."

How does the FAIR Act distribute its benefits among farmers? With respect to AMTA's production flexibility contract payments, which account for about $11 billion of the producer benefits shown in table 2.3, they were constructed to provide payments to the same set of people who received payments in 1991–95. The sugar, tobacco, peanut, and dairy programs function similarly under the FAIR Act and pre-1996 programs, so the structure of benefits for producers of those commodities is not expected to have changed substantially under the FAIR Act. Critics point to large farms getting a more than proportional share of program payments. For example, farms with more than $1 million in sales received 5.3 percent of all government payments in 1994 even though they accounted for less than 1 percent of all farms. But at the same time those farms accounted for 25 percent of farm product sales (data from USDA 1997, 83–85). Therefore, government payments played a substantially less important role in the economic situation of large farms than of smaller ones; and it seems apparent that the comparative advantage of large farms would be greater in the absence of commodity programs.

The story might be different for the new development under the FAIR Act, the increased role of loan deficiency payments for grains, oilseeds, and cotton. The roughly $7 billion annually of marketing loan benefits in 1999–2000 are received nearly proportionally to production, and benefits go to producers of soybeans who received substantially less benefits under pre-1996 programs. Also, soybeans shared in market loss payments in the emergency legislation of 1999 and 2000, even though they did not formerly receive deficiency payments or AMTA payments. Therefore, the benefits of FAIR Act programs will be distributed somewhat more to larger farms and to soybean producers, as compared to the 1987 programs (see top half of table 2.3). It still seems likely, however, that commodity programs, as compared to a free market situation with no programs, improve the competitive position of smaller as compared to large farm operations.[12]

The policy implications of this situation are unclear to me. I don't see a market failure that would warrant attempts to redirect program benefits. On the other hand, if commodity programs are to be considered a welfare program, with most benefits directed at the lowest-income people, then the whole market-oriented approach that is common to both the FAIR Act and its predecessor programs for grains and cotton should be reconsidered. Some apparently want to allocate rights to produce to small farms, with an approach similar to what has been done in the tobacco and peanut programs. The efficiency losses of this approach are likely to be enormous, however. For example, McBride (1994) estimates that variable costs of producing corn are $0.28 per bushel higher on farms that grew less than fifty acres of corn than on the average of all farms. This suggests that if we shifted eight billion bushels of corn to such smaller operations, the cost of producing the U.S. corn crop would rise by about $2 billion annually.

Possibly a return to smaller farms would be worth the cost. But the idea cannot turn upon a desire to achieve an adequate income level among households that live on small farms. Even the smallest size category, those with farm sales less than $50,000, have average household incomes within 15 percent of the average income of nonfarm households. Substantial off-farm income permits such an income level to be attained despite losses from the farming operation.

Social scientists who aren't economists draw attention to broader indicators that are neglected in looking at income data only. Some political advocates of family farms appear to believe the source of income matters, and see income from farming as more valuable than off-farm income: "Of those small farmers still on the land, 80 percent have farm income below the poverty line. They've had to make most of their income off the farm" (Turning Point Project 2000, A9). A more general line of argument is that nonpecuniary aspects of rural life have taken a hit: "It is questionable whether the quality of life in rural areas has improved over the past one hundred years," says Cochrane (1999, 3). However, most indicators of nonpecuniary well-being have improved—life expectancy, incidence of injuries or disability, educational attainment, ability to express oneself and pursue opportunities, availability of a wide range of cultural experience, and conveniences like electrical appliances, greater mobility, quicker communication, mechanization of formerly onerous chores, air-conditioned tractor cabs, and the like. There are of course social indicators, seemingly ones associated with unruly behavior by youths (but is this new?) that go the other way. Nonetheless, I feel secure in predicting that if reverse migration to farm communities of one-hundred or fifty years ago were possible, many would visit but few would stay, and not just because of income levels being lower then.

We can still ask if the emerging economic organization of agriculture might not be less desirable than a more traditional family farm structure. To put the matter in stark and artificial terms, suppose we have two counties with the same volume of production and aggregate economic activity but different structures as follows: (A) a county of three thousand equally sized, diversified family farms each with one hired hand; or (B) a county of one thousand farms of which one hundred were large, specialized farmers each employing ten workers, and nine hundred were small, part-time farmers who employed no one and earned 90 percent of their incomes off the farm? County B could have evolved from county A by losing two-thirds of its farmers and hired workers, and moving toward greater concentration of production, and indeed this is an only slightly exaggerated depiction of what has happened in U.S. agriculture since 1940. I can see some reasons for preferring the situation A. But if the move to B results in the erasure of a sixty-year shortfall of average farm household incomes compared to nonfarm households, which is what has happened in U.S. agriculture, the preference for maintaining traditional small farms becomes dubious.

Turning to larger farm operators, do we really know that they are economically well off? The great bulk of our quantitative information about the economic situation in U.S. agriculture comes from the U.S. Department of Agriculture, specifically the Economic Research Service (ERS) building upon data collected principally by the National Agricultural Statistics Service (NASS) and the periodic censuses of agriculture (until the 1990s conducted by the Commerce Department's Bureau of the Census). Particularly important are surveys of production expenditures and costs of production that have been conducted over a long period of time and currently are carried out in the Agricultural Resource Management Study that samples about ten thousand farms each year. From these data, USDA economists estimate net income from farming, the net worth of farms, rates of return to farm assets, and other financial indicators for different types and sizes of farms. Such data are the basis for Tweeten's quite optimistic assessment of the economic situation of commercial agriculture at least since 1990. I agree with his assessment, which says that on average, including both small and large farms, average household income is above the nonfarm average, and on farms with over $250,000 in sales, net income from farming alone exceeds average nonfarm income, and net worth is over $1,000,000 per farm and has been increasing even in the 1997–2000 period when commodity prices have been low.

Why then have end-of-century assessments in the national media been so gloomy? Consider a sample from the usually hype-resistant *New York Times*. That paper placed on page 1 of its "Money and Business" section, on November 28, 1999, a story headlined: "Is the Sun Setting on Farmers? Many Can't Survive the 'New Agriculture'. What facts are cited in support of this assessment? A farmer interviewed says he expects to lose money for the second consecutive year, but more systematic evidence is lacking.

The main sectorwide economic fact the *Times* adduces is that "net farm income has fallen more than 38 percent since 1997." Actually, USDA's data estimate $48.6 billion of net farm income in 1997, $44.2 billion in 1998, and $43.5 billion in 1999, for a reduction of 10 percent, from a 1997 level that was near the peak of the 1990s. More generally, referring back to longer-term trends will indicate how misleading it is to cite the 1997–99 experience as a portent of economic doom. And indeed the article itself shows a graph of real income per farm during 1980–1998 with a clear *upward* trend!

Willard Cochrane (1999) gives oxygen to similarly pessimistic views about the economic situation in U.S. agriculture, stating that "we now have a farm structure comprised of 575,000 small to medium-sized family farms struggling to survive in the midst of a farm depression and 163,000 large farms also caught in a sharp price decline."

How do we reconcile these conflicting views? Two areas are worth exploring for pinning down the differences: descriptive facts and analysis of the current situation. The issue about the facts is whether the USDA-based statistics might be incorrect or misleading. USDA data have been criticized on several counts. Some object to USDA's inclusion of the service value of farm housing in net farm income (which is not done for nonfarm households). ERS has confronted this and other technical and accounting criticisms by providing a variety of net income measures. The comparisons according to which the average farm household now has a higher income than the average nonfarm household are based on the same income concept for farm and nonfarm people.

The analytical point is that farm income is holding up only because of government payments, while the underlying situation is much more grim. Close to half of net farm income in the last three years can be accounted for by government payments; and in 1999 and 2000, livestock incomes have not been unusually low. Therefore the net income from market returns for the major crops must be extremely low. Nonetheless, it is quite misleading to focus on payments as a share of net income. Payments are a receipt to producers, as are receipts from sales of commodities. Payments in the last three years have been a little below 10 percent of gross farm receipts.[13] And, one has to consider that without those payments, as discussed earlier, farm output would have been lower and prices higher. Table 2.3 suggests that the programs as a whole generate producer gains of $19 billion. This is a lot, but the gains are short-run quasi-rents (adjustment period of two years in the data taken from Westcott and Price 2000) for a depressed-price period, and reflect quite inelastic resource supply to crop production. In the long run we would expect resource adjustment to mitigate the losses and only land rents to contribute to inelasticity of supply, so the losers would be current landowners and the losses would be considerably less than table 2.3 indicates.

PRACTICAL FOOD AND FARM POLICY

The preceding analysis may be seen as economically optimistic in the sense that the deadweight losses of current policies, estimated roughly in the neighborhood of $1 billion annually, are small compared to those of pre-1990 programs. In the arena of practical politics, while farm interests may be prudently concerned about the prospect of a reduction in governmental support, political action by the Clinton administration and Congress in 1998, 1999, and 2000 confirms the remoteness of that prospect coming to pass. Moreover, neither the two major-party candidates nor the fringe parties of left and right offered serious alternatives to current policies in the 2000 election campaign. So, do we have both economically benign and politically stable farm programs?

First, the economic story of the preceding analysis looks more benign than it really is, because the focus has been on a comparison of post- and pre-1996 programs, and has not dwelled on costs common to both regimes. One common feature is that the scale of governmental activity both before and after 1996 has been so large that a costly administrative and regulatory bureaucracy has been necessary to carry out the legal requirements of farm legislation. The Farm Service Agency has personnel and other costs of about $600 million annually and other parts of USDA are involved also. Equally important, farmers themselves have to devote a lot of time and effort to paperwork and activities in "farming the programs" that have become economically necessary. Both the public and private resources, conservatively another $1 billion annually, could be devoted to actual economic production. Even more important are the consequences of the taxes that must be collected to carry out the transfers and other costs of the programs, roughly $20 billion annually in the late 1990s. The evaluation here ties agricultural policy into the overall national economic policy picture. The most straightforward and appropriate analytical approach is to ask what this $20 billion adds to the deadweight losses imposed by the nation's tax system. If we consider these funds as generated at the margins of income tax and other tax revenue sources, current estimates of the marginal deadweight losses cover a wide range. Using the estimate of Fullerton (1991), which still appears reasonable, we would conclude that $20 billion annually creates a deadweight loss in the nonfarm economy of 25 percent of this amount, or $5 billion annually. These economywide costs, common to both pre-1996 and post-1996 programs, constitute a good reason for economists to question U.S. farm programs generally.

NOTES

1. I prefer to refer to old and new theories, as alternative explanatory hypotheses within a common broader conceptual framework that would constitute a paradigm. Alternative paradigms have such fundamental differences, and even separate vocabularies, that one can only with difficulty be understood from the standpoint of the other. An example of a different paradigm is "bank hegemony theory" in the rural sociology literature, in which we explain events in the economy of agriculture by what banks want, and market forces play only a bit part or none (see Zey-Ferrell and McIntosh 1987).

2. Neither do I object to using a term like "positivistic" to fuzz up the borderline between positive and normative. This is a way of implementing the ideas of the philosopher John Searle, who makes a good case that insisting on a sharp separation between positive and normative is slicing methodological baloney too thin. He proceeds by deriving a normative conclusion from a series of factual statements that crosses over an only quite innocuous line of demarcation. For example, factual statements about the probabilities of various outcomes of driving at night on a mountain road without headlights in a rainstorm lead to the recommendation that you should not drive without headlights in a rainstorm. That kind of argument is convincing to me that insisting on eschewal of recommendations generates no analytical gain while foreclosing real possibilities of practical welfare improvement. However, to put a starkly opposing view on the table, "Professionally responsible economists should continually remind themselves, as well as policymakers who may be listening, that their analysis holds no magic about what is good or best policy for society. And their prescription or recommendation to society to eliminate or dismantle governmental partnership with the agricultural sector is of little value, and furthermore, hardly creditable to their professionalism" (Spitze 1983, 240).

3. The parenthetical point refers to the fact that the 1985 and 1990 farm bills stated criteria for acreage reduction percentages based on USDA's projected carryover stocks for the harvest period that followed the winter period in which required ARPs were announced. The projection could be mistaken. The last ARP for corn, for example, was a 7.5 percent reduction for the 1995 crop based on a projection of supplies that would likely be price depressing in the absence of the ARP. But 1995 turned out to be a high-price, excess-demand year in which prices rose to unsustainable heights and the ostensibly stabilizing ARP rules were in fact destabilizing. This episode did a lot to damage the reputation of the pre-1996 approach, especially among a group wider than commercial commodity buyers who had generally disliked it anyway.

4. Costs of administering the program, several hundred million dollars annually, and deadweight losses from raising tax revenues for the payments, perhaps 20–30 percent of the amount raised, are properly counted as deadweight losses, but they are roughly the same in both pre-1996 and FAIR Act programs.

5. The marketing loan benefit is loan deficiency payments plus market gain on quantities placed under loan as reported in "Loan Deficiency Payment and Price Support Cumulative Activity" available on USDA's FSA website http://www.fsa.usda.gov/. The average benefit per unit output is calculated as loan benefits divided by total production of each crop in 1999 as estimated by USDA. This counts in the average a zero benefit for quantities that did not participate in the loan program. The average benefit for those who did participate is slightly higher.

6. This estimate compares USDA-FAS reports of bushels that participated in marketing loan programs with USDA-NASS crop production estimates. There are several reasons these figures might not match exactly even if all 1999 production entered the programs. For example, a farmer might have placed some grain under loan in the 1999-crop program that actually was harvested in 1998 or earlier.

7. This is a simplification, but indicates the easily available maximum. For details see USDA-FSA, 1999.

8. The 1997 Census found 199 farms with 5,000 acres or more of wheat, averaging 7,260 acres of wheat and producing 53 million bushels in aggregate, or 2.3 percent of the 1999 crop.

9. The calculation just applies the formula for the area of a triangle, 0.5 times base times height, where the base is the percentage change in Q and the height the percentage change in P (the price wedge) due to the program, so we have a numerical measure of the vertically hatched area of figure 2.1, as a percentage of PQ. The elasticities were chosen such that they are most likely to err on the high side (for the short run). If the true elasticities are lower, deadweight losses are less.

10. Effects of dairy and other marketing order programs are omitted. They are estimated to have increased the average price of milk 2 to 5 percent as of about 1980, and to have had substantial deadweight losses of about $100 million annually because of pricing milk for fluid use higher than milk used in manufactured dairy products (see review in AAEA 1986). Despite reform efforts in the 1990 and 1996 farm legislation, there does not appear to have been sufficient change in the operation of marketing orders to alter these estimates significantly, so the assessment of current relative to pre-1996 programs is unchanged.

11. The GAO sugar analysis predates USDA's purchase, in June 2000, of $54 million worth of refined sugar, and subsequent announcement of a sugar payment-in-kind program under which beet growers would have the option of plowing up and not harvesting part of the 2000 sugar beet crop in exchange for rights to the government-held sugar that had been acquired. This loss of the value of the planted crop will give a substantial boost to the deadweight cost of the sugar program, but nonetheless the acreage involved will be far less than under the grain programs of the 1980s.

12. This is not to deny that research, technical education, and public

infrastructure investment activities may well foster the development of larger farm operations.

13. That 10 percent figure is also misleading in that it aggregates commodities that get a lot of support with others that get little or no support.

ACKNOWLEDGMENT
Helpful comments were received from Keith Collins, Joe Glauber, Andy Morton, and Paul Westcott.

REFERENCES
AAEA (American Agricultural Economics Association). "Federal Milk Marketing Orders: A Review of Research on their Economic Consequences." Ames, IA: AAEA Occasional Paper No. 3, June 1986.

Babcock, B. "Agricultural Policy Update: Are FAIR's Payments Fair?" *Iowa Ag Review* 6 (Summer 2000): 1–3.

Cochrane, W. W. "A Food and Agricultural Policy For the 21st Century." Institute for Agriculture and Trade Policy, November 1999 (available online at website http://www.iatp.org).

Farm Aid. "Farmers to Rally on Capitol Hill." Press Release, September 12, 1999 (available at website http://www.farmaid.com/press/).

Fullerton, D. "Reconciling Recent Estimates of the Marginal Cost of Taxation." *American Economic Review* 81 (1991): 302–8.

Gardner, B. L. "The United States." In F. Sanderson, ed., *Agricultural Protectionism in the Industrial World,* 19–63. Washington, DC: Resources for the Future, 1990.

Glickman, D. "Statement on Signing of Emergency Farm Assistance and Crop Insurance Reform Bill." USDA New Release 0201.00, June 20, 2000 (available at http://usda.gov/news//releases/).

Gore-Lieberman Campaign. *Prosperity for America's Families: The Gore-Lieberman Economic Plan.* Nashville, September 2000.

Harberger, A. C. "The Measurement of Waste." *American Economic Review* 43 (May 1964): 58–76.

Johnson, D. G. "Labor Mobility and Agricultural Adjustment." E. O. Heady et al., eds., *Agricultural Adjustment Problems in a Growing Economy,* 163–72. Ames: Iowa State University Press, 1958.

———. *World Agriculture in Disarray.* London: Macmillan, 1973.

McBride, W. D. "Characteristics and Production Costs of U.S. Corn Farms, 1991." Agr. Inf. Bul. No. 691. Washington, DC: Economic Research Service, USDA, January 1994.

Searle, J. "Deriving 'Ought' From 'Is'." In *Speech Acts,* 175–98. Cambridge: Cambridge University Press, 1969.

Spitze, R. G. F. "Comments for the Panel." In Hamming and Harris, eds., *Farm and Food Policy,* 237–40. Clemson University, 1983.

Turning Point Project. "America's Last Family Farms?" *New York Times*, February 28, 2000, p. A9. (available on the internet at http://www.turnpoint.org).

Tweeten, L. G. "World Trade and Food Security for the Twenty-First Century." Bilkent University-FAO Conference, Ankara, Turkey. *Yapi Kredi Economic Review* 7 (December 1996): 15–25.

U.S. Department of Agriculture. "Structural and Financial Characteristics of U.S. Farms, 1994." 19th Annual Family Farm Report to Congress, Agricultural Handbook No. 712. Washington, DC: Economic Research Service, July 1997.

———. "Payment Eligibility and Limitations." Fact Sheet, Washington, DC: Farm Service Agency, November 1999.

———. "Agricultural Cash Rents, 2000 Summary." Washington, DC: National Agricultural Statistics Service, July 7, 2000.

———. "Agricultural Outlook AGO-284." Washington, DC: Economic Research Service, September 2001.

U.S. General Accounting Office. "Sugar Program." GAO/RCED/00-126. Washington, DC, June 2000.

Westcott, P., and M. Price. "Analysis of the U.S. Marketing Loan Program." Mimeo draft. Washington, DC: Economic Research Service, USDA, August 2000.

Zey-Ferrell, M., and W. A. McIntosh. "Agricultural Lending Policies of Commercial Banks." *Rural Sociology* 52 (1987): 187–207.

3

The Content of Farm Policy in the 21st Century

David Orden

This chapter examines uncertainty about future farm policy and the challenges agricultural economists face in becoming convinced (and being convincing) that the equilibrium economic paradigm applies well to the agricultural sector. Despite much that has changed in agriculture over the past seven decades, the policy reforms undertaken so far face a conundrum of being too costly to be appealing as a long-term outcome, while return to past interventions is unattractive. Luther Tweeten has performed a valuable service in making the case for markets. More such voices will need to be heard to achieve the integrated global food system necessary for bulk-product American agriculture to prosper on commercial terms in the twenty-first century.

INTRODUCTION

Desirable farm and food sector outcomes have always required supportive government institutions and foresight in the design of agricultural policy. The conditions that shaped both the mission of American agriculture and the definition of good farm policy at the beginning of the last century arose from constraints on the availability of land and labor. The continental frontier had closed, new federal immigration restrictions had reduced the influx of unskilled workers, and urban industries lured a growing share of farm people into town, where they could earn a better living working in factories and urban services. The task of agriculture was thus to produce more food without more acreage, and with much less farm labor. Good farm policy had to support the outmigration of farmworkers by promoting technologies capable of replacing unskilled laborers on the land, and by extending new educational and infrastructural assets of those who remained on farms to ensure the most productive use of the new technologies.

This production efficiency challenge was met in the twentieth century with astonishing success. Public investments in roads, electrification, education, and rural health played a central role. Yet in some respects public policy toward agriculture impeded progress. Following the economywide macroeconomic collapse of the 1930s, commodity price-support and supply-control programs were set in place that idled farm resources and distorted incentives, reducing agricultural production efficiency. Initially designed as temporary emergency measures, these market interventions became entrenched when farm lobby groups with an incentive to subsidize income over the long run gained policy control.

The depression-era commodity policy misstep is still being corrected at the turn of the twenty-first century. Completing this task is necessary to ensure the efficiency of agriculture, domestically and globally. Largely as a result of intrusive farm-commodity policies in the United States and abroad, agriculture lags behind most other industries in achieving global market integration. Closing this gap is one of the challenges facing farm policy.

WILL COMMODITY PROGRAM REFORM BE COMPLETED?
American agriculture today scarcely resembles the troubled and impoverished sector in need of emergency assistance some seventy years earlier. In 1933, the average income among six million farmers was less than one-half the national average. This income gap has now been eliminated, despite a trend decline in real farm-commodity prices, because productivity has improved for more than a half-century through technological advances, capital investments, farm consolidation, and labor outmigration. Today's farmers earn more on average than nonfarmers, and manage to do so in several different ways. Farm assets and net income are substantial for large-sized production units (just 150,000 commercial farms now produce nearly three-quarters of the nation's agricultural output). On smaller farms, the income received from nonfarm sources is more important than farm-derived income.

As the production of food and the income and employment of farmers have undergone these dramatic changes, adjustments have been made to farm policy. Orden, Paarlberg, and Roe (1999) characterize these adjustments as a "cash out" of market interventions. The rigid commodity price-support programs that in the early 1960s were idling one-fifth of cropland were partially reformed when price-support levels were lowered for corn, wheat, and cotton to enhance U.S. competitiveness in world markets, and farmers were offered payments as compensation. A substantial further advance came in the mid-1980s, when price supports, having been set too high in anticipation of inflation that did not materialize, were dropped nearly 25 percent, with direct payments once again offered to participating farmers to provide a higher nominal return for their output.

The 1996 farm bill marked a third stage in the progression of farm policy toward direct payments instead of market interventions. The election in 1994 of a Republican majority in the House of Representatives (the first in forty years) ruled out a continuation of annual acreage reduction programs (ARPs), which Republicans had always opposed. Republicans were more comfortable with existing direct payments that were partly coupled to planting decisions and market conditions, but when market prices shot up unexpectedly in 1995–96, the only way to keep the government payments coming was to decouple them from those high prices.

The 1996 farm bill has achieved some reform successes. Annual land set-asides were terminated. Farmers have responded to the planting flexibility the new bill provided with substantial shifts in acreage among crops, and the delivery of decoupled cash payments has been prompt and efficient, which farmers have come to appreciate. Adjustments were also eased because the high market prices of 1995–97 did not automatically push up commodity loan rates (producer price guarantees) under the new farm bill. When market prices subsequently fell from temporarily high levels after 1997, price supports were not so badly out of line. And for the first time since the depression, a damaging reversion to annual acreage reduction programs was avoided (in 1998, 1999, and 2000) as prices weakened. As a result, foreign competitors, who had previously been able to count on unilateral production controls in the United States to cushion their own farmers from periodic market downturns, now had to share more equitably in the adjustment burden.

The progressive shift away from intrusive market supports and toward decoupled cash payments in the last half of the twentieth century also has limitations from a policy reform perspective. This approach has purchased greater market efficiency for American agriculture at a high cost to taxpayers. It has not brought into view any date-certain for an end to farm support programs, nor has it achieved any significant income- or asset-based targeting—payments continue to be made to farmers because they are farmers, without regard to their actual need. A termination of payments to even the most prosperous commercial farm operators, the sort of "graduation" seen as desirable in most other social policies, has proven elusive in farm policy. Yet this costly and untargeted movement toward direct payments has been the only policy reform approach that has been politically acceptable to the farm program beneficiaries, who continue to exercise a dominating influence over the policy-making agriculture committees of Congress.

Continued progress along the farm policy reform path of direct payments is not guaranteed into the twenty-first century. With low farm prices, even the loan-rate caps in the 1996 farm bill have proven too high, stimulating production. Farmers have shifted among crops, but they have cut back little on total output despite low price levels for all crops. In part this has resulted from improvised political decisions: in each of the last three consecutive years (1998, 1999, 2000) of low prices, Congress has responded with temporary legislation to spend more than originally scheduled on direct payments. With a budget surplus in hand, Republicans and Democrats in

Congress have not been able to resist bidding against each other to vote emergency relief to farmers. The basic features of the 1996 farm bill have been retained, as Republicans have retained a slim congressional majority through 2000. But a new Congress will have to revisit all of the farm policy issues in 2002. Without any semblance of budget restraint, the latest phase of direct payment reform has been costly enough to lose some of its appeal as a long-term farm policy solution.

Much is at stake in how this conundrum is resolved—not only the efficacy of domestic policy but the efficiency of the global food system as well. An open global food system is necessary for bulk-product American agriculture to prosper on commercial terms in the coming century. Yet today's global food markets remain tarnished by high tariffs and nontariff border restrictions, by domestic support policies that limit market access, and by trade-distorting export subsidies. A fundamental challenge for American farm policy is to reduce these trade distortions and thus foster more global integration for agriculture in the twenty-first century.

Agriculture was originally kept out of post–World War II multilateral trade liberalization at the behest of the United States, and has been slow to come under international policy disciplines ever since. The 1986–1994 GATT negotiations finally achieved some credible rules, by establishing a framework to constrain tariffs, domestic supports, and export subsidies, and to ensure minimal international market-access levels. Yet protection for agriculture remains much higher than for other industries under the 1994 GATT/WTO agreements, and U.S. farm policy has been disciplined little so far by these agreements.

Can less-fettered world agricultural trade, essentially free trade, emerge from WTO negotiations by 2035 or 2040, roughly one hundred years after depression-era farm policy interventions were initiated in the United States? This is a worthy goal, one from which the world as a whole would reap the great benefit of a secure and low-cost basic food supply. But the limited results and fourteen-year duration of the last GATT round (eight years of negotiations and approvals, followed by six years of implementation) is a sobering reminder of the difficulty of achieving farm policy reform internationally. At this pace, only three more global agreements on agriculture may be available over the next forty years. To achieve essentially free trade in agriculture in just three rounds of negotiations will take substantial progress under each agreement: curtailing the highest tariff rates and lowering tariffs by an average of ten percentage points, doubling low-tariff market access and cutting allowed export subsidies by one-third, and limiting domestic support levels by billions of dollars. These are tall orders for accomplishment.

In the political dynamic between the setting of domestic policy and the locking-down of policy restraints through international negotiations, domestic reforms need to pave the way for international agreements, since they are seldom pushed by such agreements. In its "blue box" classification for policies that combine output expansion incentives with output constraints, the 1994 GATT/WTO agreements enshrined as exempt from international disciplines the very form of commodity-specific partially coupled payments (deficiency payments) that the 1996 U.S. farm bill revoked. The United States must stay on this course of unilateral movement toward direct payments if it hopes to achieve the open trading system in which American agriculture can best prosper in the twenty-first century. If American farm policy stays on course toward less market interventions, the United States gains increased competitiveness in world markets and can more easily press others in the WTO for simplification of trade and support policy rules, and greater trade openings.

In pursuing this purpose, there are two major risks. The first will be the temptation, so apparent over the past three years, to spend too much on production-enhancing interventions and poorly targeted direct payments to farmers. The domestic bidding war that has taken place recently between political parties easily can escalate into an international bidding war among nations. Even if the payments at issue are as decoupled as some U.S. payments have become, it is naive to assume that production and trade will not be affected. At some point the strategic advantage in international negotiations (from increased competitiveness of our own subsidized farmers) will be offset by the retarding effect that mutually large subsidies will have on the pace of progress toward more open markets.

A second risk at the turn of the twenty-first century is that the United States will be too slow to secure reforms for those few import-competing commodities that still receive significant border protection: in particular dairy, sugar, and peanuts, but also some fruits and vegetables subject to high peak-season tariffs. Such programs have been insulated from reform pressure until now because the import restrictions employed to make them work are primarily damaging to foreign farmers, while the costs borne by domestic consumers are obfuscated. Yet there is a limit to the use of this policy option. Once imports cannot be reduced further, then border measures alone will not sustain high domestic prices.

Recent stress on the U.S. sugar program illustrates this point. Until 2000, it was possible to maintain domestic prices for sugar at levels set by loan rates (about 50 percent above world price levels) without forfeitures of sugar stocks, mostly by squeezing down on imports that exceeded the U.S. minimum international market-access commitments. In 2000, domestic supply increased relative to demand, putting downward pressure on prices, even with imports at the lowest levels international agreements would permit. The USDA announced a partial plow-down of the sugar-beet crop and still purchased one million tons of sugar to prop up prices, showing vividly the efficiency loss implied by the sugar program. More such waste is in the offing if the sugar program is not changed. The United States is committed to access of additional sugar imports from Mexico by 2008 under forward-looking NAFTA provisions designed to eliminate all of the agricultural trade barriers between the two countries. This will put still more pressure on the sugar program.

One unattractive option for sugar policy in the face of these pressures will be to hold up the level of U.S. prices by maintaining current loan rates and increasing constraints on domestic supply. Stocks can be accumulated by the Commodity Credit Corporation, and if that is not enough, crops can be plowed down in the field, or marketing allotments or acreage restrictions can be re-legislated, or paid land diversions can be adopted. But these are the type of government storage and supply-control measures that Congress has progressively abolished for other crops. Their adoption now for sugar would damage the industry and weaken U.S. negotiating leverage in the WTO.

An alternative to the current program offered by some critics of sugar policy is to eliminate all domestic support unilaterally and simultaneously increase imports until U.S. prices fall to world price levels. This is too extreme a policy change to be politically viable, particularly since world prices are depressed by production and export subsidies of the European Union and other high-cost sugar producers.

To resolve this impasse, a new policy is called for that progressively converts the sugar program to one based on direct payments (Orden 2000). This policy would reduce government entanglement in the sugar market and provide more price flexibility in the short run, allowing multilateral trade policy reform that includes sugar to be sought in the long run. As a first move in this direction, the United States might adopt marketing loans that would free up sugar prices on the consumption side, while retaining current price guarantees to producers. A more dramatic option is to implement fixed direct payments and lower loan rates. Neither step would accomplish full-scale reform, but either step would move sugar policy at long last down the reform path seen for other field crops since the mid-1960s.

The policy challenges confronted for sugar illustrate that the issues to be faced in the reform of farm-commodity policy are political more than ideological. They are best understood in materialist and institutional terms, as a long twilight effort to withdraw lucrative policy entitlements from well-organized farm lobbies both at home and abroad. The beneficiaries of farm-commodity programs at home (especially those with farm asset values to protect) will not give up their benefits without a fight. Nor will the congressional committees and executive branch institutions that have built their power around the periodic renewal of farm-commodity programs be eager to relinquish their influence.

HAS THE AGRICULTURAL POLICY PARADIGM CHANGED?

In contrast to emphasizing entrenched material interests and institutional rigidities, the argument has been advanced by Luther Tweeten and Carl Zulauf (as well as William Coleman, Grace Skogstad, and Michael Atkinson) that agricultural policy has undergone a change in "paradigm" or, at a minimum, that the views of policy analysts have undergone such a change. This language is misleading to the extent that it implies that a new viewpoint has emerged to fundamentally alter either the practice of commodity policy analysis by agricultural economists or the conduct of agricultural policy as legislated. Paradigm change is possible in agricultural policy analysis or policy making, but it is questionable whether a change in paradigm has occurred in either since the New Deal.

The term "paradigm" as a description of scientific views was introduced by Thomas Kuhn in his 1962 essay *The Structure of Scientific Revolutions*. Kuhn wrote in his preface that

> I was struck by the number and extent of the overt disagreements between social scientists about the nature of legitimate scientific problems and methods. Both history and acquaintance made me doubt that practitioners of the natural sciences possess firmer or more permanent answers to such questions than their colleagues in social sciences. Yet, somehow, the practice of astronomy, physics, chemistry, or biology normally fail to evoke the controversies that today often seem endemic among, say, psychologists or sociologists. Attempting to discover the source of that difference led me to recognize the role in scientific research of what I have since called "paradigms." These I take to be universally recognized scientific accomplishments that for a time provide model problems and solutions to a community of practitioners.

Kuhn goes on to exposit the character of "normal" puzzle-solving science, the role of anomaly and new discovery in creating a crisis within a discipline, and the scientific revolution that results in a paradigm change when a new theory comes to dominate an old one because it more adequately explains a growing set of anomalies. After the scientific revolution, the new paradigm provides the "model problems and solutions" to which Kuhn refers.

In a clarification he provides in a 1969 postscript to his essay, Kuhn acknowledges that through much of his book, "the term 'paradigm' is used in two different senses." In one usage, it stands for "the entire constellation of beliefs, values, techniques, and so on shared by members of a given community." In the second usage, it denotes "one sort of element in that constellation, the concrete puzzle-solutions which, employed as models or examples, can replace explicit rules as a basis for the solution of the remaining puzzles of normal science." Kuhn asserts that the second sense is the deeper of the two.

To keep these concepts separate, Kuhn suggests that a different term be given to the first usage, and the term he provides is the (less-than-catchy) phrase "disciplinary matrix." The elements that he had previously called paradigms or paradigmatic he now argues were all constituents of the disciplinary matrix but "are no longer to be discussed as though they were all of a piece." Kuhn describes four components to his disciplinary matrix: symbolic generalizations, shared commitments to certain basic beliefs in particular models, shared values that emerge as important when a crisis must be identified and choices made between incompatible ways of practicing a discipline (of which he identifies values concerning prediction as among the most important within scientific communities), and finally "exemplars," which he describes as "the concrete problem-solutions that students encounter from the start of their scientific education." Kuhn argues that differences between sets of exemplars provide "the community fine-structure of science," and are the elements of the disciplinary matrix to which the term paradigm is most appropriate. This brings him back to his "universally recognized scientific accomplishments that for a time provide model problems and solutions to a community of practitioners" as defining a paradigm.

During the past few years I have informally compared views of paradigm change with a few colleagues in agricultural economics. It strikes me that a common paradigm is widely held about paradigm change, even (as the case under paradigms) without formalization by each of those holding the common notion. This view centers on Kuhn's idea that revolutions in the natural sciences occur when an accumulation of anomalies under one theory are better explained by a new theory. Everyone nods to the examples of Copernicus, Newton, or Einstein.

Now consider recent attempts within political science and agricultural economics to apply the term "paradigm change" to observations on developments of policy views or policy events. The political scientist Peter Hall (1993) argues that policy making involves three central variables: overarching goals, policy instruments, and precise settings of the instruments. Within this framework, he describes "first order" reforms as those in which only the settings of policy instruments change, while the instruments themselves remain in place. A slightly more ambitious "second order" reform takes place when policy instruments are modified. The only way to get a powerful "third order" reform—one that changes goals—is to adopt an entirely new theoretical or ideological vantage point that changes for policymakers even "the very nature of the problems they are meant to be addressing." Hall termed this a change in policy paradigm, analogous to Kuhn's scientific revolution. In Hall's view, first- and second-order policy reforms can be relatively easy to achieve, but third-order reforms will be difficult, because existing institutions and long-held ideas will have to be challenged. Movement toward this third order of change must come from more than just a routine political contest between organized political interests. Third-order change requires some form of societywide contestation over ideas, and is likely to be proceeded by a significant shift in the locus of authority over policy. This in turn may not happen until existing policies have failed so completely and so conspicuously as to render the existing policy set untenable: there will be an accumulation of policy failures that are seen as anomalies within the prevailing policy paradigm. Hall's (1993) comparison to the original Kuhn ideas is that a paradigm change occurs when anomalies accumulate, old policies fail, and a fundamentally new approach to policy goals emerges as dominant.

Luther Tweeten has adopted the term "new paradigm" to describe what he perceives to be changed analytic views or policies implemented with respect to the agricultural sector. In "Public Policy for Agriculture after Commodity Programs," he and Carl Zulauf (1997) write that "changes inside and outside agriculture during recent decades have markedly shifted the paradigm determining the way society pursues its goals for agriculture." Because Tweeten and Zulauf write of a constant set of goals (among which they argue the relative emphasis has changed) they might be suggesting that a shift in paradigm is less marked than the scientific or policy revolutions of Kuhn or Hall. One needs to tread carefully here, because one treads on hallowed ground when appropriating Kuhn's term "paradigm." Are the changes to which Tweeten and Zulauf refer tantamount to Kuhn's "universally recognized scientific accomplishments" to which paradigm applies?

Tweeten and Zulauf distinguish between their old and new paradigms on four classes of criteria: central economic concepts, underlying beliefs, political situation, and resulting policy prescriptions. They argue that under the "old paradigm," agriculture was characterized by a chronic economic disequilibrium of persistent excess production capacity, lack of labor out-mobility, and consequent low returns. Given positive public sentiment toward agriculture and the farm lobby's activism, the resulting policy prescription (which was followed) focused on supply control and government involvement in commodity stock adjustments. In contrast, Tweeten and Zulauf argue, under the "new paradigm" agriculture is viewed as being in long-term equilibrium with adequate returns at comparable levels with other sectors over time, and with industrialization of agricultural production creating "a belief that agriculture is no different from other industries." These economic views are held in the context of broader political perceptions arising from the failure of communism, the relative success of democratic market economies, and the increased reliance of a numerically smaller farm lobby on monetary contributions to assert influence within the political process, which takes on a negative connotation when coupled with attention to government policy failure as opposed to market failure. Under this new paradigm, Tweeten and Zulauf assert, the public policy prescription for agriculture emphasizes market efficiency.

In their 1997 article, Tweeten and Zulauf express confidence that the change in views they describe as a paradigm change within analysis of how the agricultural sector operates has also been realized in the application of agricultural policy. They write that the emphasis of the new paradigm on market efficiency raises doubts about the value of an agricultural policy focused on commodity programs, with the result that "eventually the key policy prescription, supply control, was terminated." A dramatic change is then claimed; that "the demise of commodity programs as we have known them can reinvigorate the economic equity debate and resulting policies, which most likely will be national instead of sectoral in scope." Market efficiency has become more acceptable, they argue, with governments "providing public goods and correcting externalities," and with the private sector "allocating most agricultural goods and services." Tweeten and Zulauf conclude that implementing the new paradigm raises the issue of removing market barriers, and wonder for intervention programs in sugar, tobacco, and peanuts "how long these distortions will last beyond the current farm bill, given the increasing national attention to economic efficiency and growth." Despite their concession to a constant set of goals, the paradigm change to which Tweeten and Zulauf refer in 1997 appears to be a change in policy paradigm as described by Hall (1993), not one only of changed viewpoints accepted within the scientific community of agricultural policy analysts.

In his draft "Agricultural Policy" foreword for this conference, Tweeten (2000) strikes a substantially different tone. He argues with equal force to 1997 that agriculture operates under a "new economic paradigm," in which the sector is in equilibrium with returns equivalent to other economic activities, and that this new paradigm "teaches that rates of return on agricultural resources and household incomes are determined . . . by agriculture's capacity to adjust to change." Yet instead of asserting that this viewpoint has come to dominate agricultural policy, Tweeten is pensive about why this perspective about the economic operation of the agricultural sector has not coalesced into a new policy paradigm. He writes that the new view of agricultural policy "seems not to be known by the public, farmers, or even many economists," and that "a major restructuring of public policy is long overdue."

Let me try to disentangle Tweeten's various concepts, particularly his use of the word "paradigm." Application of Kuhn's idea of paradigm change to economics (and other social sciences) faces several challenges. To start, Kuhn's original use of this word was in a context in which the basic phenomenon being studied is unchanged, even if its full character is unknown, and (as Kuhn asserts) cannot even be conceived without reliance on some theory. For this unchanged basic phenomenon, there can exist no single dominant explanation (preparadigm science); or there can be one dominant theory of explanation (a state of normal science, within an existing paradigm); or there can be competition between theories until one replaces another as dominant (Kuhn's scientific revolution and paradigm change).

This basic distinction holds up, but is more complicated to analyze, when the subject under study is a social phenomenon that is itself undergoing change. Disciplinary practitioners, who like their colleagues in natural sciences are trained to watch for and are rewarded for detecting anomalies, now must ask themselves whether the observed changes are adequately explained within the existing paradigm, or whether they throw up anomalies, which then may lead to a scientific revolution. This is inherently a difficult task.

Application of the concept of a paradigm change to policy implementation—Hall's policy paradigm change—faces yet another hurdle. To actualize a policy paradigm change, the "community of practitioners" for which the new paradigm becomes dominant must itself dominate policy decision making. Hall argues such a case of policy paradigm change occurred in Great Britain in the 1980s when those favoring monetarist views of macroeconomic policy goals and instruments came to dominate earlier Keynesian practitioners after significant failures of attempted Keyensian policy prescriptions. In contrast, various other first- and second-order policy changes induced by experience and changing circumstances did not imply a new policy paradigm over the twenty years Hall investigated.

How does Tweeten's use of "new paradigm" hold up in light of these considerations? To start, one can ask whether there is a well-defined paradigm of the economics discipline relevant to study of a sector of the economy. I believe there is a strong paradigm of equilibrium economics, and that Luther would heartily concur. The economic paradigm has all of the elements Kuhn ascribes to a disciplinary matrix, thus perhaps we should say instead that such a "matrix" exists within equilibrium economics. It includes accomplishments in specific areas that provide the exemplars (textbooks are full of them) to which Kuhn views the word "paradigm" as best applying. And there are plenty of "normal economics" puzzles to address, both assessing where markets work and where they do not. This is puzzle-solving because economists agree on how to approach these studies, even if they reach different conclusions in specific cases. There are not separate schools of market economists and market-failure economists, although schools may differ on whether markets fail in certain cases.

One of the puzzles for equilibrium economics is assessing the performance of the agricultural sector. In the context of equilibrium economic exemplars illustrating factor return equalization across industries, for example, agriculture has posed several potential anomalies. Particularly, as incomes in agriculture lagged behind nonfarm incomes, was a new paradigm required to explain this as a permanent gap, or could plausible frictions impeding factor price equalization be described adequately without stretching the equilibrium framework too far? Tweeten cites the treadmill theory of Willard Cochrane, and Glenn Johnson's fixed asset theory, as paradigms that arose in direct challenge to equilibrium theory in agricultural economic analysis (paradigms in Kuhn's deeper sense—not as a challenge to the full disciplinary matrix of equilibrium economics, which Cochrane and Johnson themselves would surely reject).

Suppose that we accept the (arguable) premise that disequilibrium theories became a paradigm for a community of policy analysts. As Luther observes, there have subsequently been indisputable changes in the demographic characteristics of the farm sector: there are fewer farms, but more importantly incomes of those living on farms have risen to be comparable with those of nonfarmers. Bruce Gardner (1992) writes that "no one questions the measured farm income trends on the grounds that with declining demand for labor in agriculture, less than perfectly elastic supply of hired labor to agriculture, and immobility due to adjustment costs or asset fixity, a chronic farm income problem has to exist." On this score, Gardner argues the data dealt a resounding blow to some dimensions of farm-sector disequilibrium theories, at least as explanations for a long-run phenomena of low labor returns. This blow can be interpreted as equivalent to a "universally recognized scientific accomplishment." The fact of change in the phenomena under investigation resolved a choice between alternative ways to practice a discipline— that is, between competing paradigms. But in this case, it is not that anomalies arose, leading to the emergence of a new paradigm, but instead that anomalies disappeared, restoring the attractiveness of the initial paradigm. Tweeten's new paradigm is essentially the assertion that the basic exemplars of equilibrium economics do apply after all to the agricultural sector! No wonder some readers are suspicious of this being something "new."

Moreover, does the claim of labor market equilibrium mean that Tweeten's equilibrium view of how agriculture operates has become the paradigm (new or not) in agricultural policy analysis. While the labor market has proven able to equilibrate, making obsolete some of the early models of disequilibrium, other characteristics of the farm sector that have long held influence on policy analysis have not changed, or at least not changed in as convincing ways. The argument that agriculture should be analyzed in open-economy instead of closed-economy models is itself a candidate for paradigm change in Kuhn's deep sense, but this has not changed the widely held perception that the farm sector is characterized as a component of the economy with relatively inelastic short-run supply and demand and subject to relatively sharp supply and demand shocks (see, for example, Kliesen and Poole 2000). By itself such characteristics do not challenge the equilibrium economic paradigm, and it is certainly a prediction of equilibrium theory that assets in agriculture earn returns equivalent to similar assets in other sectors over time, but the data have not provided evidence as unambiguous as trends in farm income. The profession remains divided on the ability of markets to equilibrate farm assets returns to other returns. Luther's own arguments (1970, 1989) on decreasing costs and increasing returns to farm size explain that returns will be low on all but the most efficient largest farms at any point in time, and that this outcome is not disappearing. And even if returns in agriculture equilibrate to returns in other sectors over long time periods, the conventional characterization of agricultural supply and demand still results in substantial instability of farm prices, stocks and asset values—not necessarily more instability than all other sectors, but substantial instability nonetheless. There does not yet seem to be a consensus among agricultural policy analysts that private markets are resilient enough to absorb this instability in a politically acceptable fashion without substantial government intervention.

In the areas described above, what is missing to assert a paradigm change is the distinguishing accomplishment that redefines the exemplars. Farm-policy problem solving still takes place on much the same set of examples and techniques (now more complex) as it long has [see, for example, the excellent paper by David Bullock and Klaus Salhofer (2000)]. Thus, Tweeten's view that agriculture operates in new ways has not established itself as a dominant new paradigm in the agricultural policy analysis community in as many ways as he implies. Instead, it remains only a coherent set of hypotheses held by a subset of analysts. The controversy goes on—seemingly endemic (so far) in Kuhn's terms.

With labor market disequilibrium theories discredited, it is now useful to concentrate discussion of the traditional farm policy controversy on land values. Unlike labor, land has not moved out of agricultural production in aggregate over the past seventy years. The land input has decreased in relative contribution as an input compared to other forms of capital and intermediate goods, but the amount of land in production has a significant effect on the level of farm product supplies at any moment in time. The policy question is whether to let markets determine land use patterns and land values around the nation without government involvement. On this score, there is no evidence of a change in basic views among policy analysts over recent decades. There remain about thirty million acres idled in the CRP in 2000, more land than was idled in the late 1950s and about half the amount idled in the 1960s when the policy movement toward less market intervention and more cash payments began. Admittedly, there has been progress made in giving land idling an environmental protection focus, but the agricultural economics profession hardly rises up in unanimity to argue that such management of land use is antiquated and conceptually wrong unless it can clearly be shown to be the most efficient approach to addressing environmental externalities.

Thus, neither in terms of the stability or the level of land asset values is there much evidence of a new paradigm within the policy analysis community. I have heard Luther argue that it would be amazing how well land markets would work and the agricultural sector absorb adjustments if only governments truly ceased their interventions. That experiment, which might prove as decisive against government intervention as labor mobility has proven to disequilibrium theories of farm labor markets, is not likely to be run any time soon. The scientific community might welcome the chance to fully test its theories, but the farm asset holders who are the beneficiaries of farm programs are not going to tolerate the launch of such a test.

CONCLUSION

This chapter has examined the uncertain future direction of farm policy and the continuing challenge the agricultural economics profession faces in becoming convinced (and being convincing) that the equilibrium economic paradigm applies well to the agricultural sector. Luther Tweeten has performed a valuable service in making the case for markets. More such voices will need to be heard if the partial reform of farm policy that has gone some distance in the direction of market foundations is to firmly repudiate agriculture-targeted depression-era interventions. Despite much that has changed, this outcome is not assured even in the twenty-first century. The partial reforms of domestic agricultural policy undertaken so far today face a conundrum of being too costly to be appealing as a long-term policy solution, while return to past interventions is unattractive. Internationally, agricultural policies among nations remain disharmonious.

It may yet turn out that we will look back from the future and assert the dominance of equilibrium economics in the analysis and implementation of farm policy. Until such time is more certain, caution is appropriate in describing shifting or evolving farm policy assessment or implementation in the language of changes in paradigms. There is an ongoing battle for the content of farm policy, but it is one mainly among organized interests and within existing institutions, not one driven by universally recognized scientific accomplishments or their equivalents.

REFERENCES

Bullock, D., and K. Salhofer. "Judging Agricultural Policies: A Framework for Understanding How We Have Done in the Past and Suggestions as to How We Should Do It in the Future." Presented at the conference Challenging the Agricultural Economics Paradigm, Columbus, Ohio, September 10–11, 2000.

Coleman, W., G. Skogstad, and M. Atkinson. "Paradigm Shifts and Policy Networks: Cumulative Change in Agriculture." *Journal of Public Policy* 16 (1997): 273–301.

Gardner, B. L. "Changing Economic Perspectives on the Farm Problem." *Journal of Economic Literature* 30,1 (1992): 62–101.

Hall, P. "Policy Paradigms, Social Learning and the State: The Case of Economic Policymaking in Britain." *Comparative Politics* 25 (1993): 275–97.

Klieson, K. L., and W. Poole. "Agriculture Outcomes and Monetary Policy Actions: Kissin' Cousins?" Federal Reserve Bank of St. Louis *Review* 82:3 (2000): 1–12.

Kuhn, T. S. *The Structure of Scientific Revolutions*. 2nd ed. Chicago: University of Chicago Press, 1970.

Orden, D. "It Is Time for Domestic Sugar Policy Reform." Testimony presented to the U.S. Senate Committee on Agriculture, Nutrition and Forestry, July 26, 2000.

Orden, D., R. Paarlberg, and T. Roe. *Policy Reform in American Agriculture: Analysis and Prognosis*. Chicago: University of Chicago Press, 1999.

Tweeten, L. "Agricultural Policy: Foreword." Presented at the conference Challenging the Agricultural Economics Paradigm, Columbus, Ohio, September 10–11, 2000.

———. 1989. *Farm Policy Analysis*. Boulder: Westview Press.

———. 1970. *Foundations of Farm Policy*. Lincoln: University of Nebraska Press.

Tweeten, L., and C. Zulauf. 1997. "Public Policy for Agriculture after Commodity Programs." *Review of Agricultural Economics* 19:2 (Fall/Winter): 263–79.

4

An Empirical Analysis of the Farm Problem: Comparability in Rates of Return

Jeffrey W. Hopkins and Mitchell Morehart

A national comparison shows that the median rate of return for nonfarm businesses exceeds farm businesses by approximately three percentage points, but that large farms are four percentage points higher than nonfarm businesses at the median. Distributional analysis reveals that nonfarm businesses lagged farm business below the median but exceeded them above the median of the distribution. Decomposition methods show that asset turnover rather than differences in operating profit margins explains most of the divergence in rates of profit to entrepreneurs in the farm and nonfarm sectors.

INTRODUCTION

The *farm problem* refers to economic difficulties of the farm sector. While perhaps intentionally vague so as to fit a wide variety of ills, the list of difficulties always includes low returns, either to the farm household in the form of low incomes or to the farm business in the form of low return on assets. Most researchers extend the farm problem's symptoms to include variability in the stream of returns over time. In developed economies, the farm problem has been traditionally addressed with price and supply controls, but in the United States these were substantially altered with the Federal Agricultural Improvement and Reform (FAIR) Act of 1996, which suspended supply control mechanisms for many commodities, including feed grains, wheat, cotton, and rice.

Passage of the FAIR Act followed more than a decade of relative income parity between farm and nonfarm households, and occurred in a sophisticated financial environment that allows farmers to manage production and price variability. Observers, including agricultural economists, felt safe to consider the farm problem a historic phenomenon, if they thought of it at all, and many envisioned an eventual end to direct farm budget subsidies (Bonnen and Schweikhardt 1998; Tweeten and Zulauf 1997).

However, emergency legislation appropriated $25.6 billion from 1998 to 2000 to compensate farm businesses for low output prices, in addition to the scheduled payments to farmers that totaled $36.5 billion over the same period. In light of the level of supplemental assistance awarded to farmers since 1997, the FAIR Act is seen by many to provide insufficient support for adverse market outcomes, and the search for post-FAIR legislation with safety-net characteristics now appears high on the agenda of policymakers (Harwood and Jagger 1999). However, it is too early to say whether a safety net for agriculture would compensate all farmers for low prices, or only those farmers with low incomes regardless of market prices. It does seem desirable to judge the health of the farm sector relative to the nonfarm economy, rather than in the ad hoc manner used over the past few years. The purpose of this chapter is to facilitate such a comparison.

The most popular way to examine the symptoms of the farm problem has been to compare the income of households that meet the U.S. Bureau of the Census definition of a farm (more than $1,000 in agricultural sales over a year) with income for all U.S. households. One such approach is to compare average farm household income from the annual Agricultural Resource Management Study (ARMS) survey to the U.S. average household income from the Bureau of the Census's Current Population Survey (CPS). This type of comparison can be found in a table entitled *Average Income to Farm Operator Households* found in the Economic Research Service's *Agricultural Outlook* published ten times a year. Another approach to assess low returns is to compare household incomes, expenditures, and net worth to a minimum threshold or poverty line. Hopkins and Morehart (2000) provide an example of this approach to measure the incidence, intensity, and inequality of poverty in agriculture.

Results could be different when using a business rather than a household as the unit of analysis, because households derive their well-being from both the farm and nonfarm sector. Many farm households, for instance, supplement their incomes through off-farm employment. Of interest is whether comparability between farm and nonfarm households in income and wealth could coexist with low rates of return to investments in agriculture (Gardner 1992) compared to the rate of return to investments in nonfarm businesses. While the methods employed to compare returns at the household level are fairly well established, comparisons of returns at the business level are incomplete. Within the agricultural sector, business returns have been compared over time (Melichar 1979) between different farm types (U.S. Department of Agriculture 1997) and regions (Angirasa, Davis, and Banker 1993). Outside of agriculture, farm returns have been compared to stock market returns (Hepp 1996; Irwin et al. 1988) but not to other businesses themselves. It is likely to be more helpful, however, to compare farm returns with returns attained on similar, family-owned and -operated nonfarm businesses. The small-business comparison is useful because households that operate a nonfarm proprietorship (such as a restaurant, dry cleaner, or store) are potentially exposed to the same macroeconomic shocks, types of risk, and asset immobility that affect farm businesses. All family-owned businesses, regardless of whether they are a farm or nonfarm proprietorship, can add to as well as drain a significant proportion of family income and wealth. Using the entrepreneurial class as the proper reference group for farm businesses, our analysis significantly adds to the empirical assessment of the magnitude of the farm problem.

In the section that follows we present an ethical and economic rationale for finding comparability in rates of return, and explain the ratio-based approach used to measure business returns. Then, we discuss the USDA and Federal Reserve microdata used to model the farm and nonfarm sector. Finally, we present our results and summarize the major findings and policy implications.

FIRM FINANCIAL PERFORMANCE

The desire for comparable returns for farm and nonfarm businesses is often motivated by an ethical concern for both distributive justice, that is, that individuals and businesses receive fair treatment in the distribution of material resources, and commutative justice, that is, that individuals and businesses receive the value of their contributions to output of goods and services (Bonnen and Schweikhardt 1998). In addition to ethical concerns, persistent gaps in relative returns are an indication of potential market inefficiency, as economic arbitrage theory predicts that capital will seek its highest return, hence returns will equalize over time among investments of comparable risk.

Economic theory predicts that a well-functioning market will align *economic* returns among sectors, where economic returns include adjustments for risk, psychic costs and benefits, and other nonmonetary factors among sectors. Bidding need not be direct between farm and nonfarm sectors in the sense that low returns on a farm should motivate a change in the business to become a grocery store. Asset markets, however, will reward those who take advantage of opportunities to enter into high-return sectors and exit from low-return sectors. Thus, the accounting (observed actual) rates of return measured in this study may differ from economic returns among sectors if capital gain, risk, and taxation liability are different among sectors. Returns to scale need not be constant within a sector for equality between sectors (Bliss 1987).

Stock price and dividend values can be compared across firms to measure corporate firm profitability under the assumption that stock prices reflect all information related to the profitability of a corporation. Sole proprietorships, however, have no outside shareholders by definition, so rates of return must be inferred through accounting methods used to measure profitability.

 Dupont analysis yields the return on asset *(ROA)* measure of firm profitability, reflecting the profitability per dollar of assets. The accrual measure of profitability combines information from both the income statement and balance sheet of a business. The popularity of ROA is in part due to its ability to compare firms of different sizes, compare returns for firms from different sectors of the economy, and compare business performance with financial instrument performance. Dupont analysis allows for the relationship between sales, costs, and assets to be examined. These relationships will not be as readily apparent when looking at the components of ROA individually.

 Decomposition of ROA into analytic components can also yield insight into the determinants of financial performance (Featherstone, Schroeder, and Burton 1988). We decompose ROA into two other common ratio measures—operating profit margin (OPM) and asset turnover (AT)—to illustrate differences in profitability between firms of different sizes and between firms in the farm and nonfarm sector. Operating profit margin is a common measure of profitability, although constructed entirely from income statement information. Operating profits reflect the ability to generate revenues and control costs in such a way as to generate a profit. Asset turnover is a ratio constructed from both income and balance sheet information, reflecting the degree to which assets generate revenues. Table 4.1 provides a useful definition and interpretation of the three measures as used in the present study.

Table 4.1. Key Financial Measures

Measure	Definition	Interpretation
Return on assets (ROA)	(Net cash income generated by all assets - labor costs - management costs + taxes + interest payments) / total business assets.	Average interest earned on all investment.
Operating profit margin (OPM)	(Net cash income generated by all assets - labor costs - management costs + taxes + interest payments) / gross revenues.	The proportion of earnings or revenue available to compensate debt and equity capital.
Asset turnover (AT)	Gross revenues / total business assets	The volume of business flowing through the asset base.

DATA

Many of the early studies of farm returns on assets, reviewed by Tweeten (1989), presented interpretation problems because their rates of return were imputed from sector data rather than disaggregated firm data. Returns gathered from systematic surveys allow returns to be treated as a distribution, that is, explicitly showing low-return firms, high-return firms, and all firms in between. This treatment of returns can be contrasted with studies that put forth a *representative firm* that shows only a single point within the entire distribution, in most cases the average firm. Heterogeneity of sector performance is important, particularly at the lower end of the distribution of firms, where low productivity (and rate of return) is one indicator of possible early exit. While ROA from surveys can be compared to financial instruments, surveys often contain other financial and management characteristics that indicate why certain types of firms experience high or low returns.

Farm business data used in the analysis come from the 1997 Agricultural Resource Management Study (ARMS) entailing an annual survey administered and maintained by the National Agricultural Statistics Service and the Economic Research Service, both USDA agencies. The ARMS survey contains over ten thousand observations, stratified into thirteen sales classes for each of the forty-eight contiguous states. Because the ARMS survey is multiphase rather than a simple random sample, it requires the use of a complex weighting strategy in order to aggregate at the state, regional, or national level. Responses in ARMS are expanded according to the probability of being selected, so that each response represents the surveyed firm and other farm households that are like it. The data set used in this study was restricted to those farm businesses that had over $50,000 in sales over the course of a year, and contained 5,480 observations. This restriction, which only embraces a third of the total number of farms on a national scale, does nevertheless include over 90 percent of the value of production of the entire agricultural sector. Due to the many important differences between small farms and large farms, we treat agricultural businesses with sales less than $250,000 as a separate group from those with sales greater than $250,000. The $250,000 breakpoint corresponds to the recommendation of the National Commission on Small Farms.

Information on nonfarm businesses was obtained from the Federal Reserve Board's Survey of Consumer Finance (SCF) for 1998 (covering the 1997 calendar year). This nationwide triennial survey collects detailed information on all household assets (including residences, other real estate, businesses, all types of financial assets, pensions, and other assets) and liabilities (including mortgages, installment loans, credit card debt, pension loans, and other debts) along with information to analyze wealth (income, demographics, marital history, employment history, attitudes). The SCF uses a dual-frame sample consisting of 3,000 households from a standard representative sample and 1,500 households drawn from a special high-wealth oversample.

In order to most closely match the farm businesses modeled, the data set in this analysis was restricted to households with sole proprietorship businesses with more than $50,000 in annual sales, and contained 245 observations. Also, data contained in the SCF did not identify whether the family dwelling was considered a business asset or was separate from the business, although in practice a dwelling could add to the overall asset level as well as provide a return in the form of housing to the household. In order to treat farm and nonfarm business returns in a parallel fashion, we exclude the market value of the family dwelling and the living space it provides throughout our reported findings.

Median values taken by components of the ROA measure are listed in table 4.2. Charges for the opportunity cost of proprietor labor were calculated using proprietor labor hour inputs and imputed wage rates. For farm businesses, unpaid labor was assumed to be compensated at $10 per hour, the rural manufacturing wage. For nonfarm businesses, unpaid labor was assumed to be compensated at the median hourly wage for the type of job performed by the proprietor, with median hourly wages estimated from wage-earning households included in the SCF data. A charge for management was calculated equal to 5 percent of net value of production for the firm. In the case of farm businesses, management charges were calculated on the value of production after first adjusting for interfarm feed and livestock purchases.

Table 4.2. Median Values of Components Used to Measure ROA, 1997

Variable	Farm	Nonfarm
	(Dollars per farm or business)	
Cash receipts	114,082	120,000
Variable costs	67,336	58,950
Unpaid labor	20,228	39,125
Management	5,262	6,000
Assets	460,177	80,000

Sources: Farm business data from ARMS; nonfarm business data from SCF.

RESULTS

Results are presented in two sections. First, we look exclusively at commercial farms and compare the returns of smaller-sized farms to larger farms. Then, we compare the two classes of farm business to all nonfarm businesses with sales greater than $50,000. Return measures are presented as distributions of ROA, OPM, and AT.

Returns of Smaller Farms and Larger Farms

Figure 4.1 shows, for smaller and larger farm businesses, return on assets using the cumulative distribution function (cdf) for 1997. In the figure, each cdf shows returns by population quantile. As an example of how to compare two cdfs, the dotted lines compare the median ROA for each group. The horizontal dotted line at the fiftieth percentile (half of all farms have higher profits and half have lower profits) read down to the horizontal axis indicates that ROA for small commercial farms is nearly zero, and for commercial farms with annual sales over $250,000 is nearly 0.07, or 7 percent. Returns at the median are higher (more than five times higher) for larger farm businesses than small businesses. Even among commercial farms, defined as those with more than $50,000 in sales, most farm businesses have sales less than $250,000 per year, hence the distribution for all commercial farms (not shown) closely resembles the distribution for small farms.

Note that while negative returns exist on both small and large farms, the incidence of negative returns is much less for larger farms. Although almost a quarter of smaller farms and more than half of the farms with sales greater than $250,000 per year had returns greater than the 6 percent yield achieved on U.S. treasury bills over most of the 1990s, many farms did not. At every point in the distribution the rate of return for large farm businesses exceeds the returns of small farm businesses, so that no matter what rate of return is desired by the producer (for example, break-even or otherwise) there are proportionately more larger-sized than smaller-sized farms that meet or exceed it.

At the bottom of figure 4.1 the probability density function (pdf) is shown. It represents the slope of the cdf at every point in the distribution. While the pdf may be more familiar than the cdf to some readers, it cannot be used to rank distributions or economic outcomes, therefore, we will present our results primarily through the cdf device.

Decomposition of ROA into AT and OPM brings additional insight into the differences by size of farm business. Figure 4.2 shows that the relative economic advantage possessed by larger farms extends to the operating profits and asset turnover distributions as well. The top panel in figure 4.2 shows that large-sized farm businesses are more cost efficient in production. The bottom panel of figure 4.2 shows that asset turnover for larger farms is greater than that for small farms over the entire distribution, indicating that large farms do not just have more assets, but use their considerable assets more efficiently than small farms in producing sales. The financial performance gap (the distance between the cdf of large and small-farm businesses) decreases in OPM but increases in AT, implying that cost control can be achieved by well-managed small farms. However, well-managed small farms did not achieve a sales volume relative to asset value equal to well-managed large firms. The steady and positive relationship between farm size and business income has been a persistent feature of the structure of agriculture (U.S. Department of Agriculture 1997).

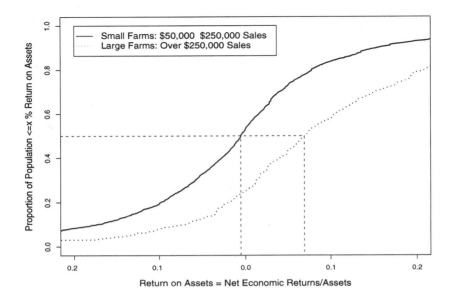

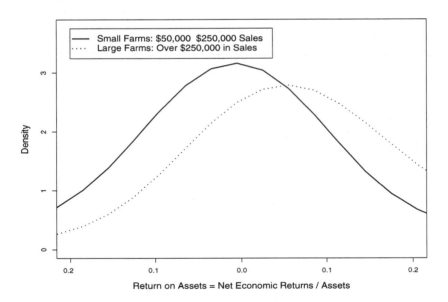

Fig. 4.1. Cumulative distribution (top) and probability density (bottom) of returns on assets to farm businesses, by sales class, 1997.

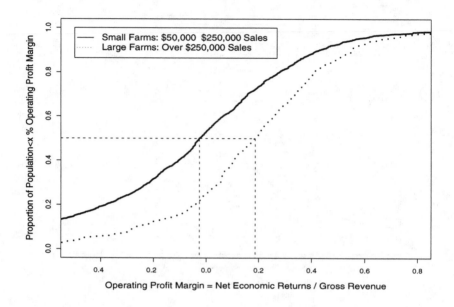

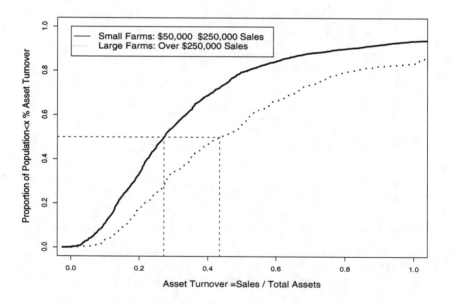

Fig. 4.2.   Cumulative distribution of operating profit margin (top) and asset turnover (bottom) for farm businesses, by sales class, 1997.

Farm and Nonfarm Returns

Figure 4.3 provides a comparison of the cdf as well as the pdf of return on assets for farm and nonfarm businesses using ARMS and SCF data for 1997. Note from the pdf that the dispersion of returns is much greater for nonfarm businesses, possibly due to the greater inherent heterogeneity of the rest of the economy relative to the farm sector. Table 4.3 shows the rate of return for three quartiles of interest. Median returns for nonfarm businesses near 3 percent compares to the small-farm rate near zero and the large-farm rate near 7 percent.

Table 4.3. Rate of Return (ROA) by Sector by Percentile, 1997

Population	25th percentile	50th percentile	75th percentile
All farms	-0.072	-0.002	0.065
Small farms	-0.076	-0.005	0.057
Large farms	-0.003	0.069	0.187
All nonfarm businesses	-0.213	0.029	0.374

Source: Author calculations.

Because the distributions cross, the ordering of farm and nonfarm businesses is reversed away from the median, so nonfarm financial performance trails small farms at the twenty-fifth percentile and exceeds large farms at the seventy-fifth percentile. Financial performance of large farms ceases to exceed that of nonfarm businesses at about the sixtieth percentile, where return on assets exceeds 10 percent, while small farm rate of return lags nonfarm returns beyond the fortieth percentile, where returns for nonfarm businesses and small farms are both negative 2 percent. Crossing distributions such as in figure 4.3 imply that identifying the dominant sector depends on the desired rate of return. If the desire is to avoid a negative return on assets, farm businesses dominate nonfarm businesses because there are proportionately fewer farms that have losses. However, if the objective is to match the treasury bill rate of return of 6 percent, nonfarm businesses are more likely to do this than farm businesses, although larger farm businesses still dominate nonfarm businesses at this higher threshold.

Decomposition of these results sheds light on the deeper question of why return measures differ between sectors. OPM distributions (top panel of fig. 4.4) show that farm and nonfarm proprietorships were much more alike in their ability to generate a profit during 1997. Large farms in particular were able to control costs relative to sales volume compared to nonfarm businesses, exceeding margins for nonfarm businesses over all but the top 20 percent of the distribution. Perhaps this should not be a surprising result, because the aspects of business management that lead to profitability are the same no matter what type of business, and have the same potential to be successfully executed. The median operating profit margins of all farm businesses and nonfarm businesses are quite close, although nonfarm businesses dominate commercial farm businesses as a group.

A comparison of farm and nonfarm business asset turnover ratios shows dramatic differences in how efficiently assets are used to generate revenue (bottom panel of fig. 4.4). Farm businesses have much lower asset turnover rates than nonfarm businesses. The inability to influence the lengthy biological processes in agricultural production and the long idle time for many agricultural assets, including land and machinery assets, could contribute to lower asset turnover relative to nonfarm businesses. Perhaps more important, the degree of asset ownership can greatly influence the asset turnover ratio. Farm or nonfarm businesses that lease the majority of their assets can achieve substantially higher turnover rates than those that own the majority of assets.

Even though nonfarm businesses dominate all farm businesses in OPM and AT, they do not dominate all farm business in the composite measure, ROA. This is because high asset turnover can be a characteristic of a business with either a positive or a negative operating profit margin, and high sales volume can amplify marginal profits in either direction. It appears that AT, in particular the absolute level of assets employed in the business, drives the wedge between farm and nonfarm financial performance, measured as ROA.

Economic theory predicts convergence in long-term returns, rather than short-term returns. Ideally, we would compare long-term returns for each firm in the survey to examine profits at more than just a single point in time, and smooth over business cycles that affect the distribution as a whole as well as firm-level dynamics of investment practices. Unfortunately, neither the SCF nor the ARMS data are structured as panel surveys, and the best that we can do is compare repeated cross-sections for additional survey years.

Figure 4.5 shows ROA for both farm and nonfarm businesses for three years (1991, 1994, and 1997) when comparable data existed for both farm and nonfarm businesses. The top panel shows the cdf and the bottom panel shows the pdf of ROA. Note that the general shape and location of the farm ROA distributions changes little over the periods examined. No first-order ranking among years is possible, as each year's distribution crosses another at some point. This is the case even for 1998 (not included), when low prices could have conceivably altered the distribution drastically.

More telling, however, are the yearly differences in nonfarm returns. While the nonfarm distributions shift widely over the years, a clear ordering exists among the years, with 1997 the best outcome, followed by 1991 and then 1994. Evidence suggests that the income of sole proprietors was strong in 1997 relative to the previous year, as aggregate proprietor incomes grew 4.7 percent at the same time that the number of those claiming self-employment as the primary occupation decreased by 1.9 percent (Headd 2000).

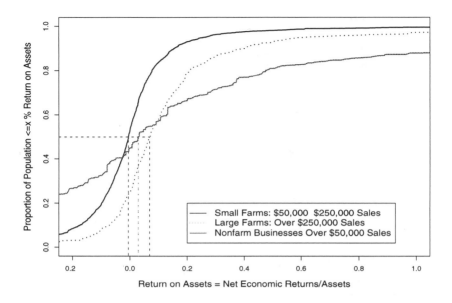

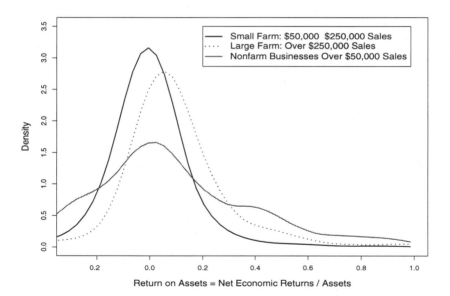

Fig. 4.3.   Cumulative distribution (top) and probability density (bottom) of returns on assets for farm and nonfarm businesses, 1997.

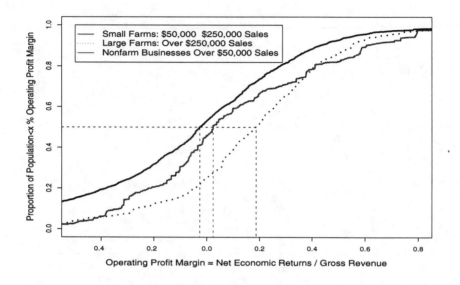

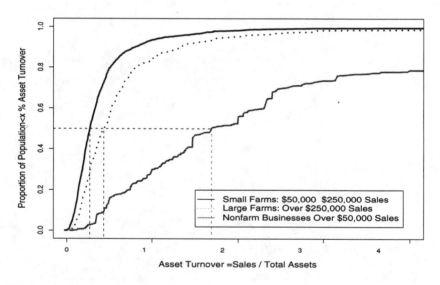

Fig. 4.4.   **Cumulative distribution of operating profit margin (top) and asset turnover (bottom) for farm and nonfarm businesses, by sales class, 1997.**

The measured difference between farm and nonfarm returns may not reveal an equilibrium if ROA as measured leaves out important sources of returns for farm firms, such as capital gains in the agricultural sector. Agricultural sector assets are dominated by agricultural real estate, which makes up over 75 percent of the overall farm asset portfolio on average. Agricultural real estate has appreciated at an average compound rate of 4 percent per year since 1987 (U.S. Department of Agriculture 2000), so farm businesses may tolerate a low current return on assets in exchange for relatively higher capital gains in the asset value. Under this strategy, the asset is held both for its annual dividend as a production input, in the form of net cash returns, as well as its investment value, in the form of a capital gain. We simulate the effect of an increase in the value of agricultural real estate by adding 5 percent of the value of real estate to the firm's current pretax cash income. The top panel of figure 4.6 demonstrates that a capital gain adjustment would be sufficient to close the gap at the median between all farm and nonfarm returns in 1997. The effect of the capital gain is to shift the crossing point of the nonfarm business distribution further to the right, so that at the same time the negative gap between farm and nonfarm returns decreases at the upper end of the distribution, the positive gap is getting greater at the lower end. Capital gains increase the proportion of farms that exceed nonfarm returns as well as increase the range over which farms dominate nonfarm businesses.

Another factor that may be responsible for differences in farm and nonfarm returns is the effect of government payments on farm returns. Direct government payments are designed to compensate for low returns (although payments are not means-tested), so that we might expect some of the gap between farm and nonfarm returns at the lower end of the distribution to disappear if not for the direct government support that exists for farms but not for other businesses. The bottom panel of figure 4.6 demonstrates the effect on ROA of an elimination of direct government payments, which consisted almost entirely of Production Flexibility Contract payments in 1997. The overall shift in the cdf appears to be slight, smaller in magnitude than the effect of capital gains, and the effects are relatively evenly distributed throughout the distribution. While the FAIR Act has been criticized for failing to provide an adequate safety net for farm households, in fact the percentage of extremely adverse outcomes realized by farms is lower than for households operating businesses in the nonfarm economy. Although direct payments enter into current returns and are capitalized into asset values, we make no attempt to model how asset values would adjust over the longer term. The simulation we show should accordingly be interpreted as the short-term effect of an unanticipated loss of income.

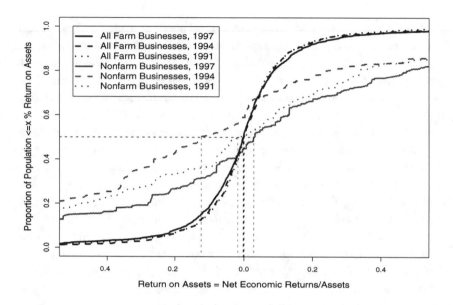

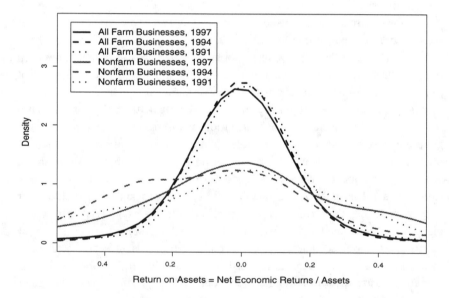

Fig. 4.5.   Cumulative distribution (top) and probability density (bottom) of returns on assets to farm and nonfarm businesses, 1991, 1994, and 1997.

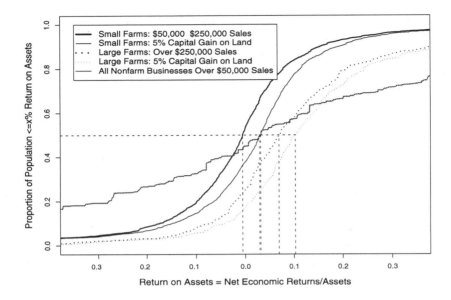

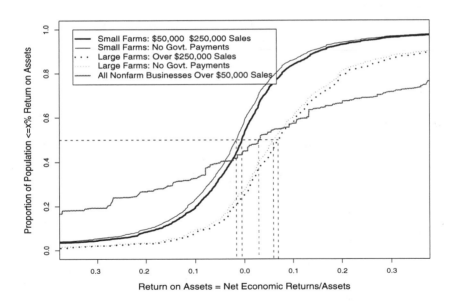

Fig. 4.6. **Cumulative distribution of returns on assets simulations with a 5 percent capital gain on farm real estate (top) and without direct government payments (bottom), 1997.**

The top panel of figure 4.7 compares the level of assets used in farm and nonfarm businesses, showing dramatically higher levels of assets used in farm than in nonfarm businesses. Smaller farms, which also employ a vastly larger set of assets than nonfarm businesses, are disadvantaged relative to nonfarm businesses regarding both their ability to control costs and their ability to efficiently produce sales from a given asset level. Larger farms, on the other hand, are highly competitive at controlling costs, but perhaps because of high asset requirements they lag high-performing nonfarm businesses in ROA.

One of the primary implications of a low return to investments is a low rate of wealth creation for farm households. In particular, this will be true if a household depends solely on agricultural sources of income and equity investments, and fails to diversify outside of its own businesses. The ARMS and SCF data can be analyzed to detect differences in net worth between households operating farm and nonfarm sole proprietorships. To carry out this comparison, we combined the business and nonbusiness assets and debt held by each household. The bottom panel of figure 4.7 shows that the net worth of households with a farm business is quite similar to the net worth of households with a nonfarm business. The median wealth of households with nonfarm businesses is slightly less than the median wealth of all households with a farm business, and much smaller than the median wealth of large farm businesses. It is also relevant to compare wealth levels of proprietorships to wealth levels of wage-earning households, also shown in the bottom panel of figure 4.7. As seen, the level of wealth found in the wage-earning population is dominated by all classes of household proprietorships.

CONCLUSIONS

Our analysis compared business returns in agriculture to returns outside of agriculture. We used a unique data set that allowed us to look at a wide range of sole proprietorship businesses to assess comparability in returns. This approach allowed for a deeper investigation of ways in which farm businesses are different from nonfarm businesses. We found three major differences.

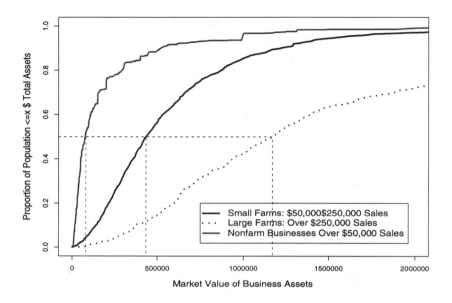

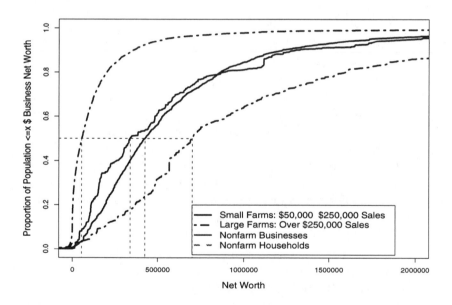

**Fig. 4.7.   Cumulative distribution of total assets for farm and nonfarm
businesses (top) and net worth for farm households, nonfarm
households, and all households (bottom), 1997.**

First, nonfarm businesses achieved a median rate of return that was slightly greater than that of all farm businesses and slightly less than that of farm businesses with sales greater than $250,000. When a capital gain was considered on farmland, the gap between farm and nonfarm business rates of return disappeared, and the gap between large farm and nonfarm businesses grew. The effect of direct government payments, on the other hand, was slight and did not change the ordering of median returns. The entire distribution of nonfarm returns was more dispersed relative to farm returns, indicating that even though the complexity and heterogeneity of farming is appropriately stressed in forming farm policy, performance of the sector as a whole is stabilizing relative to the rest of the economy. At the lower end of the distribution, for firms with negative return on assets, nonfarm businesses performed worse than farm businesses.

Second, return measures can be decomposed into a profitability measure (operating profit margin) as well as an efficiency measure (asset turnover). Large farms fare well relative to nonfarm businesses regarding their profitability, with similar to greater OPM for farm businesses over much of the distribution, although high-return large farms underperformed high-return nonfarm businesses. Smaller farms, on the other hand, have lower operating profit margins than nonfarm businesses at every point in the distribution. A hypothesis (which unfortunately cannot be tested with available data) is that smaller commercial farms accept a lower return in part because they receive part of their total return as a psychic income from the farm way of life. The most compelling difference between farm and nonfarm businesses is found in the ability of nonfarm businesses to generate much higher sales from assets relative to farm businesses.

Third, the net worth positions held by households with nonfarm businesses largely coincide with the net worth positions of households with farm businesses. In the case of large farm businesses, total household wealth is greater than wealth of households with nonfarm businesses. This suggests that despite highly variable rates of asset depletion and creation in the nonfarm proprietorship sector, similar patterns of household wealth have been attained. The level of wealth creation for nonfarm businesses is likely in part achieved through diversification of the investment portfolio of households. While farm businesses may also diversify, their wealth to a large extent seems to have come about as a result of much less diversification than nonfarm entrepreneurs.

Two main policy implications arise from the study. First, because most of the value added in agriculture occurs on farms that achieve profit levels similar to profits achieved on nonfarm businesses, the agricultural sector does not seem to be suffering from the *farm problem* as defined by low returns. Low returns, where they exist, are slightly augmented by capital gains and government payments. However, these do not appear to significantly change the overall distribution of returns, implying that their incidence does little to target low-return businesses.

Safety nets, by definition, would need to assist firms whose returns are judged to be excessively low at the household level. Diversification of farm investments outside of the sector would likely boost net returns for the overall household portfolio, particularly for small farms, who appear to have significant levels of poorly performing assets.

Second, the *farm problem* is frequently interpreted as one of excessive volatility in returns. Our analysis indicates, however, that cross-sectional if not temporal variability in the distribution of returns is much greater for the nonfarm business sector than the farm business sector. More direct testing is required here, however. Returns for a given firm may be much more variable than quantile returns if a firm changes its order within the distribution each year due to idiosyncratic shocks. One can imagine that the large variation in performance at different quantiles shown here is a lower bound on the firm-level variation in business returns. Both farm and nonfarm businesses are likely to change their ordering within the distribution from year to year. Unfortunately, collecting data required to follow returns over time for a farm or nonfarm business is not a component of the Survey of Consumer Finances or ARMS.

REFERENCES

Angirasa, A. K., B. Davis, and D. E. Banker. "Comparative Performance of Southern Plains and U.S. Farm Businesses". *Agribusiness* 9,4 (1993): 363–75.

Bliss, C. "Equal Rates of Profit." In *The New Palgrave: A Dictionary of Economics.* London: McMillan Press, 1987.

Bonnen, J. T., and D. Schweikhardt. "The Future of U.S. Agricultural Policy: Reflection on the Disappearance of the Farm Problem." *Review of Agricultural Economics* 20 (Spring/Summer 1998): 2–36,.

Featherston, A. M., T. C. Schroeder, and R. O. Burton. "Allocation of Farm Financial Stress among Income, Leverage, and Interest Rate Components: A Kansas Example." *Southern Journal of Agricultural Economics* 20 (December 1988): 15–24.

Gardner, B. L. "Changing Economic Perspectives on the Farm Problem." *Journal of Economic Perspectives* 30 (March 1992): 62–101.

Harwood, J., and C. Jagger. "Agriculture's Safety Net: Looking Back to Look Ahead." *Choices* 14 (4th Quarter 1999): 54–60.

Headd, B. "Small Business Economic Indicators, 1998." Technical report. Washington, DC: Office of Advocacy, Small Business Administration, March 2000.

Hepp, R. E. "Comparison of Michigan Farmland Returns to Nonfarm Investments." Working Paper 96-93. East Lansing: Michigan State University Department of Agricultural Economics, October 1996.

Hopkins, J. W., and M. Morehart. "Distributional Analysis of Farm Household Income." Select Paper for the Western Economic Association International Conference, Vancouver, British Columbia, 29 June–3 July 2000.

Irwin, S. H., D. L. Forster, and B. J. Sherrick. "Returns to Farm Real Estate Revisited." *American Journal of Agricultural Economics* 70 (August 1988): 580–87.

Melichar, E. "Capital Gains versus Current Income in the Farming Sector." *American Journal of Agricultural Economics* 61 (December 1979): 1085–92.

U.S. Department of Agriculture. "Financial Performance of U.S. Commercial Farms, 1991–94." Agricultural Economic Report 751. Washington, DC: Economic Research Service, USDA, June 1997.

————. "Accumulated Farm Real Estate Value will Help Farmers and Their Lenders through Period of Declining Cash Receipts." *Agricultural Income and Finance Situation* and *Outlook Report* AIS-74. Washington, DC: Economic Research Service, USDA, February 2000.

U.S. National Commission on Small Farms. "A Time to Act." Miscellaneous Publication 1545. Washington, DC: U.S. Department of Agriculture, January 1998.

Tweeten, L. *Farm Policy Analysis.* Boulder, CO: Westview Press, 1989.

Tweeten, L., and C. Zulauf. "Public Policy for Agriculture after Commodity Programs." *Review of Agricultural Economics* 19(Spring/Summer 1997): 263–80.

5

Income Variability of U.S. Crop Farms and Public Policy

Carl R. Zulauf

Three interrelated analyses concerning crop farm income variability comprise this chapter. The first two analyses find that farm program crops do not have a higher per-acre gross income risk than nonprogram crops, and that a change in average U.S. yield between planting and harvest does not create a per-acre gross income risk for corn, cotton, oats, soybeans, and wheat. Based on the latter finding, a new type of crop insurance is proposed: base indemnities on the yield of an individual farmer relative to U.S. average yield. Empirical analysis finds that this proposed new type of crop insurance may be viable without the need for federal subsidies.

INTRODUCTION

Two problems historically have been identified with U.S. farm policy: low farm income relative to the nonfarm sector and farm sector variability (Gardner 1992; Tweeten 1989). Concern about low farm income has declined as farm household income has achieved equality with nonfarm household income due to larger farms and increased off-farm income (Gardner 1992; Tweeten 1995; Zulauf 1994). In contrast, concern about farm sector variability has increased. In particular, income variability of crop farmers has attracted attention because of weather's impact on crop production.

This chapter starts with the proposition that income variability is more important to crop farmers than is variability of price and yield. The latter are concerns only to the extent that they translate into income variability. Given this proposition, three separate but interrelated analyses comprise the chapter.

The first analysis examines whether annual average gross cash income per acre has greater variability for farm program crops (specifically, oilseeds, grains, cotton) than for nonprogram crops. The analysis finds that, while the U.S. average annual yield and price of program crops have higher variability than the nonprogram crops, there is no difference in the variability of U.S. average annual gross cash income per acre. The reason for the latter finding is that, relative to nonprogram crops, program crops have a more negative correlation between changes in annual yield and annual price. Thus, to the extent that current public income support payments, such as AMTA and marketing loan payments, are based on the desire to reduce gross farm income risk, it would be appropriate to spread these payments across a broader array of crops than currently is done.

The second analysis examines whether the risk that national yield at harvest may differ from expected national yield at planting translates into a risk that national gross cash income per acre realized at harvest will differ from that expected at planting. The analysis is conducted for corn, cotton, oats, soybeans, and wheat. It finds that, at the national level, this yield risk does not on average translate into a change in gross cash income per acre. The reason is that a lower national average yield is offset by a higher national average price—that is, the elasticity between average U.S. price and average U.S. yield is –1 for these crops.

Based on this finding, a new type of crop yield insurance is proposed. It would insure the change in a farmer's yield from its expected value relative to the change in U.S. average yield from its expected value. This yield-difference insurance recognizes that a 30 percent yield shortfall for an individual farmer results in very different gross income risk when national yield is 40 percent less than expected, than when national yield is 10 percent higher than expected.

The third analysis reported in the chapter applies the proposed yield-difference insurance using county-level data. Yield-difference insurance was found to pay out approximately the same average indemnity as insurance based on shortfalls in the level of yield, the concept behind the current multiple peril crop insurance. However, the average annual indemnity for the U.S. varies much less from year to year for yield-difference insurance than for yield-level insurance, implying that the proposed yield-difference insurance may succeed in the private market.

IS INCOME PER ACRE MORE RISKY FOR FARM PROGRAM CROPS THAN OTHER CROPS?

This section addresses whether annual gross cash income per acre is more variable for farm program crops than for other crops. Variability in annual yield and cash price also are examined because they are the components of gross cash income per acre. Risk is measured as the standard deviation of $\ln X_{t+1} - \ln X_t$, where ln is the natural logarithm, X_{t+1} and X_t are annual average U.S. yield, cash price, or gross cash revenue per acre (average cash price times average yield) in years t and t + 1.

Taking the first difference of an economic variable is a common procedure for creating a stationary series, a necessary statistical property to reliably compare standard deviations. Furthermore, the first difference of the natural logarithm of a variable can be interpreted as a percent change, which facilitates comparison when a variable, such as price or yield, can be measured in different units. Last, the natural logarithm of price is commonly used because changes in price usually follow a logarithmic normal distribution and because price cannot be less than zero.

The appropriate measure of yield variability is a point of debate. A commonly used measure is the standard deviation of the residuals from a regression trend line fit to historical yields. This measure was also calculated. It generated the same qualitative results for yield variability as the first difference of the natural logarithm.

The analysis was conducted for twenty-six crops over the period from 1986 through 1999 (see table 5.1). Each of these crops had average U.S. price and yield data available over the analysis period. Nine of the crops are farm program crops: barley, corn, cotton, oats, rice, sorghum, soybeans, sunflowers, and wheat. They are highlighted in bold in table 5.1. The nonprogram crops include rye, hay, flaxseed, and various vegetables and fruits. The analysis starts with 1986 because the Food Security Act of 1985, that is, the 1985 farm bill, substantially lowered price support loan rates, thus increasing the reliance upon markets to determine price (Glasser 1986). Subsequent farm bills continued the movement to greater reliance upon markets.

Variability of average annual U.S. yield ranged from 4 percent (Irish potatoes) to 33 percent (flaxseed), while variability of average annual U.S. cash price ranged from 4 percent (cucumbers for pickles and snap beans for processing) to 30 percent (flaxseed) (see table 5.1). To statistically test whether variability differed between farm program and nonfarm program crops, a nonparametric rank sum test was conducted (Dixon and Massey 1983). A rank sum test evaluates whether the program and nonprogram crops, when ranked by level of variability, are randomly distributed or whether the two groups cluster toward opposite ends of the distribution. For example, if price variability is greater for the program crops, they would cluster toward the high end of the price variability distribution while nonprogram crops would cluster toward the low end of the distribution.

The null hypothesis is that farm program and nonfarm program crops have the same variability. The rank order test rejects this null hypothesis for annual average U.S. yield at the 98 percent confidence level and for annual average U.S. cash price at the 93 percent confidence level. Thus, from a statistical testing perspective, since 1985 the nine program crops have been characterized by greater annual yield and price variability than the seventeen nonprogram crops.[1]

Table 5.1. Variability of Average Annual Yield, Price, and Gross Cash Income Per Acre, Selected Crops, United States, 1986–1999

Crop	Annual average yield variability[a] (%)	Crop	Annual average cash price variability[a] (%)	Crop	Annual average gross cash income per acre variability[a] (%)
Flaxseed	33	Flaxseed	30	Head lettuce	30
Sunflowers	21	Oats	29	Rice	26
Corn for grain	20	Rice	27	Cotton	24
Oats	17	Head lettuce	26	Oats	22
Cotton	16	Dry edible beans	25	Cranberries	22
Green peas (processing)	15	Sweet potatoes	21	Sunflowers	21
Sorghum for grain	15	Sunflowers	20	Sweet potatoes	21
Soybeans	13	Sorghum for grain	20	Flaxseed	20
Barley	13	Wheat (all)	20	Dry edible beans	20
Cranberries	12	Rye	20	Tomatoes (fresh)	18
Sweet corn (processing)	11	Onions	20	Onions	18
Snap beans (processing)	11	Corn for grain	19	Green peas (processing)	18
Head lettuce	10	Irish potatoes	19	Irish potatoes	17
Rye	10	Barley	19	Sorghum for grain	16
Wheat (all)	10	Soybeans	17	Rye	16
Dry edible beans	10	Cranberries	17	Wheat (all)	15
Hay (all)	9	Tomatoes (fresh)	16	Corn for grain	15
Tomatoes (processing)	7	Cotton	15	Barley	14

Crop		Crop		Crop	
Sweet potatoes	7	Hay (all)	12	Sweet corn (fresh)	13
Strawberries	7	Sweet corn (fresh)	11	**Soybeans**	**13**
Cucumbers for pickles	6	Green peas (processing)	9	Tomatoes (processing)	12
Tomatoes (fresh)	6	Strawberries	8	Sweet Corn (processing)	11
Sweet corn (fresh)	6	Tomatoes (processing)	8	Snap beans (processing)	10
Onions	5	Sweet corn (processing)	5	Strawberries	10
Rice	5	Cucumbers for pickles	4	Hay (All)	10
Irish potatoes	4	Snap beans (processing)	4	Cucumbers for pickles	6

Source: Author calculations using data from the U.S. Department of Agriculture.

Note: Farm program crops appear in boldface.

[a] Standard deviation of the first difference of ln of average annual U.S. yield (average annual U.S. cash price; average annual U.S. gross cash income per acre).

In contrast, for average annual U.S. gross cash income per acre (yield times cash price), the rank order test of the null hypothesis that farm program and nonfarm program crops have the same variability has a statistical confidence level of 77 percent. This is below the commonly used statistical significance levels of 90 percent and 95 percent. Thus, from a statistical testing perspective, since 1985 gross cash income per acre has not been more variable for program crops than for nonprogram crops. This different finding occurs because, excluding rice and cotton, program crops had a -0.50 or lower correlation between changes in annual U.S. cash price and annual U.S. yield (see table 5.2). A negative correlation implies that changes in yield and changes in cash price move in opposite directions, thus stabilizing gross cash income per acre. Everything else constant, the more negative the correlation, the more stable is gross cash income per acre from year to year. A negative correlation between yield and price is often referred to as a natural hedge.

Table 5.2. Correlation between Average Annual Cash Price and Average Annual Yield, Selected Crops, United States, 1986–1999

Crop	Correlation[a]	Crop	Correlation[a]
Flaxseed	-0.80	Sweet corn (processing)	-0.33
Corn for grain	**-0.73**	Cucumbers for pickles	-0.32
Dry edible beans	-0.72	**Rice**	**-0.31**
Wheat (all)	**-0.71**	Sweet potatoes	-0.27
Soybeans	**-0.68**	Snap beans (processing)	-0.19
Barley	**-0.66**	Strawberries	-0.03
		Green peas (processing)	-0.02
Irish potatoes	-0.66		
Oats	**-0.65**	Sweet corn (fresh)	0.13
Rye	-0.63	Cranberries	0.16
Hay (all)	-0.62	**Cotton**	**0.21**
Sorghum for grain	**-0.61**	Tomatoes (fresh)	0.35
Sunflowers	**-0.50**	Head lettuce	0.36
Onions	-0.45	Tomatoes (processing)	0.41

Source: Original calculations using data from the United States Department of Agriculture.
Note: Farm program crops appear in boldface.
[a]The correlation is calculated between the first difference of the ln of average annual price and the first difference of the ln of average annual yield.

Results suggest that commodities were not selected for government price and income support based on their level of variability of receipts. However, when interpreting the results of this analysis, two caveats should be kept in mind. First, although the analysis applies to the situation experienced by nonparticipants in farm programs, it includes some years when farm programs may have influenced U.S. average yields and prices, especially of farm program crops. Thus, the existence of farm programs may have affected the results of this analysis.

Second, variability of yield increases as the level of aggregation moves from the country to the individual farm (Harwood et al. 1999, 8). The same relationship probably holds for price. These two relationships imply that, everything else constant, the smaller the total acres of a crop in the United States, the greater will be the variability of the crop's average annual U.S. price and yield. The acres planted to program crops is higher, often much higher, than the acres planted to most nonprogram crops. Thus, everything else constant, the larger acres of the program crops means less variability in their average annual U.S. yield and price than for the smaller acre nonprogram crops. This implication reinforces the finding that program crops have higher yield and price variability than nonprogram crops.

In contrast, the relationship between total U.S. acres and variability of annual per-acre income is more difficult to determine. While greater price and yield variability translates into greater per-acre income variability, this increase may be offset by a more negative correlation between price and yield. Examination of table 5.2 suggests that the larger a crop's total U.S. acres, the more negative the correlation between average annual U.S. yield and price. To examine empirically this visual observation, a correlation was calculated between the price-yield correlations presented in table 5.2 and harvested U.S. acres in 2000. The correlation was -0.41 and significant at the 95 percent confidence level. Thus, everything else constant, the correlation between average annual U.S. yield and price becomes more negative as total U.S. acres increase, making it difficult to determine the impact of the size of the U.S. production area on the variability of per-acre income.

WHAT IS THE RISK FACED BY U.S. FARMERS WHEN MAKING CROP PLANTING DECISIONS?

The negative correlation found between changes in average annual U.S. yield and average annual U.S. cash price for most crops raises a related issue: What is the correlation between change in yield and change in price between planting and harvest? Expected harvest price and expected yield are important inputs into planting decisions. Furthermore, within a rational expectation framework (Muth 1961; Peck 1975), income risk occurs when the realized income deviates from its expected value at the time the production decision was made.

This issue was investigated using data for corn, cotton, oats, soybeans, and all wheat. These crops accounted for 71 percent of the acreage planted to the principal crops[2] in 1999 (USDA 2000). Each of the five crops is traded on a futures market. Substantial evidence exists that U.S. farmers use futures prices in forming price expectations and allocating production resources (e.g., Gardner 1976; Hurt and Garcia 1982; Chavas, Pope, and Kao 1983; Eales et al. 1990). Thus, the price of the harvest futures contract at planting can be used as a proxy for expected price. The harvest futures contracts are September for oats and wheat, November for soybeans, and December for corn and cotton.

Since 1974, USDA has released its first estimate of supply and demand conditions for the forthcoming crop year during late April/early May.[3] The estimates appear in the *World Agricultural Supply and Demand Estimates* (*WASDE*), a widely followed benchmark for information concerning supply and demand. The late April/ early May release date falls before or during the planting of the crops being investigated, except for winter wheat.

The expected price at harvest during spring was assumed to be the first nonlimit close of the harvest futures after the release of the first new crop estimates in *WASDE*. Using the first nonlimit close allows the futures market to react to unanticipated information contained in a *WASDE* report (Garcia et al. 1997).

Expected U.S. average yield during spring was taken from the late April/early May *WASDE* if a yield was reported for the forthcoming new crop year. For years in which this *WASDE* release did not report an expected yield, an expected yield was estimated. Two estimation procedures were evaluated: (1) an average of yields for the previous five crop years minus the high yield and the low yield (hereafter referred to as the Olympic average) and (2) a one year out-of-sample forecast based on a linear trend line regression of yields for the previous twenty years. The two estimation procedures generated similar yield forecasts. Furthermore, the results generated by subsequent analyses that used the two yield forecasts were similar. The analyses using the Olympic average are presented.

U.S. average yield at harvest was the yield reported in the September *WASDE* for oats and wheat, and November *WASDE* for corn, cotton, and soybeans. The realized price at harvest was the first nonlimit close of the harvest futures after the release of these *WASDE* reports.

Regression analysis is used to investigate the relationship between the spring-to-harvest change in yield and the spring-to-harvest change in harvest futures price. The dependent variable was the natural logarithm (ln) of the harvest futures price at harvest minus the ln of the harvest futures price during spring. The independent variable was the ln of the yield at harvest minus the ln of expected yield during spring.

The estimated slope coefficients are reported in table 5.3 for both the period since 1973 and the period since 1985. All coefficients are negative and significantly different than zero at the 95 percent level of statistical confidence. This finding was expected because a decline in yield translates into a decline in supply, represented graphically as a leftward shift in the supply curve. From economic theory, a leftward shift in supply should cause price to increase, given a constant, positively sloped demand curve.

Table 5.3. Regression of Spring-to-Harvest Change in ln Harvest Futures Price against Spring-to Harvest Change in ln U.S. Average Yield, Selected Crops, United States, 1974–1999

Crop	1974–1999		1986–1999	
	Slope coefficient	R^2	Slope coefficient	R^2
Corn	-0.90	0.52	-1.03	0.67
Cotton	-0.59	0.17	-0.47[b]	0.12[b]
Oats	-1.67[a,b]	0.50[b]	-1.55[b]	0.47[b]
Soybeans	-1.08	0.49	-0.50	0.21
Wheat	-0.95	0.20	-1.65[d]	0.45[c]

Source: Original calculations using data from the U.S. Department of Agriculture's *World Agriculture Supply and Demand Estimates* (*WASDE*) and from the Chicago Board of Trade.

Note: Harvest futures contracts were September for oats and wheat, November for soybeans, and December for corn and cotton. Spring price was the first nonlimit futures close after the release of the first new crop estimates in *WASDE*. This occurred in late April/early May. Harvest price was the first on limit futures close after the September *WASDE* for oats and wheat and the November *WASDE* for corn, cotton, and soybeans.

[a] Coefficient is significantly different from –1 at the 5 percent level of significance.

[b] The Durbin-Watson statistic revealed positive autocorrelation. The Cochrane–Orcutt procedure was used to correct for this positive autocorrelation.

[c] White's correction was used for heteroskedasticity.

An unexpected finding was that, except for oats during the 1974–1999 period, the regression coefficients do not differ significantly from -1 using the 95 percent confidence test level. Since both variables were expressed as differences in ln values, the estimated slope coefficient is an elasticity. A slope coefficient greater than -1 (i.e., closer to zero) was expected because short-run supply and demand elasticities at the farm level are generally thought to be inelastic (Tweeten 1989, 20; Schnepf 2000).

An elasticity of -1 implies that changes in the two variables comprising the elasticity offset each other, resulting in a stable product of the two variables. Gross cash income per acre is the product of yield and price. Thus, for corn, cotton, oats, soybeans, and all wheat, the -1 elasticity found in this analysis implies that average gross cash income per acre for the United States on average does not change when average U.S. yield changes between spring and harvest. In other words, the risk that U.S. average yield changes between spring and harvest for these five crops on average does not translate into a gross income risk.

IMPLICATIONS FOR CROP INSURANCE

The finding of the previous section implies that if an individual farmer's yield changes by the same proportion as the U.S. average yield between spring and harvest and assuming no basis risk, then the farmer on average experiences no per-acre gross income risk. In other words, on average his/her gross income per acre will be the same at harvest as expected during the previous spring. By extension, an individual farmer experiences a per-acre gross income risk when his/her yield changes proportionally less from spring to harvest than does the U.S. average yield. To illustrate, if a farmer's harvested yield equals his/her yield expected during the spring, the farmer still incurs a per-acre gross cash income risk if U.S. average yield at harvest is 10 percent higher than U.S. expected yield during spring. A crop insurance program that protects against this risk would need to pay an indemnity based on the difference in the change of a farmer's yield and the change of U.S. average yield between spring and harvest. This program will be referred to as yield-difference insurance.

To provide an initial evaluation of the properties of this proposed new type of crop insurance, a comparative analysis was conducted against a program that insures against declines in an individual farmer's yield level relative to expected yield. This is the type of insurance provided by Multiple Peril Crop Insurance (MPCI). It will be referred to as yield-level insurance.

The data for this analysis are county-level yields of corn, cotton, oats, soybeans, and winter wheat from the 1972 through 1998 crop years. Following existing insurance products, winter wheat is divided into three separate programs, depending on the geographical area of the country. A county's expected yield was the Olympic average yield for the five previous years.

For yield-level insurance, an indemnity was paid if the county's yield for a year was less than the insured amount (coverage level times the county's expected yield). Three insurance coverage levels were evaluated: 50 percent, 75 percent, and 90 percent. Any shortfall in yield relative to the insured amount is valued at the average price of the harvest futures contract during the period used to establish price guarantees under the MPCI contracts in force for the 2000 crop year. To illustrate the yield-level insurance modeled in this chapter, assume a county's harvested corn yield is eighty bushels per acre and the county's Olympic average is one hundred bushels per acre. Thus, a 20 percent yield shortfall occurs relative to expected yield. An insurance indemnity would be collected at the 90 percent coverage level, but not at the 50 percent and 75 percent coverage levels. Amount of indemnity paid at the 90 percent coverage level is ten bushels per acre [(90 percent - 80 percent) * 100 bushels] times the established price guarantee.

For yield-difference insurance, an indemnity was paid if the percentage ratio of the county's harvested yield to the county's expected yield was lower than the percentage ratio of the U.S. harvested average yield to the U.S. expected yield by more than the coverage level. The coverage levels are ten percentage points, twenty-five percentage points, and fifty percentage points. These coverage levels are analogous to the 90 percent, 75 percent, and 50 percent coverage levels, respectively, for yield-level insurance. The indemnity equals the percentage point yield shortfall times the expected yield times the payout price. To illustrate the yield-difference insurance modeled in this chapter, assume the same county situation as in the previous illustration: a harvested yield that is 20 percent less than the county's expected yield. Also assume that national average U.S. corn yield was 10 percent above its Olympic average. The differential shortfall of the county yield relative to the U.S. average yield is thirty percentage points (-20 percent - +10 percent). Thus, an indemnity would be paid at the ten-percentage point and twenty-five-percentage point coverage levels, but not at the fifty-percentage point coverage level. Subtracting the ten-percentage point coverage level from the thirty-percentage point shortfall yields a twenty-percentage point payout rate. Multiplying this rate by one hundred bushels per acre yields a twenty-bushel per acre payout, which is multiplied by the established price guarantee to generate the indemnity payout.

Results of the analysis are reported in table 5.4. As expected, average annual indemnity increases as the insurance coverage level increases. Furthermore, because yield-difference insurance removes the marketwide or systemic effect of large-scale weather events, the standard deviation of annual indemnities was substantially less. In most cases, the standard deviation of annual indemnity under yield-difference insurance was at least 50 percent lower than the standard deviation of annual indemnity under yield-level insurance.[4] Compared with yield-level insurance, substantially fewer counties collected under yield-difference insurance during years with large-scale weather events. On the other hand, substantially more counties collected when U.S. average yield was above average. To illustrate using corn, 1,041 counties collected an average indemnity of $32 per acre under the 90 percent yield-level insurance during the severe 1988 drought, while only 426 counties collected an average indemnity of $8 per acre under the ten-percentage point yield-difference insurance. In contrast, during the record yield year of 1994, 48 counties collected an average indemnity of $1 per acre under the 90 percent yield-level insurance, while 494 counties collected an average indemnity of $9 per acre under the ten-percentage point yield-difference insurance.

Table 5.4. Average and Standard Deviation of Annual Insurance Payouts for Selected Crops, County Observation Unit, United States, 1972–1998

Crop	50% yield level[a]	50 yield difference[b]	75% yield level[a]	25 yield difference[b]	90% yield level[a]	10 yield difference[b]
	Average annual insurance payout ($ per acre per county)					
Corn	0.57	0.36	4.16	3.00	10.14	9.63
Cotton	0.83	0.72	6.48	6.76	17.37	18.84
Oats	0.18	0.04	1.35	0.85	3.62	3.17
Soybeans	0.12	0.07	1.93	1.65	6.10	6.32
CBOT winter wheat[c]	0.03	0.01	0.83	0.72	3.41	3.91
KCBOT winter wheat north[d]	0.16	0.16	1.54	1.91	4.73	6.17
KCBOT winter wheat south[e]	0.15	0.03	1.64	1.67	4.87	5.61
	Standard deviation of annual insurance payout ($ per acre per county)					
Corn	0.86	0.50	5.56	2.25	11.32	3.77
Cotton	1.50	0.82	6.98	3.44	14.20	5.22
Oats	0.68	0.08	2.90	0.68	5.16	1.44
Soybeans	0.26	0.09	2.38	0.93	5.87	1.90
CBOT winter	0.13	0.02	1.69	0.66	4.29	1.79

KCBOT winter wheat north[d]	0.43	0.29	1.85	1.39	3.56	2.31
KCBOT winter wheat south[e]	0.30	0.03	1.92	0.79	4.17	1.84

Source: Original calculations using data from the U.S. Department of Agriculture.

Note: Number of counties used as observations per crop: corn, 1,679; cotton, 299; oats, 517; soybeans, 1,131; CBOT winter wheat, 680; KCBOT winter wheat north, 252; and KCBOT winter wheat south, 480.

[a] For yield level insurance, a payout was made if the county's harvested yield for the year was less than the coverage percent (50, 75, 90) times the county's expected yield (Olympic average yield for the five previous years).

[b] For yield-difference insurance, an indemnity was paid if the percentage ratio of the county's harvested yield to the county's expected yield was lower than the percentage ratio of the US harvested average yield to the US expected yield by more than the coverage level. The coverage levels are ten percentage points, twenty-five percentage points, and fifty percentage points.

[c] Delaware, Georgia, Illinois, Indiana, Kentucky, Louisiana, Maryland, Michigan, North Carolina, New Jersey, New York, Ohio, Pennsylvania, Tennessee, Virginia, West Virginia, and Wisconsin.

[d] Iowa, Idaho, Minnesota, Montana, North Dakota, Nebraska, Oregon, South Dakota, Utah, Washington, and Wyoming.

[e] Arkansas, Arizona, Colorado, Kansas, Missouri, New Mexico, Oklahoma, and Texas.

Given the substantial difference in the standard deviation of annual insurance payout between yield-difference and yield-level insurance, it was surprising that the average indemnity of the two insurance programs was almost identical for a given crop. This held true even though average indemnity varied substantially by crop. Using the highest level of coverage, cotton had the highest average annual indemnity at $17 to $19 per county observation while oats and Chicago Board of Trade (CBOT) winter wheat had the lowest average annual indemnity of $3 to $4 per county observation. The similar level of performance of the two types of insurance also extends to the minimum level of gross cash income when insurance is purchased. Due to space constraints, this analysis is not presented here but is available from the author.

CONCLUSIONS AND IMPLICATIONS

Three separate but interrelated analyses comprise this chapter. Each addresses some aspect of crop income variability. In the first analysis, farm program crops were found to have higher yield and cash price variability than nonfarm program crops. However, a high negative correlation between yield and price for most program crops resulted in no statistical difference in the variability of gross cash income per acre for program and nonprogram crops. This finding suggests that it is not appropriate to single out the current farm program crops for special treatment when public policy seeks to mitigate gross farm income risk. Thus, to the extent that current farm income payments, such as AMTA and marketing loan payments, are based on a desire to reduce gross farm income risk, it would be appropriate to spread these payments across a broader array of crops than currently is done.

In the second analysis, the spring-to-harvest change in average U.S. price of corn, cotton, oats, soybeans, and wheat was found to have an elasticity of -1 with respect to the spring-to-harvest change in U.S. average yield. Thus, changes in U.S. average yield between spring and harvest do not on average translate into a gross cash income risk. Stated alternatively, for these crops, changes in national yield between spring and harvest do not on average create a systematic or marketwide risk in gross income per acre.

Before proceeding to a discussion of the implications that a unitary price-yield elasticity has for crop insurance, it is worth noting that this finding also raises questions about the ability of countercyclical price policy, such as marketing loan rates and target prices, to stabilize income. Specifically, stabilizing on price rather than on receipts or net income increases instability in years when a systemic yield-reducing weather event occurs and price is below the price support level when planting decisions are made.

This unitary elasticity finding implies that if an individual farmer's yield changes by the same proportion as the U.S. average yield between planting and harvest and assuming no basis risk, then the farmer on average experiences no gross income risk. On average, his/her gross income will be the same at harvest as expected during the previous spring. By extension, an individual farmer experiences a gross income risk when his/her yield changes proportionally less from spring to harvest than does the U.S. average yield. To illustrate, even though a farmer's expected and realized yield are the same, he/she still incurs a gross income risk if U.S. harvested yield turns out to be 10 percent higher than U.S. expected yield at planting time.

Based on this finding, a new crop insurance program is proposed, specifically one that pays an indemnity based on the difference in the change of a farmer's yield and the change of U.S. average yield between planting and harvest. This yield-difference insurance was operationized in this study as: pay an indemnity whenever the percentage ratio of harvested yield to expected yield was lower than the percentage ratio of U.S. harvested average yield to U.S. expected yield by more than the coverage level. The coverage levels were ten percentage points, twenty-five percentage points, and fifty percentage points. These coverage levels are analogous to the 90 percent, 75 percent, and 50 percent coverage levels, respectively, for yield-level insurance (i.e., the concept underlying Multiple Peril Crop Insurance). The indemnity equals the percentage point yield-difference shortfall times the expected yield times the payout price.

Using county-level data from 1972 through 1998, the yield-difference insurance program resulted in essentially the same annual average indemnity payout as an insurance program based on yield levels. However, the number and level of payouts differ dramatically by year. Compared with yield-level insurance, yield-difference insurance pays lower indemnities in years when a market-level yield event, such as a drought, occurs but pays higher indemnities in years when national yields are average to above average. Thus, in most of the scenarios examined, year-to-year variation in indemnity payment under yield-difference insurance was at least 50 percent less than the year-to-year variation in indemnity payment under yield-level insurance.

Research on crop insurance has focused on moral hazard and adverse selection as reasons for the failure of private sector insurance (Knight and Coble 1997). While these problems, which are caused by inadequate information, undoubtedly exist for crop insurance, they also exist for all other types of insurance. Thus, the question is whether crop insurance offered by the private market would be more susceptible to these two common insurance problems than other types of insurance offered by the private market. This author is not aware of any evidence to support an affirmative answer. Furthermore, both moral hazard and adverse selection can be dealt with in crop insurance by using a moving average of past yields over a reasonably short period of time to represent expected yield. If a farmer does "farm" the insurance program one year, chances of collecting in a subsequent year are reduced.

On the other hand, inclusion of a systemic risk is a reason for failure of private market insurance (Goodwin 2001; Mahul 2001; Miranda and Glauber 1997). A private insurance company can easily go bankrupt when the systemic risk occurs and, thus, many policyholders collect. In particular, yield-level insurance includes the systemic risk associated with a national drought. The solution is not, as the United States currently does, to support subsidized public insurance, especially when the systemic risk does not carry an associated income risk. Instead, the response should be to encourage the development of a private market insurance contract that removes the systemic risk. Yield-difference crop insurance appears to provide such an option, thus allowing the federal government to eliminate its subsidies to crop insurance without destroying the insurance industry.

The specific yield-difference insurance investigated in this analysis could be implemented for farm program crops using individual farm-level yield data available from the Farm Service Agency. Whether the conclusions obtained from the county-level data used in this study hold at the individual farm level is an issue that needs investigation. Variability is likely to be greater at the individual farm level. Another issue is whether the empirical findings for corn, cotton, oats, soybeans, and wheat hold for other crops. The significant negative correlation found between total U.S. harvested acres and the correlation between average annual U.S. yield and price suggests it may not. Nevertheless, this analysis suggests that for the five large-acreage crops examined in this chapter, a yield-difference insurance program has the potential to substitute the private market for current government subsidies.

NOTES

1. Data on annual average U.S. cash price existed for thirteen other nonfarm program crops. They were agaricus mushrooms, almonds (California), apples, avocados (California), English walnuts (California), grapefruit, grapes, hazelnuts (Oregon), nectarines (California), oranges, peaches, pears, and sweet cherries. The null hypothesis of equal price variability with the nine farm program crops could not be rejected, as the confidence level was only 40 percent.

2. Principal crops include almost all crops except fruits and vegetables (USDA 2000, 5).

3. For oats, the first year that *WASDE* reported supply and demand numbers for the forthcoming crop year during late April/early May was 1976.

4. Given the negative correlation between changes in average U.S. yield and price between spring and harvest, how does the year-to-year variation in average indemnity payout of revenue insurance compare with yield-level insurance? To investigate, a gross revenue insurance program was modeled. Expected gross revenue was set equal to a county's Olympic average of per-acre gross revenue. In general, standard deviation of the yearly payout was greater under the gross income insurance program. The greater variation in annual payout that resulted from insuring

against demand and nonyield supply factors more than offset the reduction in the annual variation of indemnity payouts that result from the negative correlation between changes in national price and national yields. Details of this analysis are available from the author.

ACKNOWLEDGMENTS
The author thanks Luther Tweeten and Roy Black for their helpful comments on an earlier version of this chapter. The author also thanks Matthew Pullins, Xudong Feng, Sharon Baxter, and Tim Cordonnier for their assistance in collecting and analyzing the data.

REFERENCES
Chavas, J., R. Pope, and R. Kao. "An Analysis of the Role of Futures Prices, Cash Prices, and Government Programs in Acreage Response." *Western Journal of Agricultural Economics* 8 (July 1983): 27–33.

Dixon, W. J., and F. J. Massey. *Introduction to Statistical Analysis.* 4th ed. New York: McGraw-Hill, Inc. 1983.

Eales, J. S., B. K. Engel, R. J. Hauser, and S. R. Thompson. "Grain Price Expectations of Illinois Farmers and Grain Merchandisers." *American Journal of Agricultural Economics* 72 (August 1990): 701–8.

Garcia, P., S. H. Irwin, R. M. Leuthold, and L. Yang. "The Value of Public Information in Commodity Futures Markets." *Journal of Economic Behavior and Organization* 32 (April 1997): 559–70.

Gardner, B. L. "Futures Prices in Supply Analysis." *American Journal of Agricultural Economics* 58 (February 1976): 81–84.

———. "Changing Economic Perspectives on the Farm Problem." *Journal of Economic Literature* 30 (March 1992): 62–101.

Glasser, L. K. *Provisions of the Food Security Act of 1985.* Agriculture Information Bulletin No. 498. Washington, DC: National Economics Division, Economic Research Service, USDA. April 1986.

Goodwin, B. K. "Problems with Market Insurance in Agriculture." *American Journal of Agricultural Economics* 83(August 2001): 643–49.

Harwood, J., R. Heifner, K. Coble, J. Perry, and A. Somwaru. *Managing Risks in Farming: Concepts, Research, and Analysis.* Agricultural Economic Report No. 774. Washington, DC: Economic Research Service, Market and Trade Economics Division and Resource Economics Division, USDA. March 1999.

Hurt, C. A., and P. Garcia. "The Impact of Price Risk on Sow Farrowings, 1967–78." *American Journal of Agricultural Economics* 64 (August 1982): 565–68.

Knight, T. O., and K. H. Coble. "Survey of U.S. Multiple Peril Crop Insurance Literature Since 1980." *Review of Agricultural Economics* 19 (Spring/Summer 1997): 128–56.

Mahul, O. "Managing Catastrophic Risk Through Insurance and Securitization." *American Journal of Agricultural Economics* 83(August 2001): 656–61.

Miranda, M. J., and J. W. Glauber. "Systemic Risk, Reinsurance, and the Failure of Crop Insurance Markets." *American Journal of Agricultural Economics* 79 (February 1997): 206–15.

Muth, J. "Rational Expectations and the Theory of Price Movements." *Econometrica* 29 (April 1961): 315–35.

Peck, A. E. "Hedging and Income Stability: Concepts, Implications, and an Example." *American Journal of Agricultural Economics* 70 (March 1975): 410–19.

Schnepf, R. " Assessing Agricultural Commodity Price Variability." *Agricultural Outlook*, 8–13. Reprint titled, *Managing Farm Risk: Issues and Strategies*. Economic Research Service. February 2000, USDA.

Tweeten, L. *Farm Policy Analysis*. Boulder, CO: Westview Press. 1989.

———. "The Twelve Best Reasons for Commodity Programs: Why None Stand Scrutiny." *Choices* 10 (Second Quarter, 1995): 4–7, 43–44.

U.S. Department of Agriculture, *Acreage*. Cr Pr 2-5 (6-00)a. Washington, DC: National Agricultural Statistics Service, USDA, June 30, 2000.

Zulauf, C. R. "U.S. Farm and Food Policy: Evolution of a New Covenant." *Choices* 9 (Fourth Quarter 1994): 35–36.

6

Crop Insurance: Inherent Problems and Innovative Solutions

Shiva S. Makki

Crop insurance programs have taken the center stage in the U.S. farm policy arena following the 1996 farm bill, which eliminated deficiency payments and other support structure that protected farmers against uncertain production and markets. The federal crop insurance programs have grown rapidly since then and their use as risk management tools has increased, costing taxpayers millions of dollars. This chapter examines causes for the absence of private crop insurance and presents several innovative proposals to make insurance programs more efficient.

INTRODUCTION

Crop yield insurance has been a part of the government's farm program for over a half century, but was not considered to be a major farm risk management component until the 1990s.[1] The crop insurance program received a major impetus with the passage of the Crop Insurance Reform Act in 1994. The act paved the way for new insurance products that provide protection against both yield and price risks. Since the passage of the 1996 FAIR Act, which was designed to reduce the government's intervention in agricultural markets, the policy focus has shifted toward greater emphasis on crop insurance. The Agricultural Risk Protection Act of 2000, which allocated $8.2 billion for crop insurance over the next five years, will increase premium subsidies for all coverage levels in an attempt to bring more farmers into the program and lessen the need for ad hoc disaster assistance.[2] The Risk Protection Act also aims at popularizing higher coverage levels and revenue insurance products.

Despite widespread efforts to promote crop insurance, the program historically has been plagued by poor actuarial performance. It may be impossible to fully evaluate whether the markets for crop insurance work in light of the fact that government was always involved in providing crop insurance. This is, of course, true for any analysis of agricultural markets where government plays a significant role. At issue is whether government insurance exists because private insurance markets have failed, or whether the lack of private insurance markets are due to crowding out by government involvement.

Theoretical and empirical literature on crop insurance suggests two major reasons for the failure of crop insurance markets. First, systemic, nondiversifiable risk in crop yields poses a pervasive problem for the efficient functioning of crop insurance markets and, in particular, may pose more serious obstacles to the emergence of an independent private crop insurance market (Miranda and Glauber 1997). Second, the presence of asymmetric information, which can lead to adverse selection and moral hazard problems, is the major cause for inequities in rating and the failure of crop insurance markets (Goodwin and Smith 1995; Just, Calvin, and Quiggin 1999; Makki and Somwaru 2001a).

Given the inherent problems associated with insurance markets, and the changing policy and market environment, several crucial questions emerge concerning the role of crop insurance in farm risk management. Why have the private crop insurance markets failed in agriculture? Is the absence of private crop insurance a case of market failure or a case of government failure? What are the consequences of market or government failure on resource allocation in agriculture? What reform of the crop insurance program could achieve an expanding list of risk management objectives without discouraging the development of private insurance markets? Finally, what, if any, type of programs can efficiently and effectively provide the desired economic support to the farm sector without encouraging farm production above competitive market levels, especially in marginal, high-risk areas?

This chapter is organized as follows: The next section examines possible reasons offered for the absence of a market-based crop insurance programs in agriculture. The discussion also considers whether the poor actuarial performance of crop insurance is caused by market failures or policy/government failures. The third section presents a brief discussion of the consequences of subsidized federal crop insurance followed by a section presenting policies and programs to improve the efficiency of the crop insurance program and facilitate the allocation of resources in agriculture. The final section presents conclusions.

LACK OF MARKET-BASED PRIVATE INSURANCE IN AGRICULTURE: SOME POSSIBLE REASONS

The insurability of a loss by the private sector depends on two crucial conditions—the independence of losses across insured individuals and the symmetric information between insurer and the insured regarding the probability distribution of the underlying risk. Neither of the two conditions is satisfied in the crop insurance market. Hence, private insurance markets presumably fail in the absence of public subsidies. Of course, government may be no better than the private sector in dealing with these two conditions. Another possible explanation for the absence of private crop insurance markets may be that government intervention crowds out private insurers. The final reason may be that farmers are not very risk averse or are unwilling to pay actuarially sound premiums because they find their own less costly methods to protect against loss, including savings, futures markets, and off-farm incomes. This important issue, however, is beyond the scope of this chapter.

As is true with most arguments for intervention, government involvement is usually justified on the basis of market failure. Given such arguments, it is useful to review in greater depth the reasons offered for the failure of private crop insurance markets and implications of government intervention on private insurers.

Systemic Risk
Systemic risk refers to the likelihood of widespread losses across the insurance pool due to a high degree of correlation among individual losses. Crop losses are often driven by natural disasters such as drought or floods, which systematically affect a large number of farms over widespread areas. The correlation among individual losses by these systemic weather events raises the risk associated with holding a portfolio of individual crop insurance contract liabilities above what it would otherwise be if individual losses were independent of each other. For example, Miranda and Glauber (1997) estimate that a portfolio of geographically diversified crop insuranc contracts is as much as twenty times riskier than a portfolio of conventional health or automobile insurance contracts.

Systemic risk undermines a crop insurer's ability to diversify risks across farmers, crops, or even regions, and prevents the company from performing the essential function of an insurance intermediary: the pooling of risk across individuals. It also creates reinsurance problems for crop insurance. The cost of maintaining adequate reserves to cover catastrophic losses associated with widespread natural disasters could be prohibitive. In addition, reinsurance rates are often difficult to calculate for low probability and high loss events such as floods and droughts (Kunreuther 1977).

Asymmetric Information
Farmers choose an insurance product based on expected benefits derived from the product. It is often argued that farmers, who produce under uncertainty, know more about their own expected losses (benefits) than can be discerned by the insurer. Such asymmetry of information can give rise to the problem of adverse selection, with negative consequences for the efficient functioning of the crop insurance market. *Adverse selection* in insurance markets refers to a situation in which the insurance provider is unable to accurately assess the risk of loss and, therefore, fails to set premiums commensurate with risk (Makki and Somwaru 2001a; Miranda

1991; Goodwin and Smith 1995). Asymmetric information exists in agriculture because of differences in inherent farm risks arising from factors such as the farm's location characteristics and farmers' managerial abilities (Knight and Coble 1997).

Recent empirical evidence indicates that asymmetric information can lead to rating problems. Using individual insurance records for 1997, Makki and Somwaru (2001a) show that, for Iowa corn and soybean farmers, risk levels might not have been assessed appropriately. They show that, in the case of individual yield and revenue insurance products, the actual premium rates paid by high-risk farmers are lower than their competitive rates and vice versa, while in the case of area (county) yield insurance, the actual premium rates are comparable to the competitive rates across all risk types. They also show that the disparity between actual and competitive rates is higher at extreme risk levels, implying the underlying difficulty in assessing individual farmers' risks. Using nationwide farm-level data on corn and soybeans, Just et al. (1999) also conclude that asymmetric information is a major problem in crop insurance markets.

Another major problem associated with asymmetric information is that of *moral hazard*. It is a situation where information asymmetries caused by the unobservability of the actions by the insured results in them gaining more program benefits than they would otherwise (under full information). That is, moral hazard occurs when producers, after purchasing insurance, alter their production practices in a manner that increases their chances of collecting indemnities (Chambers 1989). Chambers shows that moral hazard decreases the insurability of a loss because the market is impaired by opportunistic actions of the insured.

The empirical evidence regarding moral hazard in crop insurance has been mixed. Several studies have attempted to link farmers' use of inputs to the purchase of crop insurance and the consequent expected indemnities. Some studies showed evidence for increased use of inputs (Vercammen and van Kooten 1994; Horowitz and Lichtenberg 1993), while others showed evidence for decreased use of inputs (Babcock and Hennessey 1996; Goodwin and Smith 1995; Quiggin, Karagiannis, and Stanton 1993).

Coble et al. (1997) investigated the effects of moral hazard on insurance indemnities to examine whether moral hazard affects the actuarial soundness of the multiple peril crop yield insurance. Using a panel data of Kansas wheat farms for the period 1986 through 1990, Coble et al. show that moral hazard is likely to be a significant cause of excess losses in the crop insurance program. Furthermore, they argue that the insured change their behavior only in years when losses appear imminent, and conclude that better monitoring in regions where substantial losses appear likely in a particular season or year could substantially reduce losses.

However, many of these studies of moral hazard in crop insurance markets are based on data prior to 1996, when commodity programs including deficiency payments, supply controls, and loan rates dominated American farm policy. It is quite possible that producers' decisions prior to 1996 were influenced by programs other than crop insurance. Therefore, one needs to be careful in relating farmers' behavior to the performance of crop insurance markets based on old data and a different policy structure.

Asymmetric information is not unique to agriculture. Automobile and health insurance markets deal with asymmetric information problems through using observable information in rating risk. Obtaining information is neither easy nor cheap, but private insurance markets in certain sectors have found ways to operate efficiently. However, in crop insurance markets there is little incentive for private insurance companies to address this important problem. I discuss this issue next.

Market Failure or Government Failure?

Asymmetric information is often blamed for the failure of private crop insurance markets (Makki and Somwaru 2001a; Goodwin and Smith 1995). However, whether the problems of adverse selection and moral hazard are intrinsic to crop insurance, or whether they are consequences of program design, is not clear. The fact that private crop insurance markets have not developed is not *prima facie* evidence of a market failure. Rather it could also be construed as evidence of government intervention crowding out private insurers, lack of risk aversion by producers, or ingenuity of producers in pursuing alternative risk management strategies. Even though one of the objectives of government intervention in the crop insurance is to facilitate the development of an independent private crop insurance market, in reality, it seems to have had the opposite effect. Next, I discuss some of the actions of government that may affect the efficient operation of the crop insurance programs and the development of private insurance markets in agriculture.

Assessment of Risk

The most critical issue for any effective insurance program is to be able to assess the risk appropriately and set premium rates commensurate with that risk. Under current crop insurance programs, premium rates are driven by a producer's average yield, where producers with higher average yields are assessed lower premium rates on the assumption that expected losses decrease as the expected yield increases. The variance of yield is not taken into consideration. Under the current rate structure, farmers with low expected yields have more incentive to purchase insurance because of their higher likelihood of collecting indemnities. It is, however, quite possible that some producers with higher average yields may have a higher likelihood of collecting indemnities. As Skees and Reed (1986) argued, it is possible that two farmers with different expected yields have similar yield distributions and vice versa. Both situations can lead to inappropriate insurance rates and adverse selection.

If premium rates are determined according to average risks, the insurer will be left with an adversely selected pool composed of more high-risk individuals for whom insurance is underpriced and fewer low-risk individuals for whom insurance is overpriced (Rothschild and Stiglitz 1976). Therefore, inaccurate premium rates can seriously jeopardize the viability of the insurance program. It is often the common practice in crop insurance that premium rates are adjusted for county and statewide losses to eliminate large differences in premium rates across counties. Such smoothing and loss spreading could induce adverse selection by distorting insurance rates.

Another shortcoming of current risk assessment approaches is that the rate-making procedure allows farmers to substitute the lowest yield record on the farm with the county average yield to obtain the average production history, commonly referred to as the APH yield. Since the coverage levels are tied to APH yields, this procedure of estimating premium rates understates the level of risk by artificially reducing the variance of yield and, therefore, the level of risk.

Reinsurance Arrangements
Another weakness with the current crop insurance program design is the reinsurance arrangements between private insurers and the government. Although crop insurance contracts are delivered by private insurance companies, the government plays a significant role in the design and implementation of the crop insurance program. The federal government subsidizes farmers' premiums and pays private insurance companies most of the administrative and loss adjustment costs. Since the 1980s, the Federal Crop Insurance Corporation (FCIC) has also been the primary reinsurer for crop insurers. FCIC establishes the terms and conditions for reinsurance. Under the current reinsurance agreement, companies can cede premiums to one of the three reinsurance funds—assigned risk, developmental, and commercial—with each fund offering different levels of risk sharing between the government and companies. Under the assigned risk fund, the company must retain 20 percent of the net book premium and associated liability for ultimate net losses. That is, companies cede up to 80 percent of risk to FCIC on policies in the assigned risk fund. The developmental fund allows more flexible risk sharing, where companies retain 35 to 50 percent of the risk. Companies retain the greatest share (up to 100 percent) of risks on policies in the commercial fund (Risk Management Agency 2000).

The major limitation with such an arrangement is that companies can place risky policies in the assigned risk fund backed by FCIC, while retaining low-risk policies in the commercial fund where companies can retain the majority of profits. In addition, since FCIC assumes most of the risk, companies have little incentive to monitor adverse selection and moral hazard problems at the farm level (Miranda and Glauber 1995).

Disaster Assistance

Crop insurance and disaster assistance are often viewed as competing programs. The purpose of disaster assistance is to compensate farmers who suffer catastrophic losses due to floods, droughts, fire, earthquake, or any other natural disaster. In recent years, disaster assistance has become institutionalized, providing a safety net to farmers whenever widespread losses occur due to natural disasters (Lee, Harwood, and Somwaru 1997). Disaster assistance is often blamed for low participation rates in the crop insurance program (Goodwin and Smith 1995; Miranda, Skees, and Hazell 2000). This is because the provision of free disaster aid may eliminate the need for individual farmers to purchase crop insurance. Even though there have been several attempts to make crop insurance a substitute program for disaster assistance, such attempts have not been successful. A reason is because disaster assistance brings high political rewards. Despite repeated commitments by Congress to replace disaster assistance with crop yield and revenue insurance, in 1999 and 2000 the government responded with disaster aid to farmers who experienced low incomes due to extreme weather conditions and low market prices. Such ad hoc disaster assistance payments not only reduce the incentive to participate but also affect the development of private risk insurance markets.

Political Nature of the Insurance Program

Yet another reason for the absence of private insurance markets is the political nature of the insurance program. Statutory restrictions limit how much a rate can be adjusted from year to year or for different risk types. Crop insurance, as a government program, cannot be discriminatory. Unlike private insurance in which actuarially sound premium rates are likely to be of paramount importance, crop insurance as a government policy instrument reflects goals and objectives of elected officials. Any attempt to make premium rates actuarially sound may be perceived as discriminatory against high-risk farmers (who could be family farmers, minority farmers, or marginal-land farmers). Insurers seem reluctant to vary premiums with many of the characteristics they can observe that might indicate the extent of risk. Nonetheless, adverse selection or moral hazard arising because an observable risk indicator is ignored does not constitute market failure.

UNINTENDED CONSEQUENCES OF GOVERNMENT INVOLVEMENT

Though the stated purpose of the U.S. crop insurance program is to provide economic stability to agriculture in an efficient manner, the current program seems to have had several unintended consequences. As a public program, crop insurance may never attain the economic efficiency of a market-based private insurance. However, it is important to know what those unintended consequences are and understand the short- versus long-run consequences on the efficiency of the program as well as on the efficiency of resource allocation. The unintended consequences of the federal crop insurance program may include high loss ratios, acreage response, changes in production practice, and increased land values.

Table 6.1. Loss Ratios for U.S. Crop Insurance,1981–1999

Selected states and crops	Loss Ratio		
	1981–1989	1990–1999	1981–1999
Selected states			
Arkansas	3.09	2.94	2.97
California	1.57	1.79	1.71
Florida	2.06	2.17	2.12
Georgia	2.36	2.88	2.68
Illinois	1.84	0.91	1.12
Indiana	1.70	1.40	1.48
Iowa	1.13	0.95	1.01
Kansas	2.06	1.44	1.62
Minnesota	1.21	1.47	1.40
Nebraska	1.20	1.08	1.11
North Carolina	1.63	2.89	2.40
North Dakota	2.18	2.07	2.16
Ohio	1.79	1.59	1.62
South Dakota	1.55	1.83	1.79
Texas	2.69	2.73	2.72
Washington	1.22	1.34	1.28
Wisconsin	1.40	1.54	1.48
United States	1.99	1.84	1.88
Crops			
Corn	1.36	1.28	1.33
Soybeans	2.28	1.28	1.65
Wheat	2.27	2.03	2.14
Cotton	2.05	2.67	2.55
Tobacco	1.61	3.17	2.41
Peanuts	1.97	2.77	2.49
Sorghum	2.30	2.02	2.13

Source: Participation data (Risk Management Agency 2000).
Loss ratios estimated by dividing indemnity by farmer paid premium.

Loss ratio is defined as indemnity paid out per dollar of premium collected from farmers. Table 6.1 presents the loss ratios for the 1981–99 period across different states and crops. Over the 1981–99 period, the average loss ratio for the nation was about 1.88, which implies that on average $1.88 is paid out in indemnities for every dollar collected in premiums. The loss ratios for selected states vary from a low of 1.01 for Iowa to 2.97 in Arkansas during the 1981–99 period. There also is significant variation in loss ratios across different crops. Corn, which is by far the most widely insured crop, has the lowest loss ratio at 1.33, while cotton has a loss ratio of 2.55.

Even though the loss ratio for the United States declined from 1.99 in 1981–89 to 1.84 in 1990–99, the loss ratio increased in several states (table 6.1). Similarly, loss ratios decreased for field grains and increased for cotton, tobacco, and peanuts between the two time periods. High loss ratios in the U.S. crop insurance markets are due partly to asymmetric information problems and partly to government regulations that restrict changes in premium rates from year to year or among contiguous counties. Private insurance may not be able to operate with such high loss ratios.

The loss ratios presented in table 6.1 imply that riskier crops and regions benefit more from crop insurance. This leads to the argument that the subsidized crop insurance program distorts production patterns and allocation of resources in agriculture. Skees (2000), for example, indicates that crop insurance subsidies coupled with disaster assistance payments encourage farmers to take more risk and affect land-use patterns. He estimates that every ten-percentage-point increase in participation in crop insurance brings an additional 5.9 million acres to crop production. The major limitation of his study is that the analysis fails to account for other commodity programs that dominated the study periods 1978–82 and 1988–92. In other words, the study attributes all changes in acreage to the subsidized crop insurance program, which became the predominant farm program only in the late 1990s.

Goodwin and Vandeveer (2000) indicate that the effects of increased premium subsidy levels on acreage response is fairly modest. For example, their results show that a 50 percent decrease in insurance premium rates results in only a 2 to 3 percent increase in planted acres to corn and soybeans. In another study, Young et al. (2000) also conclude that crop insurance premium subsidies generate small shifts in aggregate planting and that the impact on net farm returns is small. Using the POLYSIS-ERS simulation model, they show that the net effect of a $1.4 billion payout in annual crop insurance subsidies increases net farm income from crop production by about $1.2 billion annually, resulting in a loss of $0.2 billion to the society.

Even though the aggregate response seems to be moderate, there are concerns that the subsidized crop insurance program may cause cropping to be extended to environmentally fragile lands, including pasture and marginal areas. As a result, output will exceed economically efficient levels and suppress prices for all producers. In other words, such a response would increase both production and price risks to producers, undermining the very purpose of the program.

Like any other government commodity programs, subsidized crop insurance also has the potential to land values in the long run. That can increase debt service cost and the unit cost of production. High land values can create barriers to entry for new farmers—the very thing that government wants to prevent.

Given these inherent problems, one should still ask whether market-based insurance in agriculture works optimally. As argued elsewhere, at least the intention of government intervention is to make markets work for farmers. Assuming that the government will continue to be involved in the crop insurance markets, the question is how to establish a crop insurance program that is actuarially efficient. Also at issue is how to design a program that will permit maximum flexibility for producers and will not distort market incentives. In the next section, I discuss a few ideas aimed toward reforming crop insurance programs in that direction.

REFORMING CROP INSURANCE: SOME INNOVATIVE PROPOSALS

Although welfare assessment of public programs such as crop insurance is often difficult, one can at least conclude that there is no presumption for optimal equilibrium in the presence of government subsidies. That is, subsidized crop insurance programs are not justified on economic or equity grounds. Market failure arguments are rather tenuous for crop insurance because the long history of government involvement seems to have discouraged the development of independent private insurance markets. The evidence shows that the private sector does provide numerous risk management tools including forward contracting, futures, options, and hail insurance (Harwood et al. 1997, 1999). In addition, farmers use self-insurance, crop and geographic diversification, and spreading sales to mitigate production and market risks (Makki and Miranda, Forthcoming). Subsidized crop insurance undermines such private risk management strategies. Federal crop insurance needs reforms not only to reduce market inefficiencies and public costs, but also to prevent crowding out of private insurance markets.

While insured acres increased in the 1990s, less than one-third of farm producers participate in the crop insurance program (Makki and Somwaru 2001b). This is because the current government-assisted insurance program is mostly field-crop based. The program is not as widely available for specialty crops and noncrop farm enterprises. In addition, the availability of crop insurance products varies among geographical areas. If the policy intention is to provide an equitable support to all farmers, then the insurance program must include both crop and noncrop farming in all geographic areas.

The proposals to reform the U.S. crop insurance program revolve around three basic themes: actuarial soundness, market efficiency, and equity. For example, the proposal to include the coefficient of variation of yield (revenue) in assessing risk could improve the actuarial soundness of the crop insurance program. Area income insurance and weather index insurance have the potential to mitigate both adverse selection and moral hazard problems. I also discuss whole-farm insurance and tax-deferred savings accounts that have the potential to provide a broad safety net to all rural households including farmers and nonfarmers who depend on agriculture for their livelihoods. These proposals along with a publicly supported area reinsurance could facilitate the development of independent private crop insurance markets.

Accounting for Variance of Yield in Assessing Risk

Setting insurance rates to reflect the risks of different farmers is necessary to reduce the potential for adverse selection. Under the current system, individual risk is represented by the yield span ratio (ratio of farm expected yield to the county average yield). If the expected yield is less than the county average yield, then such a farm is considered as high-risk and vice versa (Goodwin 1994; Makki and Somwaru 2001a; Skees and Reed 1986). This construction is based on the dubious assumption that expected losses decrease as the expected yield increases. As indicated earlier, this method of assessing risk can lead to inappropriate rates and adverse selection. Explicit accounting for both the mean and the variance of yield in assessing risk could reduce such ambiguities. Also, the assessment of risk based on individual yield or revenue volatility could allow farmers with different expected yields to obtain similar levels of protection. To attract new farmers and retain low-risk farmers, the present premium rates need to reflect farmers' risks appropriately.

Area Revenue Insurance

Studies show that area insurance plans suffer less from adverse selection and moral hazard problems because their guarantee levels and indemnity payments are based on county-level yields (Mahul 1999; Makki and Somwaru 2001a; Miranda 1991). For example, Group Risk Plan (GRP), which provides indemnity payments when the actual county average yield drops below a preset critical or guaranteed county-based level regardless of the yield of the individual farmer, is known to have little or no adverse selection or moral hazard problem. However, participation rates have been extremely low for GRP. The new plan, Group Risk Income Protection (GRIP), which adds a revenue component to GRP, may attract more participation because of its income protection component. Coverage is based on county-level income, calculated as the product of the county yield and the harvesttime futures market price. GRIP is available for corn and soybeans under a pilot program in selected counties in Iowa, Illinois, and Indiana where GRP is offered. Area or group insurance, however, can fail to provide protection to farmers suffering unique individual losses (Miranda 1991; Goodwin and Smith 1995).

Regional Weather Index Insurance
In their report to the World Bank, Miranda, Skees, and Hazell (2000) explore the option of insurance contracts based on a regional weather index. Even though the proposal is for developing countries, it does carry some useful implications for the U.S. crop insurance program. Insurance contracts based on area yield, rainfall, soil moisture, temperature, or a combination of these could effectively reduce the problems of asymmetric information and facilitate the development of an independent private crop insurance market. In principle, a contract could be written against specific events defined and recorded at the regional level.

Miranda et al. propose insurance contracts to be sold in standardized units with a fixed indemnity per contract if the insured event occurs. Buyers could purchase as many units of the insurance as they wish and it is not restricted to farmers only. That is, regional index insurance could be sold to anyone in the region whose income or livelihood is correlated with the insured event. These insurance contracts are relatively inexpensive and largely free of adverse selection and moral hazard problems because indemnity payments are not based on individual losses. Regional index insurance could effectively protect farmers against catastrophic events that affect the region but not the losses caused by idiosyncratic events specific to a farm. Therefore, this program may have all the problems associated with area insurance, particularly low participation.

Whole Farm Insurance
Since the introduction of revenue insurance, more areas have been added, and revenue insurance has come to cover a substantial portion of insured acreage in some areas. Revenue insurance products continue to expand, with several new products introduced on a pilot basis in selected counties in 1999. For example, Adjusted Gross Revenue (AGR) is a new revenue insurance product offering coverage on a whole-farm rather than on a crop-by-crop basis. AGR bases insurance coverage on revenue from agricultural commodities reported on Schedule F of the grower's federal income tax return. AGR protection is calculated by multiplying the approved gross income (based on Schedule F) and the percentage coverage level selected by the farmer. Loss payments are triggered when the adjusted gross income for the insured for the insured year falls below the guaranteed level. AGR provides an income safety net for multiple commodities in one insurance product (Dismukes 1999).

Whole-farm insurance potentially can protect gross farm income at low cost because it minimizes moral hazard (Mahul and Wright 2000). Such insurance is less likely to distort markets because whole farm insurance is less likely to influence farmers' planting and other management decisions compared to some other insurance plans. By using coverage levels based on gross sales receipts reported on Schedule F of income tax returns, the costs of administration and delivery of the programs could be greatly reduced.

Tax-Deferred Savings Accounts for Farmers
A program of tax-deferred savings accounts for farmers is another proposal to help farm operators manage their year-to-year income variability. This program would build a cash reserve to be available for risk management. By depositing income into accounts during years of high net farm income, farmers could build a fund to draw on during low-income years. Federal income taxes on eligible contributions could be deferred until withdrawal. Under the Farm and Ranch Risk Management (FARRM) account proposal, for example, farmers could take a federal income tax deduction for deposits of no more than 20 percent of eligible farm income. Deposits would be made into interest-bearing accounts at approved financial institutions, and interest earnings would be distributed and taxable to the farmer annually. Withdrawals from principal would be at the farmer's discretion (no price or income triggers for withdrawal), and taxable in the year withdrawn. Deposits could stay in the account for up to five years, with new amounts added on a first-in first-out basis. Deposits not withdrawn after five years would incur a 10 percentpenalty (see Monke and Durst [1999]for more details on FARRM accounts).

FARRM accounts are likely to cause little distortions in farmers' decisions on acreage allocations or on levels of inputs to apply and, therefore, are less market distorting. This program is simple and relatively less expensive (tax dollars and administrative resources) in providing an income safety net to farmers. Another advantage with FARRM accounts is that they could include all farmers, even those for whom insurance is not available currently such as livestock, dairy, poultry, and fishery farmers. Tax-deferred savings plans might be attractive to farmers who use self-insurance to manage risks. In addition, the program could be extended to other rural households whose income depends on agriculture. The major weaknesses of the FARRM program is that risks are not pooled and time is required to build up the account. That is, a producer who is poor, has participated in the program only a few years, or who incurs several low-revenue years in a row may receive little or no risk protection from the program.

Area Yield Reinsurance
One of the arguments for government reinsurance of agricultural risk markets is that no private insurance pool is likely to have sufficient reserves to cover the systemic risk. Thus, the government becomes the reinsurer of last resort. This argument may have some merit, though it should also be noted that the international

reinsurance market is both deep and wide. Reinsurance is likely available, though it may be the case that the premium necessary to obtain such coverage may result in crop insurance premium rates that exceed farmers' willingness to pay for insurance (Goodwin and Smith 1995).

Miranda and Glauber (1997) argue that systemic risk poses a serious obstacle to the emergence of an independent crop insurance market. According to their findings, some changes to the existing reinsurance agreements would be worth considering. They propose area-yield reinsurance contracts that would indemnify the insurer based on shortfalls in regional yields, offering private insurance companies protection against catastrophic losses arising from widespread natural disasters. In a simulated scenario, Miranda and Glauber (1997)demonstrate significant risk reduction to the insurer when area-yield reinsurance was available at the state level. More important, their results show that state area-yield reinsurance contracts would allow crop insurers to reduce their portfolio risk to levels comparable to those enjoyed by automobile and health insurers.

Area-yield reinsurance could be provided by the government at a low cost. County-yield estimates have been compiled by the U.S. government since 1963 for most major crops, while state-yield estimates are available back to the turn of the century for many crops and states. Area yields cannot be manipulated by crop insurers. Moreover, under an area-yield reinsurance program, insurance companies could be tasked with setting premium rates for crop insurance contracts sold to producers. This structural change would also restore a strong incentive for crop insurance companies to improve actuarial performance by closely monitoring the adverse selection and moral hazard problems that can arise between the insurance company and the individual farmer. Government can also encourage private insurers to hedge their risk with yield futures contracts sold at the Chicago Board of Trade (CBOT). Currently, the volume of futures contracts sold at CBOT is rather low, perhaps due to the relative attractiveness of subsidized government reinsurance.

Mandated Disaster Aid
The last proposal is to eliminate crop insurance and make disaster aid a mandated program responding to catastrophic events wherever and whenever they occur. This program, however, has the potential to be misused because the amount of assistance is often decided by elected officials on a year-to-year basis, and may not provide predictable risk protection to farmers. Payment triggers and amounts, however, need not be ad hoc. In the 1970s a mandatory disaster assistance program, with fixed trigger levels and per-unit-of-loss payment amounts, was in effect. Due to large outlays, however, the program was replaced in 1980 with a reformed crop insurance program where farmers shouldered a portion of the costs through payment of premiums.

CONCLUSIONS

The major purpose of this chapter is to examine some of the inherent problems associated with crop insurance markets and explore innovative ideas to reform the existing programs—some of which have already been initiated. As policymakers consider ideas to reform the federal crop insurance program and improve its actuarial performance, the issues and some of the innovative proposals made in this chapter will be important.

The current crop insurance program, which includes yield and revenue insurance products, is designed to mitigate yield and price risks in agriculture. The program, however, has been plagued by poor actuarial performance and low private market participation. Independent private markets have not developed for crop insurance (except hail insurance) because of nondiversifiable risks, asymmetric information problems, high administrative costs, low risk aversion by farmers, and self-insurance by farmers. As a result, to increase participation the government has been heavily involved in designing, implementing, and financially assisting programs. As a public program, crop insurance is subject to political influences and policies that are detrimental to the efficient functioning of the market.

In this chapter, I discuss several innovative ideas that may improve the actuarial soundness of the crop insurance program and facilitate efficient allocation of resources. Programs can be made more actuarially sound by aligning premiums with the level of risk each farmer represents. To do that, the assessment of risk must account for the variance of yield (revenue), which could reduce rating inequities and adverse selection problems in the market for yield (revenue) insurance products. Area revenue insurance and whole-farm insurance programs may also help alleviate some of the inherent problems associated with crop insurance markets. Tax-deferred savings accounts could help farmers build cash reserves to draw on during low-income years. Whole-farm insurance and tax-deferred savings account programs would cost less to taxpayers and could include other sectors within agriculture for which insurance is not available currently. These initiatives provide ways to improve the efficiency of the crop insurance program and the allocation of resources in the agricultural sector.

Sustained government involvement in agricultural risk insurance would crowd out private risk management tools (e.g., futures markets, self-insurance, diversification) and would adversely affect the development of independent private insurance markets. With some changes to the existing program design and limiting the government to provide only broad safety-net-type insurance products and area reinsurance, crop insurance programs could be made more efficient and less market distorting.

NOTES

1. The 1980 crop insurance act was the first attempt to make crop insurance a major farm program, but participation remained low through the 1980s and early 1990s.

2. The level of subsidy varies by the level of coverage chosen by the producer. The following table, reproduced from the Agricultural Risk Protection Act (ARPA) of 2000, compares the subsidy rates (in percentages) under the current and the proposed systems for alternative coverage levels:

Coverage Level (%)	50	55	60	65	70	75	80	85
Current System (%)	57	48	39	42	32	24	17	13
ARPA 2000 (%)	67	64	64	59	59	55	48	38

ACKNOWLEDGMENTS

I gratefully acknowledge helpful comments received from Joy Harwood, Agapi Somwaru, Luther Tweeten, Robert Dismukes, and Monte Vandeveer. The views expressed herein are those of the author and not those of the Economic Research Service or U.S. Department of Agriculture.

REFERENCES

Babcock, B. A., and D. Hennessy. "Input Demand Under Yield and Revenue Insurance." *American Journal of Agricultural Economics* 78(May 1996): 416–27.

Chambers, R. G. "Insurability and Moral Hazard in Agricultural Insurance Markets." *American Journal of Agricultural Economics* 71(August 1989): 604–16.

Coble, K. H., T. O. Knight, R. D. Pope, and J. R. Williams. "An Expected Indemnity Approach to the Measurement of Moral Hazard in Crop Insurance." *American Journal of Agricultural Economics* 79(February 1997): 216–26.

Dismukes, R. "Recent Developments in Crop and Revenue Insurance." *Agricultural Outlook*, May 1999, pp. 16–21.

Goodwin, B. K. "Premium Rate Determination in the Federal Crop Insurance Program: What Do Averages Have to Say About Risk?" *Journal of Agricultural and Resource Economics* 19(December 1994): 382–95.

Goodwin, B. K., and M. Vandeveer. "An Empirical Analysis of Acreage Distortions and Participation in the Federal Crop Insurance Program." Washington DC: U.S. Department of Agriculture, 2000.

Goodwin, B. K., and V. H. Smith. *The Economics of Crop Insurance and Disaster Aid*. Washington, DC: American Enterprise Institute, 1995.

Harwood, J., D. Heifner, K. Coble, and J. Perry. "Alternatives for Producer Risk Management." *Agricultural Outlook Forum*. Washington, DC: U.S. Department of Agriculture, February 1997.

Harwood, J., D. Heifner, K. Coble, J. Perry, and A. Somwaru. *Managing Risk in Farming: Concepts, Research, and Analysis*. Agricultural Economics Report Number 774. Washington, DC: Economic Research Service, U.S. Department of Agriculture, 1999.

Horowitz, J. K., and E. Lichtenberg. "Insurance, Moral Hazard, and Chemical Use in Agriculture." *American Journal of Agricultural Economics* 75(November 1993): 926–35.

Just, R. E., L. Calvin, and J. Quiggin. "Adverse Selection in Crop Insurance: Actuarial and Asymmetric Information Incentives." *American Journal of Agricultural Economics* 81(November 1999): 834–49.

Knight, T. O., and K. H. Coble. "Survey of U.S. Multiple Peril Crop Insurance Literature Since 1980." *Review of Agricultural Economics* 19(Spring/Summer 1997): 128–56.

Kunreuther, H. *Disaster Insurance Protection: Public Policy Lessons*. New York: John Wiley and Sons, 1977.

Lee, H., J. Harwood, and A. Somwaru. "Implications of Disaster Assistance Reform for Non-Insured Crops." *American Journal of Agricultural Economics* 79(May 1997): 419–29.

Mahul, O. "Optimum Area Yield Crop Insurance." *American Journal of Agricultural Economics* 81(February 1999): 75–82.

Mahul, O., and B. Wright. "Designing Optimal Crop Revenue Insurance." *Working Paper*, Department of Economics, INRA, France, 2000.

Makki, S. S., and A. Somwaru. "Asymmetric Information in the Market for Yield and Revenue Insurance Products." *ERS Technical Bulletin*. Washington, DC: Economic Research Service, U.S. Department of Agriculture, March 2001a.

———. "Farmers' Participation in Crop Insurance Markets: Creating the Right Incentives." *American Journal of Agricultural Economics* 83(August 2001b): 662–67.

Makki, S. S., and M. J. Miranda. *Self-Insurance and the Utility of Conventional Agricultural Risk Management*. Washington, DC: World Bank Report on Risk Management, Forthcoming.

Miranda, M. J. "Area-Yield Crop Insurance Reconsidered." *American Journal of Agricultural Economics* 73(May 1991): 233–42.

Miranda, M. J., and J. W. Glauber. "Systemic Risk, Reinsurance, and the Failure of Crop Insurance Markets." *American Journal of Agricultural Economics* 79(February 1997): 206–15.

Miranda, M. J., J. Skees, and P. Hazell. "Innovations in Agricultural and Natural Disaster Insurance for Developing Countries." *Working Paper.* Columbus: Ohio State University, June 2000.

Monke, J., and R. Durst. "Tax Deferred Savings Accounts for Farmers: A Potential Risk Management Tool." *Agricultural Outlook*, May 1999, pp. 22–24.

Quiggin, J., G. Karagiannis, and J. Stanton. "Crop Insurance and Crop Production: An Empirical Study of Moral Hazard and Adverse Selection." *Australian Journal of Agricultural Economics* 37(August 1993): 95–113.

Risk Management Agency. Website, *www.rma.usda.gov*, 2000.

Rothschild, M., and J. E. Stiglitz. "Equilibrium in Competitive Insurance Markets: An Essay on the Economics of Imperfect Information." *Quarterly Journal of Economics* 90(November 1976): 629–49.

Skees, J. R. "The Potential Influence of Risk Management Programs on Cropping Decisions." Lexington: Department of Agricultural Economics, University of Kentucky, 2000.

Skees, J. R., and M. R. Reed. "Rate Making for Farm-Level Crop Insurance: Implications for Adverse Selection." *American Journal of Agricultural Economics* 68(August 1986): 653–59.

Vercammen, J., and G. C. van Kooten. "Moral Hazard Cycles in Individual Coverage Crop Insurance." *American Journal of Agricultural Economics* 76(May 1994): 250–61.

Young, E. C., R. D. Schnepf, J. R. Skees, and W. W. Lin. "Production and Price Impacts of U.S. Crop Insurance Subsidies: Some Preliminary Results." Washington, DC: Economic Research Service, U.S. Department of Agriculture, 2000.

Impact of Agribusiness Market Power on Farmers

Suresh Persaud and Luther Tweeten

Economic studies indicate that agribusiness industries supplying farm inputs and processing and marketing farm output are imperfectly competitive. The economic efficiency gains attending economies of size attained through agribusiness concentration and firm growth have overshadowed any losses from agribusiness market power. Thus, increasing concentration of agribusiness has benefited the nation. Gains have gone to consumers and to farmers as consumers. An agribusiness of small competitive firms would not have created the innovative input technologies nor advertised to create the food demand that have raised American farm households to unprecedented levels of prosperity. Because of high development costs and economies of size in agribusiness, imperfect competition will become the rule rather than the exception. Farmers will need to adapt to (and can continue to prosper) operating in that environment that benefits the nation.

INTRODUCTION

Mergers and acquisitions (M & A) leading to fewer firms in an industry frequently encounter opposition based on the assumption that greater concentration reduces the economic welfare of society. A perennial concern voiced by farmers is that they are exploited by the market power wielded by imperfectly competitive and highly concentrated input supply and product marketing agribusiness firms. Often underlying the controversy surrounding M & A activity is a presumption that structural characteristics of an industry (e.g., firm size, numbers, concentration) are inexorably tied to industry conduct (e.g., predatory, exclusionary behavior) and performance (efficiency, innovation, marketing margin behavior).

The junior author of this chapter noted prices of Coke and Pepsi in twelve-ounce cans for several weeks while grocery shopping. Of interest is that the price per ounce of these soft drinks is typically one-fifth the price of carbonated water in similar size containers. It is well known that Coke and Pepsi dominate the soft drink industry, while the bottled water industry is less concentrated and characterized by easy entry and widely available technology. The very important lesson for the unwary from this admittedly small and unscientific sample is that market structure alone is an elusive and unreliable guide to industry performance.

A large number of studies cited in this chapter (see Heffernan 1999; Purcell 1990) and elsewhere document the structure of agribusiness industries. We target the relationship between agribusiness industry structure and performance in this chapter. We address agribusiness commodity marketing and processing before briefly examining performance of farm input supply industries.

This chapter assesses the current state of knowledge of the impacts of concentration of agribusiness industries into few firms buying farm products and selling inputs to farmers. The objective is to address the following questions: (1) Does increasing concentration in agribusiness reduce income of farmers and consumers? (2) Who gains and who loses from greater concentration?

Any attempt to relate structural characteristics of an industry to performance, whether through a single comprehensive study or through the integration of different case studies, must acknowledge the impact of industrial concentration on (1) market power and/or (2) economic efficiency. Greater industry concentration enhances market power and hence raises profits and marketing margins, other things equal. Greater industry concentration usually means fewer and larger firms, and hence reduces costs and marketing margins through realizing economies of size, other things equal. If the efficiency effects of greater concentration outweigh the market power effects, then mergers and acquisitions leading to a smaller number of larger firms is beneficial for society. The opposite is true if the market power effects dominate the efficiency gains. This chapter attempts to answer the question of which effect predominates.

MARKETS FOR FARM COMMODITIES

This section reviews studies of the economic impacts of concentration on livestock marketing and processing. As shown in figure 7.1, the share of cattle (hog) slaughter accounted for by the top four firms increased to approximately 68 percent (45 percent) in 1994 from about 30 percent (35 percent) in 1979. These increases in the four-firm concentration ratios (proportion of industry output from four biggest firms) have amplified fears of exploitative behavior by the marketing sector.

However, figure 7.2 shows that the real (CPI-adjusted) retail price of beef has fallen at a faster rate than the real (PPI-adjusted) farm price, especially from 1979 to 1985, leading to a decline in the real marketing margin. The real pork margin (fig. 7.3) also exhibits a moderate declining trend. The absence of an upward trend in the real marketing margins (despite the demand by consumers for greater processing) may indicate that economic efficiency dominates market power impacts of greater concentration. With these empirical observations in mind, we now turn to a selected review of in-depth analytical studies of the behavior of firms in the marketing sector.

Studies prior to 1990 gave conflicting results regarding the impact of meat packing industry concentration on farm prices. Ward (1988, 198) found two studies of the beef packing industry that found lower state or regional fed cattle prices associated with the high levels of industry concentration. Two other studies of the beef packing industry reviewed by Ward found no significant relationship between market share and prices paid for fed cattle in local or regional markets.

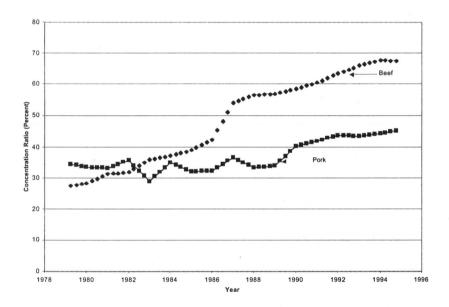

Fig. 7.1.   Concentration ratios for the beef and pork Sectors, 1979.1–1994.3. (Source: U.S. Department of Agriculture)

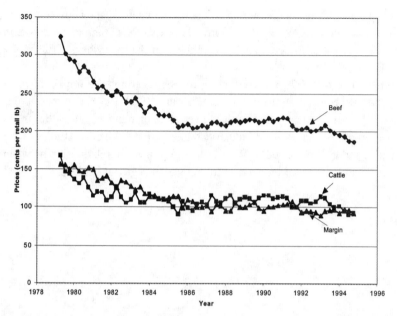

Fig. 7.2.   **Retail and farm prices for beef sector (retail weight basis),**
                1979.1–1994.3. (Source: U.S. Department of Agriculture)

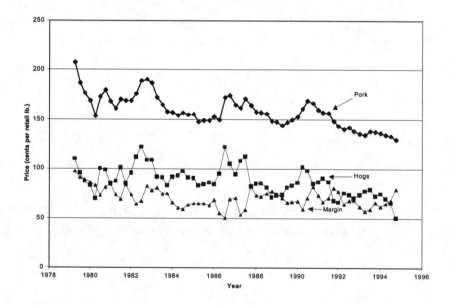

Fig. 7.3.   **Retail and farm prices for pork sector (retail weight basis),**
                1979.1–1994.3. (Source: U.S. Department of Agriculture)

In a 1986 study, Quail et al. (p. 55) estimated that slaughter cattle prices would have been 24 to 47 cents higher per hundredweight in four U.S. regions if the regions would have had lower beef-packer firm concentration ratios and hence more competition. Critics strenuously objected to these findings, however. This price increment is probably statistically insignificant and is dwarfed by cattle cycle price swings of thirty dollars or more per hundredweight. Ward and Bullock rejected the conceptual and statistical models used by Quail et al. Ward (1986) accused the authors of underestimating economies of size—less concentration could have increased packer costs and reduced beef prices even more. Bullock (1986)noted that transportation costs and whether regions are surplus or deficit in beef production relative to consumption were not adequately accounted for by Quail et al. These factors according to Bullock might better explain price differences attributed to concentration and could support the conclusion that the beef packing industry is competitive.

The Grain Inspection, Packers, and Stockyards Administration (GIPSA) of the U.S. Department of Agriculture (USDA) used weekly cost and revenue plant-level data from the forty-three largest steer and heifer slaughter plants to examine the effects of concentration on prices paid for cattle. Nonparametric tests indicated that the data used by this study were not consistent with profit maximizing behavior. Because the underlying profit maximizing assumptions of the model may not have been valid, the statistically significant findings of market power were considered questionable by GIPSA. The agency stated that "the analysis did not support any conclusions about the exercise of market power by beef packers" (GIPSA 1996).

The weekly plant-level data collected by GIPSA and the scholarship exemplified by its study were in themselves strong contributions to the literature, as was the finding that fed-cattle prices in all geographic regions are linked. The findings of strong regional price linkages and rapid price adjustments imply that markets behave as if slaughter hogs and cattle are bought and sold in a single national geographic market. These results provide "strong evidence that measurement and analyses of concentration in beef packing need to focus on relatively broad geographic markets." That is, prices in local areas are affected very little by differences in concentration in those regions due to strong, rapid regional price adjustments. Thus, the results of studies examining the margin-concentration relationship using national measures of concentration would apply to regional markets as well.

Whereas an oligopoly (oligopsony) describes an industry of a few firms, such that the activity of any one seller (buyer) affects other firms in the industry and may influence their behavior, monopoly (monopsony) describes an industry with one seller (buyer). In a competitive sector comprised of many firms, output for any one firm does not influence output of other firms. In an imperfectly competitive sector such as oligopoly (few sellers) or oligopsony (few buyers), however, output of one firm influences profits and hence the behavior and output of other firms in the sector. The Quail et al. (1982) study did not account for this impact on other firms. Many of the studies examined below test for oligopoly/oligopsony behavior using the method of conjectural variations.

A conjectural variation of firm 1 about firm 2 indicates firm 1's expectation of how firm 2 responds to firm 1's choice of output (Varian 1993). The conjectural variations method allows the degree of cooperation among marketing firms to be quantified (Tweeten 1992, ch. 8). For example, if conjectural variations are such that a given firm believes that a decrease in its output will be offset by increases in its rivals' output (leading to an unchanged level of industry output), then no firm has an incentive to cut back production with the objective of raising its prices and profit levels. If these conjectural variations are found to exist (even in a highly concentrated industry), then the behavior of firms is indistinguishable from perfect competition.

Schroeter (1988) extended Appelbaum's (1982) conjectural approach to allow the estimation of monopoly and monopsony power in the beef packing industry. Using annual data from 1951 to 1983, this system was estimated using full information maximum likelihood. The results indicate modest, statistically significant degrees of market power in the input and output markets. However, these distortions have been relatively stable since 1970, despite the post-1977 increase in industry concentration.

Azzam and Pagoulatos (1990) tested for oligopolistic and oligopsonistic behavior in the U.S. meat packing industry using annual aggregate time series data for beef, pork, sheep, and lamb over a sample period from 1959 to 1982. The model was estimated using iterative nonlinear three-stage least squares, correcting for first-order serial correlation. Likelihood ratio tests indicated a rejection of the null hypothesis of price-taking behavior in the input and output markets. Additionally, it was found that the degree of oligopsony power in input purchases exceeded the degree of market power in the output market.

It is important to note that the *structure* (e.g., size and number of firms) and the *conduct* (e.g., how one firm reacts to actions of another) do not reveal "bad" performance of the sector measured by lack of innovation, rates of return that are too high or low, or exploitation of consumers. For example, the above study by Azzam and Pagoulatos did not address whether increasing concentration in the meatpacking industry is harmful or beneficial to society. They did not quantify the possible efficiency gains resulting from a smaller number of larger firms. Findings of market power based on conjectural variations do not provide a complete picture of the impacts of concentration. Cost-savings due to economies of size may indeed outweigh market power distortions, implying the need to extend Azzam and Pagoulatos's study.

Additionally, Azzam and Pagoulatos (1990) combined the time series data for beef, pork, sheep, and lamb. This aggregate approach, however, does not allow conclusions about any particular industry. For example, the behavior of the sheep and lamb packing industry may differ from that of the beef packing sector.

Brester and Musick (1995) modeled the impact of the four-firm concentration ratio on the farm-to-wholesale and wholesale-to-retail price spreads for lamb, while also controlling for the impacts of uncertainty and changes in the prices of lamb pelts (a byproduct of lamb slaughter). Their results indicate that a 10 percent increase in the concentration ratio increased the farm-to-wholesale and wholesale-to-retail price spreads by 4.8 percent and 4.6 percent, respectively. These elasticities were statistically significant at the 5 percent level. Although Brester and Musick's approach does not constitute a test of market power (as in Schroeter or Azzam and Pagoulatos), it does capture the net effect on margins of increasing concentration in lamb packing. Consequently, Brester and Musick's (1995) findings are more appropriate for assessing whether industry consolidation is socially harmful or beneficial.

One limitation of their study is that Brester and Musick could not obtain a data set for the retail price of lamb (which is required to compute the margins). This time series was constructed using the retail price for one month and the CPI for lamb and organ meat. Brester and Musick's findings may be sensitive to their method of imputing the retail price of lamb, because the CPI for lamb and organ meat reflects the price behavior of other meats besides lamb. For example, if the retail price of organ meat was rising faster than the actual retail price of lamb, then the aggregate CPI for organ meat and lamb would tend to overstate any retail price increases in the latter (or understate the retail price decreases). Consequently, marketing margin behavior would be biased upward as well with rising lamb prices.

The limitation noted above applies to the retail price, and hence to the findings of a positive relationship between the wholesale-to-retail price spread and concentration. The unavailability of a data set for the retail price most likely does not nullify Brester and Musick's finding that higher concentration tends to raise the farm-to-wholesale price spread. However, even if their results hold, sheep and lamb expenditures are of relatively minor importance compared to beef and pork, which dominate red meat consumption.

Schroeter and Azzam (1991) examined the impact on marketing margins for the pork sector of output price uncertainty with a model using conjectural variations to account for imperfect competition among risk-averse marketing firms. The implied oligopoly and oligopsony distortions were generally small and insignificant (though not jointly), while the coefficient for output price risk was positive, larger in magnitude, and highly significant. The authors' findings were consistent with competitive behavior in the 1980s, though less so in the early 1970s, implying a trend towards more competitive conduct over time. Ignoring the impact of output price uncertainty would give the incorrect conclusion that output market power exists. This conclusion was evident when Schroeter and Azzam reestimated the model without the price risk term, as the model gave larger, statistically significant estimates for oligopoly components of the margin.

Wohlgenant and Haidacher (1989) developed an econometric model allowing variable factor proportions in their estimates of the impacts on retail food prices and farm prices of demand, supply, and marketing margins. Wohlgenant and Haidacher used reduced-form specifications for retail and farm prices, using extraneous (prior) estimates for retail demand elasticities to overcome identification problems. Using annual time series data from 1956 to 1983, Wohlgenant and Haidacher estimated the model for eight food commodities (beef and veal, pork, poultry, eggs, dairy products, processed fruits and vegetables, fresh fruit, fresh vegetables) using the joint generalized least squares technique.

With the exception of fresh fruit, neither the symmetry nor the constant-returns-to-scale restriction (given the former) could be rejected, consistent with a constant-returns-to-scale industry production function for a competitive food-marketing sector.

Muth and Wohlgenant (1999) tested for oligopoly and oligopsony power in the beef processing industry with a model allowing for variable proportions but not requiring estimation of output demand and factor supply schedules. Output supply and input demand functions were estimated using nonlinear three-stage least squares based on restricted and unrestricted specifications. When the coefficients of the model were assumed to be constant (variable) over time, the results were consistent with imperfect (perfect) competition. Additionally, constant returns for the industry were rejected.

Holloway (1991) extended Gardner's (1975) model of perfect competition to provide a conceptual framework that allowed for oligopoly with endogenous entry. The author's comparative statics were based on the assumptions of Cournot competition. The model was specified using Wohlgenant and Haidacher's (1989) approach and data. The null hypothesis of perfect competition could not be rejected for poultry, eggs, dairy, processed fruits and vegetables, fresh fruits, and fresh vegetables. Note that the sufficiency of Holloway's test of perfect competition implicitly assumes variable proportions. However, the author could not reject the null hypothesis that the elasticity of substitution between farm and nonfarm inputs is zero in the production of beef, veal, and pork. Although Holloway's test was (strictly speaking) not applicable to the beef and pork sectors, the author maintains that any departures from competition in these sectors have been relatively insignificant.

Matthews et al. (1999) tested for asymmetry in the beef and pork sectors and the impacts of concentration in the beef sector by estimating an endogenous switching model using monthly data from 1979 to 1996. The model allowed farm-to-wholesale-to-retail and retail-to-wholesale-to-farm price transmission effects to be computed. The authors found that beef and pork prices responded asymmetrically, making more rapid upward than downward adjustments. The impacts of greater packer concentration in the beef sector were examined using the Herfindahl-Hirshman Index (HHI)[2] as a measure of concentration. The results indicated that greater concentration leads to higher farm prices and lower farm-to-wholesale price spreads.

Using annual data from 1955–81, Thurman found, based on the Wu-Hausman test, that the real retail price of poultry meat is predetermined while quantity is not. Thurman maintained that these findings suggest either (1) competitive markets with constant returns to scale (CRS) technology and horizontal factor supplies or (2) market structures where prices are sluggish. If the latter is true, then demand shocks have significant but delayed impacts on price, implying that the former does not necessarily hold.

To test for significant effects of demand shocks, the author regressed the poultry price on lagged demand residuals, costs of production (feed, fuel, and labor costs), and the grain-to-meat conversion efficiency. Testing the null hypothesis of competitive markets amounts to testing the null that the lagged residuals have no impact on the poultry price. The results indicated a competitive market structure for poultry and constant returns to scale, consistent with the 1989 findings of Wohlgenant and Haidacher. Using annual data for 1970–92, Azzam (1997) found that the cost-efficiency effects of concentration are twice the market power effects in the U.S. beef packing industry. The model assumed fixed (variable) proportions between the farm (nonfarm) inputs and the output, and price-taking behavior in the acquisition of nonfarm inputs and in the output market, though not necessarily in the purchase of farm ingredients. A Diewert cost function was substituted into the first-order condition for profit maximization, eventually yielding an expression relating the agricultural input margin to the sum of a market power component and a marginal processing cost component. This equation was estimated along with an equation for the supply schedule of grain-fed cattle facing packers, using a nonlinear three-stage least squares statistical model.

In their empirical specification, concentration was represented by the four-firm concentration ratio. Besides cattle, the packing industry was assumed to use labor, energy, and transportation, while in the production of grain-fed cattle, the major inputs were corn, calves, and transportation services. In the preliminary steps of the empirical work, the test of the firms' conjectural variations yielded results consistent with the hypothesis that each firm expects a change in its farm input purchases to be offset by a change in its rivals', so that the farm input price is unchanged. Although the hypothesis that packers are price takers in the cattle market was rejected, the decomposition of the market power effects into its cost and efficiency components indicated that the latter component was twice that of the former.

However, Azzam's (1997) research showing that unit costs decline with greater concentration begs an important question: who benefits from a more concentrated meatpacking industry—farmers, consumers, or both? Azzam computed the margin as the difference between the wholesale and the farm prices. The inverse margin-concentration relationship quantified by Azzam may involve retail prices decreasing relative to farm prices (benefiting consumers), or farm prices rising relative to retail prices (benefiting farmers). A contribution of Persaud (2000) is that the impacts on the margins of concentration are decomposed into the impacts on retail and farm prices, in addition to separating the market power and cost efficiency effects of changing concentration.

To capture the incidence of the impacts of changing concentration (consumers versus farmers' benefits), Persaud developed a nonlinear structural model allowing price transmission effects to be quantified. In this study, price transmission effects are defined as (1) the impact on retail food prices of changes in farm prices (e.g., the extent to which a change in the farm price of beef cattle is passed through to the retail price of beef) and (2) the impact on farm prices of changes in retail prices.

Although research on price transmission effects is generally conducted as an issue separate from the impacts on margins of concentration, Persaud maintains that these issues are inseparable. For example, mergers and acquisitions affecting the degree of concentration in the beef packing industry may have different impacts on retail and farm prices. However, those changes in farm (retail) prices are transmitted to the retail (farm) level. In equilibrium, after the price transmission effects have occurred, the result may be a change in the marketing margins. If higher concentration leads to lower retail prices due to economic efficiency gains while farm prices are unaffected in equilibrium, then this would be an example of an inverse margin-concentration relationship.

Persaud (2000) treats the margin as an equilibrium entity computed as the difference between the retail and farm prices (both on a retail equivalent basis), taking into account the interaction between these two prices referred to as price transmission effects. Analyzing the farm and retail prices using a descriptive approach shows the close linkage between these time series. Figure 7.2, showing the farm and retail prices for the beef sector, suggested symmetry in that changes in retail (farm) prices tend to be mirrored by changes in farm (retail) prices.

Although descriptive analysis can be revealing, this approach does not allow us to assign a causal role to either farm prices or retail prices, or to both. Based on causality tests, previous studies have tended to find that retail price changes originate from movements in farm prices, and not vice versa. But, causality tests are flawed. For example, the restricted and unrestricted regressions for testing the direction of causality use a lagged structure of an arbitrary length.

Properly functioning markets allow retail price movements to feed back to the farm level, in addition to permitting the commonly documented pass-through of farm price movements to the retail level. The market system achieves vertical coordination through mutual (perhaps even simultaneous) farm and retail price adjustments. A maintained hypothesis in Persaud (2000) is that the close farm-retail price linkages shown in figures 7.2 and 7.3 result from bidirectional pass-through effects.

Rather than relying on flawed tests of causality to rule out the existence of the retail-to-farm pass-through effects, the current study develops a structural model flexible enough to quantify both price transmission effects so that their magnitudes, sign, and statistical significance can be examined. Although this approach is sensitive to the specification of the equations, causality tests are also sensitive to specification (Pindyck and Rubinfeld 1991).

Persaud assumed that (1) a fixed yield of the final homogenous product is obtained from each unit of livestock, that is, the fixed proportions assumption; (2) the retail market is perfectly competitive, though not necessarily the input market for the farm ingredient; and (3) the meat packing sector exhibits price-taking behavior in the market for the nonfarm inputs (labor and capital). The study assumed that the profit of a particular marketing firm (engaged in processing, transport, and storage activities) is its payoff and its strategy options are the possible output levels it can produce.

To compute the price transmission effects, the behavior of farmers and consumers must also be considered, in addition to the behavior of the marketing sector, discussed above. Thus, there are three equations (the optimizing condition from applying the profit maximizing condition, the farm supply, and the retail demand) and three endogenous variables (the retail price, farm price, and quantity).

Persaud's model is an improvement on Azzam (1997) because the former allows estimation of the price transmission effects with both farm and retail prices endogenous. Azzam's specification ignored the equilibrium relationship between retail and cattle prices. Properly functioning markets should allow retail price changes to be transmitted back to the wholesale and farm levels through the marketing sector. If cattle prices are indeed affected by retail price movements, the presence of the former on both sides of Azzam's (1997) optimizing equation will lead to simultaneous equation bias. Additionally, unlike Persaud's model, Azzam's model unrealistically assumed that the farm supply is static.

Persaud's analysis proceeds on four levels: (1) The Broyden-Fletcher-Goldfarb-Shanno nonlinear three-stage least squares algorithm was applied to the system, yielding estimates of the weighted average of the conjectural variations. (2) Because the conjectures were found to be negative 1.0, the model was reestimated imposing the competitive assumption. (3) The structural parameters from the second step were used to compute the impacts of concentration on the retail and farm prices. (4) The price impacts were used to calculate the effect on margins of changes in concentration. Steps 1–4 described above were performed for both short-run and long-run cases, estimated in accordance with the partial adjustment hypothesis.

The model outlined above was applied to the beef, pork, and turkey sectors. Persaud's empirical results indicating competitive conjectures have two major implications in this study: (1) the market power effects of increasing concentration are zero; and (2) the price transmission effects (which are a function of the conjectures) are unity, meaning that a one cent change in the retail (farm) price is fully reflected at the farm (retail) level. Additionally, the empirical results indicated significant cost-efficiency effects of greater concentration. Specifically, a statistically significant inverse relationship is found between retail prices and concentration, in contrast to the insignificant impacts of concentration on farm prices. The net effect is that margins decrease in response to increasing concentration in the meat packing/processing sector for beef, pork, and turkeys.

EXTENDING ANALYSIS TO OTHER FARM COMMODITIES AND INPUTS

The conduct and performance of farm input supply industries have been studied less than of food processing and marketing industries. A review of farm input industrial organization studies raises no red flags, however, and many of the industries that supply American farmers are world-renowned for innovation and competitiveness (see Barse 1990; Leibenluft 1981; Tweeten 1989, ch. 8).

Economic theory holds that if farm resources are mobile, as they are today (Tweeten 1989,114–23), then farms facing imperfectly competitive agribusinesses will still earn returns as high as resources would earn elsewhere but fewer resources will be devoted to farming than under competitive agribusiness conditions. However, in the real world a case can be made that oligopoly *increases* the level of farm output and commodity prices.

Farm input supply and product marketing firms in many instances are oligopolies (few sellers) or oligopsonies (few buyers). Although it is not possible to conclude a priori that oligopolistic (oligoponistic) firms will be more or less efficient or have lower or higher prices than would an atomized (numerous firms) market structure, oligopolistic (oligoponistic) industries are known to engage in extensive product innovation and advertising. Large outlays for food advertising may be one reason why the principal malnutrition problem in the United States today is chronically eating too much rather than too little. Although persistently eating too much is socially undesirable, it benefits farmers as producers. A more perfectly functioning market providing optimal nutrition would reduce domestic demand for food by an estimated 12 percent (Finke and Tweeten 1996). An atomized food industry with less product differentiation, innovation, and promotion, and controlled to serve the public interest likely would reduce the demand and price for farm output on average. Farm receipts would be less.

Overall aggregate profits of agribusiness firms that process and market farm products average less than five cents out of each dollar spent by consumers for food in supermarkets. Stock market investors are highly perceptive, and they price food processing and marketing firms as slow-growth, low-profit businesses (Koontz 2000, 3). In recent years many high-tech firms providing genetically modified new plant varieties that might be expected to be profit leaders have merged, consolidated, or ceased operations for lack of profits. Similarly, numerous full-line farm machinery makers have ceased operations or merged with other firms because of low or negative profits. Among other constraints on excess profits, domestic and foreign farm input suppliers are eager to enter and expand in the large American market if profits beckon. They constitute what industrial organization economists call a "credible threat," encouraging competitive behavior in an imperfectly competitive industry.

If agribusiness firms are wielding market power to accrue excessive profits, then cooperatives should prosper along with private agribusiness firms. The presence of producer cooperatives reduces chances for exploitation of farmers. Producer-owned cooperatives constitute approximately one-third of farm input and product markets and are prominent in nearly all major categories of farm outputs and inputs. A number of cooperatives have integrated vertically to operate in nearly all phases of farm input supply, contracting, product processing, and product marketing. They have not prospered in competition with private firms and, indeed, many would not survive without government help. Cooperatives have consolidated at a rapid pace in recent years to compete and survive. Some cooperatives have consolidated or in other ways grown to a size providing countervailing power against large private firms. In fact, the size and predatory conduct of some large cooperatives have drawn the attention of antitrust agencies in recent decades.

William Heffernan (1999, 12, 13) expresses concern for farmers over concentration of market power in clusters of agribusiness firms, and predicts that

> ...as the food chain clusters form, with major management decisions made by a small core of firm executives, there is little room left in the global food system for independent farmers. . . . If the number of [U.S. ?] farmers is reduced to about 25,000 in the next decade, there will be many farm families who will be involuntarily removed from their land.

Heffernan's presumption of only 25,000 farms remaining in a decade is premature. The number of farms fell from 1,925,700 in census year 1992 to 1,911,859 in 1997, the latest census year, or at a rate of only 0.14 percent per year (U.S. Department of Agriculture, 10). At this rate, 3,096 years instead of Heffernan's predicted 10 years will be required to reach 25,000 farms. The rate of loss of farms, in fact, has slowed as agribusiness concentration has grown.

Rather than blame productive, output-increasing, and labor-saving technologies for lower farm prices, populists have found scapegoats in agribusiness. Neil Harl (2001, 49) states that "without much doubt the greatest economic threat to farmers as independent entrepreneurs is the deadly combination of concentration [in agribusiness] and vertical integration." He worries that farmers "without meaningful competitive options" will become "serfs" (p. 45). Such rhetoric is unjustified. The largest impact on farm size and numbers in the future as in the past will come from *exemplary* performance of agribusiness in innovating new institutions and improving technologies rather than from "unfair" pricing due to market power.

We know of no empirical study indicating that anticompetitive agribusiness behavior is causing farms to get larger and fewer (see Barse 1990; Leibenluft 1981). In fact, farms are consolidating for the same reasons that agribusiness and indeed all industries are consolidating. These reasons include availability of large, expensive, indivisible technological and human capital that reduces costs per unit of output. Costs per unit are reduced, however, only if that "lumpy" capital is spread over many units of output.

Firms are consolidating also to gain advantages of task specialization. That is, costs are lower and efficiency higher by having specialists in the respective fields of management, information systems, marketing, finance, and blue-collar activities. An all-purpose family member in a family firm performing each of these tasks will not do so very efficiently. Financing expensive research and development, advertising in national media, coping with risks, meeting government regulations (e.g., food safety and quality, environment), and obtaining access to national venture capital markets are also reasons to lower unit costs by expanding size through firm growth or consolidation.

Millions and sometimes billions of dollars are required to research and develop each new sophisticated technology through bioengineering, computer assisted design, and other means. A firm must have command over massive resources to afford the human, material, and technological capital essential for success. To survive, it must be able to engage in enough ventures so that the few successful ventures will offset the many failures.

Overhead costs are incurred whether or not firms are operating. With high overhead relative to operating costs, a premium is placed on a reliable, steady supply of raw material to keep processing plants operating near capacity and overall costs per unit low. Firms can utilize production and marketing contracts to ensure product to process and thereby keep total unit processing costs low.

Firms also utilize production and marketing contracts rather than consolidation to achieve economies of size. Koontz (2000, 5) states, "I would argue that little contract production has emerged because of power. It has emerged to produce a product more consistent with low-cost processing systems and consumer wants." Consumers are demanding foods and services tailored to their wants, needs, and lifestyles. Agribusinesses must have farm products at the right time, place, quantity, quality, and price to process and meet consumers' demand. Contracts are a way to meet this demand for products at low transaction costs.

In short, farms consolidate to achieve economies of farm size arising mainly from technology rather than from pressures of (or response to) agribusiness concentration. Even in the highly unlikely case that a less-concentrated agribusiness sector would raise farm commodity prices, family farms would not necessarily fare better. Benefits of higher prices would be bid into land prices and higher land prices would make entry impossible for some potential farm operators. Higher commodity prices would enable operators to acquire more laborsaving machinery, displacing farms through consolidation.

As noted in chapter 1 by Tweeten, very high support prices in Japan and the European Union have not saved family farms. It follows that the largest "threat" to family farm number is not the perfidy but the productivity of the agribusiness sector in developing a new generation of inputs that function much in the way that tractors, combines, milking machines, hybrids, and pesticides interacted in the past to raise the output of each farmer.

The United States has antitrust laws adequate to address anticompetitive behavior—such laws need to be enforced. However, as noted by Tweeten and Flora (2001), greater individual transparency in production contracts and market pricing could help each buyer and seller be better informed relative to other participants in markets. Such information would reduce chances for exploitation.

To summarize, farming has been influenced far more by favorable performance of agribusiness bringing increased productivity than by unfavorable conduct bringing high farm input prices or low commodity prices through market power. The major source of decline in number and increase in size of family farms has been technology, especially adoption of laborsaving farm machinery. Such technology is the result of ingenuity and is not the result of monopoly structure or subpar performance of agribusiness. Scale-influencing technologies would have caused losses in commercial farm numbers even if farm prices would have been much higher or production contracts had never been devised. Productivity gains have brought massive national income benefits to society as a whole and hence to farmers in the long run because farm income per capita has trended toward national income per capita—as predicted by the new agricultural paradigm. A major source of the spectacular rise in farm household income since the 1930s has been laborsaving technology that has freed farmers to operate larger units and work off farms.

One cannot help but be struck by the stark contrast between vilification of agribusiness industries by populists and the absence of evidence justifying such vilification through numerous in-depth economic studies of agribusiness. There is no evidence that farm problems of annual and cyclical income instability and squeezing out of commercial family farms would be any different today if the agribusiness sector acted as if it were perfectly competitive. Evidence indicates that an increasingly concentrated structure of agribusiness has maintained high performance measured by innovation, rising economic efficiency, and falling real marketing margins—corrected for additional food processing that today's affluent, time-conscious consumers demand.

COMPETITION AND THE NEW PARADIGM

The new economic paradigm for agriculture recognizes the need to accept some imperfect competition in food and agricultural sectors for the sake of economic efficiency, technological progress, and rising living standards. In postindustrial agriculture where knowledge creation, dissemination, and information technologies dominate, fixed (overhead) costs for product research and development are massive. In contrast, once technologies are developed, marginal costs of widely supplying those technologies to customers are small. Operating competitively with price equal to marginal cost, a firm will not cover overhead costs and will fail financially.

Thus, for society to reap the rewards of the cornucopia of technological progress potentially forthcoming from the private sector, the public will need to accept pricing in excess of marginal cost and hence some imperfect competition. Some agribusiness firms unable to recover development costs in a competitive economy will seek and receive intellectual property rights protection in the form of patents, copyrights, and trademarks. In other cases, farmers will need to learn to live with natural monopolies (monopsonies) because only one production contractor, for example, will be able to operate in an area at low cost and be nationally and internationally competitive. Public reporting of contract terms will be important to monitor conduct and performance of such firms.

SUMMARY AND CONCLUSIONS

The studies reviewed in this chapter that merely test for competitive behavior do not address the more relevant issue of whether or not concentration reduces the economic welfare of society. Studies do indeed reveal that agribusiness markets are imperfectly competitive, but cost efficiencies resulting from greater concentration exceed losses from market power distortions, causing a net improvement in economic welfare.

Azzam (1997), for example, found that increasing concentration in the beef packing sector may be beneficial because the cost-efficiency effects of concentration are twice the market-power effects in the beef packing industry. That is, the net effect of fewer beef packing firms is lower marketing margins.

Persaud goes beyond Azzam's study to decompose margin changes into retail and/or farm price impacts. He found that marketing margins go down as agribusiness concentration goes up. But farmers do not gain or lose from changes in concentration, unlike consumers, who gain from greater concentration. These results support the competitive notion of consumer sovereignty—benefits of market efficiency gains are passed to consumers rather than to producers. Producers will reap normal profits if farm resources adjust. Economic theory predicts that farmers will operate further down on their output supply curve (lower commodity prices and quantities) than when facing a perfectly competitive agribusiness sector. In reality however, farmers are likely to receive higher commodity prices and sell more output with concentrated agribusiness structure due to heavy advertising and innovation by oligopolistic agribusiness firms.

On the other hand, oligopsony in farm input supply industries speeding innovation and laborsaving technologies has raised farm productivity and brought fewer and larger farms. The performance of agribusiness has been exemplary rather than predatory. The public may not wish to force (as if it could) agribusiness into a competitive structure that would advertise and innovate less because national income losses would be enormous.

NOTES

1. The opinions expressed in this chapter are those of the authors and do not necessarily reflect the views of the United States Department of Agriculture.

2. The HHI, unlike the four-firm concentration ratio, gives a higher value if production of output is concentrated in the biggest firm within the four largest firms. HHI is the sum of the share proportions squared.

REFERENCES

Appelbaum, E. "The Estimation of the Degree of Oligopoly Power." *Journal of Econometrics* 19 (1982): 287–99.

Azzam, A. "Measuring Market Power and Cost-Efficiency Effects of Industrial Concentration." *Journal of Industrial Organization* 45(1997): 377–86.

Azzam, A., and E. Pagoulatos. "Testing Oligopolistic and Oligopsonistic Behavior: An Application to the U.S. Meat-Packing Industry." *Journal of Agricultural Economics* 41(1990): 362–70.

Barse, J., ed. *Seven Farm Input Industries.* Agricultural Economic Report Number 635. Washington, DC: Economic Research Service, U.S. Department of Agriculture, September 1990.

Brester, G. W., and D. C. Musick. "The Effect of Market Concentration on Lamb Marketing Margins." *Journal of Agricultural and Applied Economics* 27(1995): 172–83.

Bullock, J. B. "Evaluation of NC 117 Working Paper WP-89. " (mimeo). Columbia: Department of Agricultural Economics, University of Missouri, 1986.

Finke, M., and L. Tweeten. "Economic Impact of Proper Diets on Farm and Marketing Resources." *Journal of Agribusiness* 12(1996): 201–7.

Gardner, B. L. "The Farm-Retail Price Spread in a Competitive Food Industry." *American Journal of Agricultural Economics* 57(1975): 399–409.

GIPSA (Grain Inspection, Packers, and Stockyards Administration). *Concentration in the Red Meat Packing Industry*. Washington, DC: GIPSA, U.S. Department of Agriculture, 1996.

Harl, N. "The Structural Transformation of the Agribusiness Sector." In J. Schnittker and N. Harl, eds., *Fixing the Farm Bill*, 45–55. Proceedings of Conference held in Washington, DC. Ames: Department of Economics, Iowa State University, 2001.

Heffernan, W. *Consolidation in the Food and Agriculture System*. Report to the National Farmers Union. Columbia: Department of Rural Sociology, University of Missouri, 1999.

Holloway, G. J. "The Farm-Retail Price Spread in an Imperfectly Competitive Food Industry." *American Journal of Agricultural Economics* 73(1991): 979–89.

Koontz, S. *Concentration, Competition, and Industry Structure in Agriculture*. Testimony at Agricultural Concentration and Competition Hearing, April 27, 2000. Washington, DC: Committee on Agriculture, Nutrition, and Forestry, U.S. Senate, 2000.

Leibenluft, R. *Competition in Farm Inputs: An Examination of Four Industries*. Washington, DC: Federal Trade Commission, February 1981.

Matthews, K. H., W. F. Hahn, K. E. Nelson, L. A. Duewer, and R. A. Gustafson. "U.S. Beef Industry: Cattle Cycles, Price Spreads, and Packer Concentration." TB-1874. Economic Research Service, U.S. Department of Agriculture, 1999.

Muth, M. K., and M. K. Wohlgenant. "A Test for Market Power Using Marginal Input and Output Prices With Application to the U.S. Beef Processing Industry." *American Journal of Agricultural Economics* 81(1999): 638–43.

Persaud, S. *Investigating Market Power and Asymmetries in the Retail-to-Food and Farm-to-Retail Price Transmission Effects*. PhD Dissertation. Columbus: Department of Agricultural, Environmental, and Development Economics, Ohio State University, 2000.

Pindyck, R. and D. Rubinfeld. *Economic Models and Economic Forecasts*. Third ed. New York: McGraw-Hill, 1991.

Purcell, Wayne, ed. *Structural Change in Livestock: Causes, Implications, Alternatives*. Blacksburg, VA: Research Institute on Livestock Pricing, 1990.

Quail, G., B. Marion, F. Geithman, and J. Marquardt. "The Impact of Packer-Buyer Concentration on Live Cattle Prices." NC 117 Working Paper WP-89. Madison: Department of Agricultural Economics, University of Wisconsin, May 1986.

Schroeter, J. "Estimating the Degree of Market Power in the Beef Packing Industry. " *Review of Economics and Statistics* 70 (1988): 158–162.

Schroeter, J., and A. Azzam. "Marketing Margins, Market Power, and Price Uncertainty." *American Journal of Agricultural Economics* 73 (1991): 990–99.

Thurman, W. N. "The Poultry Market: Demand Stability and Industry Structure." *American Journal of Agricultural Economics* 69 (1987): 30–37.

Tweeten, L. *Farm Policy Analysis.* Boulder, CO: Westview Press, 1989.

———. *Agricultural Trade.* Boulder, CO: Westview Press, 1992.

Tweeten, L., and C. Flora. *Vertical Coordination of Agriculture.* Task Force Report 137. Ames, IA: Council for Agricultural Science and Technology, March 2001.

U.S. Department of Agriculture. *Red Meat Yearbook.* Washington, DC: Economic Research Service, U.S. Department of Agriculture, March 1997.

Varian, H. R. *Intermediate Microeconomics: A Modern Approach.* New York: W.W. Norton, 1993.

Ward, C. "The Impact of Packer-Buyer Concentration on Live Cattle Prices: A Review and Comments." (Mimeo) Stillwater: Department of Agricultural Economics, Oklahoma State University, 1986.

———. *Meatpacking Competition and Pricing.* Blacksburg, VA: Research Institute on Livestock Pricing, 1988.

Wohlgenant, M. K., and R. C. Haidacher. "Retail to Farm Linkage for a Complete Demand System of Food Commodities." TB-1775. Economic Research Service. U.S. Department of Agriculture, 1989.

8

Do Farmers Receive Huge Rents
for Small Lobbying Efforts?

David S. Bullock and Jay S. Coggins

It is often argued that federal U.S. agricultural program benefits are an anachronism, an unnecessary throwback that today's farmers could and should be forced to do without. Yet the programs stubbornly remain, seemingly defying gravity as they transfer tens of billions of dollars from taxpayers to relatively wealthy farmers. This would be less of a mystery if it were not for the fact that political contributions by agricultural producers are small in comparison to the benefits they produce. Ideas from political economy cannot explain this fact. We argue that the dynamic nature of lobbying is crucial to understanding why agricultural policies continue and do so with very little lobbying expenditures. Several potential explanations are given, including the possibility that program benefits are capitalized into land values to such a degree that today's farmers, having paid an artificially high price for their land, realize very little in the way of actual benefits.

INTRODUCTION AND BRIEF HISTORICAL PERSPECTIVE

In response to the unprecedented economic hardships in rural areas, in 1933, as part of F. D. Roosevelt's New Deal, Congress passed legislation that for the first time provided federal agricultural price supports, mandated reductions in planted acreage to further increase prices, and transferred income from taxpayers to those in production agriculture. The effectiveness of these programs in easing the plight of farmers remains a subject of considerable debate. Standards of living on U.S. farms remained depressed until World War II, when once again European supplies were disrupted, world commodity prices soared, and U.S. farm incomes soared along with them.

After World War II ended in 1945, concerns mounted throughout the world that a great depression might return. While the agricultural economy remained healthy in the years immediately following the war, in 1948 the United States came to an agricultural policy crossroads. The crucial decision was whether the U.S. government would step out of agricultural income support programs originated in 1933 and let markets function and farmers survive as they might or might not in those markets, or whether the New Deal philosophy and policies would be continued into the second half of the century (Cochrane and Ryan 1981, 72–76). In the end, the New Dealers won out. Congress passed the Agricultural Act of 1948 and then a related act in 1949, providing high, rigid price supports and various types of acreage and production control in an attempt to avoid surplus production.

The programs' proponents and opponents alike perceived that the agricultural programs passed in 1948 and 1949 were in some sense "permanent" in nature— that they were meant to be, or at least that political realities required that they be, maintained into the distant future. Agricultural economist George Brandow wrote approvingly, "A permanent program is definitely intended. A withdrawal of government from the farm price field will not happen" (Cochrane and Ryan 1981, 75). Agricultural economist John D. Black, who opposed such high, rigid levels of price support, wrote, "It was evident even while the Act of 1948 was being jammed through the closing hours of the 80th Congress that the proponents of rigid 90 per cent of parity intended to extend . . . the provisions . . . in the following Congress. This they have already succeeded in doing for 1950–51, and it is already apparent that they fully intend to keep the liberalizations provided for 1951–52 from going into effect when the time comes" (Cochrane and Ryan 1981, 76). Brandow's and Black's perceptions have proved correct. More than fifty years later, if Congress fails to pass a farm bill approximately every five years, then under law, by default the U.S. government's agricultural policy reverts back to the generous provisions of the 1948 Agricultural Act.

While U.S. farm policy has changed incrementally over the past five decades, by many measures recent government involvement in agriculture is as great as it has ever been (see chapter 1 by Tweeten). Agricultural policy persists, despite huge changes in the socioeconomic characteristics of U.S. farmers and farm landowners.

Relatively well-off farmers receive a large share of government payments to agriculture (U.S. Department of Agriculture 1999). Nothing about this fact is inherently surprising; tales of inequitable government payments are commonplace. What may be surprising, however, is that it appears that farmers are receiving their huge "rents" from government while making relatively little effort to lobby government. Of course, the efforts and expenditures of farm lobbyists are respected, sometimes even feared in Washington.[1] But the little-considered fact remains that while tens of billions are received in transfers, only tens of millions are spent in lobbying. Table 8.1 shows contributions by agricultural interests to all political campaigns in each of the last six two-year election cycles. The data include contributions from individuals, contributions from political action committees, and soft-money contributions.

CAN THE CURRENT POLITICAL ECONOMY LITERATURE EXPLAIN THE HISTORY OF U.S. AGRICULTURAL POLICY?

Explanations from the New Political Economy Literature
The main purpose of this chapter is to employ various political economy models to explain the continuing generous support of the farm economy by government. "New Political Economy" models attempt to explain how small, politically well-organized groups can exploit larger, less-focused groups via the political process (Olson 1965; Stigler 1971; Becker 1983; Gardner 1987). Such models help to account for many aspects of current policy, but do not explain why an event like the Great Depression should trigger agricultural transfers in the first place. Other models refer to various types of altruism to explain why farmers are more likely to receive transfers from government when they are hard-pressed by economic markets (Becker 1985; Bullock 1994). These models may explain government response to the Great Depression, but also predict that transfers stop when economic conditions improve—a prediction not borne out in the U.S. history.

Explanations from the Rent-Dissipation Literature
Any attempt to understand the historical lobbying expenditures of U.S. agricultural groups must first refer to the rent-dissipation literature. Gordon Tullock (1967, 1980) was the founder of this literature. Tullock argued that potential recipients of a transfer would spend the full amount of that transfer in lobbying to win the "prize" of the transfer. Krueger (1974) followed Tullock and simply assumed that deadweight losses due to rent seeking would be equal to the rents conferred by the transfer program: in her paper, dissipation is assumed to be "complete."

Since Krueger's work, an extensive literature on rent dissipation has arisen. A principal topic of this discussion has been whether economic theory leads us to expect dissipation to be complete, as in Krueger, or not (Tullock 1980, 1984, 1985, 1987, 1989; Coggins 1995; Bullock and Rutström 2000). Overdissipation or underdissipation occur if rent-seeking "waste" is respectively greater than or less than the rent sought. A significant portion of this literature has used static, one-period game theory to predict that we might expect rents from government programs to be nearly completely dissipated, or even overdissipated by lobbying (Millner and Pratt 1989; Ellingsen 1991; Baye, Kovenock, and de Vries 1994; Fabella 1995; Anderson, Goeree, and Holt 1996). Several economic experiments confirmed the prevalence of overdissipation of rents when subjects play one-period games (Millner and Pratt 1989; Shogren and Baik 1991; Schotter and Weigelt 1992; Davis and Reilly 1998). Important to our discussion here is that none of the rent-dissipation

Table 8.1. Political Expenditures on U.S. Agriculture by Major Commodity Groups (Nominal Dollars)

Year	1990	1992	1994	1996	1998	2000
Crop production & basic processing	4,866,651	4,990,432	7,297,078	10,317,211	8,352,573	11,256,535
Fruits/vegetables	698,795	2,098,279	1,523,396	2,000,752	1,755,375	2,136,646
Sugar	1,863,632	2,319,768	2,555,173	3,356,273	2,752,299	3,349,268
Dairy	1,918,153	2,348,741	2,157,717	2,423,212	1,846,208	2,643,331
Livestock	1,573,072	2,511,410	2,306,650	3,107,593	2,344,913	4,569,676
Poultry & eggs	540,075	1,038,996	904,756	1,476,536	1,048,087	1,460,542
Total[a]	11,462,368	15,309,618	16,746,764	22,683,573	18,101,453	25,417,998

Source: http://www.opensecrets.org.

[a] Total exceeds sum of components shown due to minor unallocated expenditures.

literature explains radical underdissipation of rents that is, how U.S. farmers could currently be obtaining large rents with relatively small lobbying expenditures.

Why the Theoretical Literature Has Missed the Mark in Explaining U.S. Agricultural Policy

Our arguments above imply that agricultural political economists over the past several decades have missed the mark. Our models of political economy and our principal ideas about political economy do not explain many basic observations about the actual U.S. agricultural political economy. In particular, the basic point that government transfers to agriculture are not primarily generated through lobbying is a new and radical one given the current state of the agricultural political economy literature, not to mention the current state of popular political sentiment. Now at the beginning of the twenty-first century, one of the most prevalent topics in political discourse is that of the power of money and lobbying in persuading government. This type of discussion is not new, but the past generation has increased the focus on government as a negative force in society. Thirty years ago, economists considered the idea that government policy might run counter to the general public interest as an interesting puzzle to be pondered. Between the New Deal and Vietnam War, "political economy" models, such as they were, reflected a general public optimism about the ability and willingness of government to implement policy for the general social good.

But as popular optimism about government "programs" waned, economists increasingly began to question the assumption that the objective of government is to increase the general good. Skeptical ideas about government being motivated by the lobbying efforts of special interest groups appeared in the literature (c.f. Olson 1965; Stigler 1871; Becker 1983) to explain empirical observations of government programs that transfer money from the relatively poor to the relatively rich at the expense of the general economy. Social waste through misallocation of resources caused by agricultural policy is perhaps the central theme of Luther Tweeten's work. This work owes an intellectual debt to the writings of Hotelling and later Harberger (cf. 1954, 1964) on deadweight losses caused by transfer policies that distort market prices. Indeed, the ideas of Hotelling (1938), Harberger (1964), and in some sense even of many classical economists, about resource allocation and deadweight loss provide intellectual backing for some of the major arguments made by neoclassical economists against monopoly and for free trade. How disappointing to many of us, then, when empirical studies of the size of Harberger's triangles find them to be relatively small compared to the sizes of the transfers!

It is perhaps partly because of this disappointment in the size of Harberger triangles in empirical work that the ideas of Gordon Tullock (1967, 1980) about rent dissipation have been so well received in the political economy and social choice literature over the past two decades. Tullock's idea that rent dissipation is large is compelling to economists because if we are allowed to call any transfer an efficiency loss, then the harm done by distortionary policy is much larger than implied by Harberger triangles alone. This gives us an additional theoretical reason to continue to follow our inclination to stand against government intervention and for free markets. In summary, to a significant degree we agricultural economists are guilty of building political economy models, not to better understand the world that we carefully observe, but rather for the purpose of upholding our prior beliefs.

EXPLANATIONS OF SMALL LOBBYING EXPENDITURES IN AGRICULTURE

It is apparent that the lobbying models prevalent in the political economy literature, tending as they do to predict complete or near-complete rent dissipation, fail to explain observed outcomes in the agricultural political economy. How can this failure be accounted for? In this section we develop a series of possible explanations, each of which, when incorporated in a more formal and complete model, might provide at least part of the answer.

A common theme runs through the list of potential explanations that we describe. Each calls into question the idea, implicit in the literature on rent seeking, that the observed transfer is the prize at issue in any rent-seeking contest. It is our contention instead that there is a great deal of inertia in the political system that would provide sizable agricultural program benefits even if no lobbying occurred in a given year or perhaps for several years. The question "What would happened if no players lobbied?" is to our knowledge never examined in the extant literature. Yet the question, concerning what we term the "no-lobbying baseline," seems essential and is central to our analysis.

We argue that a large portion of the transfers to farmers under U.S. agricultural policy is not contested in any given year. That is, the "no-lobbying baseline" transfer to farmers is large. Policymakers consider the existence and continuance of this no-lobbying baseline transfer to be "off the table"—that is, that it is a subject that need not and will not be discussed seriously during negotiations over the content of a farm bill. The question that concerns us in the remainder of the chapter is why this might be.

Farm Fundamentalism

We consider five possible explanations, all of which could be incorporated in a formal model of political economy. The first, discussed elsewhere by Luther Tweeten (1979, 7–10; 1989, 73–81), is farm fundamentalism. Farm fundamentalists agree with the idea that "farmers are better citizens, have higher morals, and are more committed to traditional American values than are other people; indeed the nation's moral and social character depends on farmers" (Tweeten 1989, 73). It follows that in the minds of farm fundamentalists it is a good thing for the government to subsidize farming to assure the financial health of the agricultural sector. To the extent that sufficient numbers of nonfarmers and farmers agree with the basic tenets of farm fundamentalism, it is likely that it will take little if any political pressure to bring about government transfers to agriculture, especially in difficult economic times for the agricultural sector.

Dynamic Rent Seeking

Our second possible explanation, upon which the remaining three draw, is the idea that actual rent-seeking interactions between agriculture interests and their opponents play out over many years. This is in contrast to most of the models in the literature on political economy, in which the game is played only once. The dynamic nature of rent-seeking contests, played out over many decades, can have a profound effect in many ways.

Political Capital

Consider, for example, our third possible explanation for the low level of rent-seeking expenditures and of rent dissipation in agriculture: the idea of political capital. Here we define political capital as the political skills and connections that a pressure group develops with experience. Farmers and farm groups have been heavily involved in politics for several generations. They have political organizations in place and running. They have personal and political relationships with people in positions in power, as well as with people in the government bureaucracy (i.e., in the USDA). With that political capital in place, it may take relatively little active politicking to maintain transfers. It may well be that even with no explicit lobbying from any groups, the political capital that farm groups have created or inherited over several generations leads to large government transfers to farmers.

A subtle but potentially powerful form of political capital is an interest group's lobbying potential. That is, it may be that a group can rely on threats of lobbying to get what it wants, so that it can achieve its goals with only minimal actual lobbying expenditures. We might follow Theodore Roosevelt in calling this the "speak softly and carry a big stick" approach to politics. For example, potential political opposition to European Union farm subsidies might be kept at bay by the knowledge that French agricultural unions can obtain their own domestic agricultural supports by marshaling huge protests in which farmers use their tractors to tie up Paris automobile traffic for days. This idea of lobbying potential is essentially dynamic. It is because political games are played over and over that French farmers are able to achieve their desired result by wreaking havoc on Paris commuting only once. Then they have gained enough credibility so that the mere threat of a similar action creates the political pressure necessary to generate large transfers for years to come. A similar, if less dramatic, force is probably at work in the formation of U.S. agricultural policy.

Voting and Other Institutions

Our fourth possible explanation concerns the many ways in which institutions may help to limit rent dissipation. This is another factor that is inherently dynamic. In a path-breaking economic treatise on institutions, Aoki (2000, 1–10) describes institutions as *"self-sustaining system[s] of shared beliefs* about a salient way in which the game is repeatedly played" [italics in original]. We note that, in a sort of grand political-economic equilibrium, people develop institutions such as voting, and respect the outcomes of elections, because this is a much cheaper way to make public decisions than to have continuing all-out fights over government transfers. The repeated nature of voting makes losers accept losses gracefully. They know that not accepting losses, and instead carrying out riots or other costly lobbying efforts, might lead other groups to carry out similar disruptive efforts in the future. The Folk theorem is the well-known result that in a noncooperative repeated game, equilibrium outcomes can take on characteristics similar to those of a cooperative game (Friedman 1990, 111). Beliefs about the threat of how opponents may play in future rounds of the game tends to sustain cooperative-like play. Thus, the major purposes of public decision-making institutions may be exactly to prevent the magnitude of rent dissipation predicted in the Nash equilibria of Tullock-type one-shot games. These institutions can exist and be maintained precisely because of the repeated nature of the actual political game being played.

Capitalization of Benefits into Land Values

Finally, our fifth possible explanation for a high level of no-lobbying baseline transfers to agriculture concerns the idea that agricultural program benefits are capitalized into agricultural land values. The claim here is that the first generation of farmers, those who owned their land when agriculture programs were first put in place, occupied a privileged position historically. The first programs were created not in response to large lobbying expenditures by farmers but because of a consensus political concern that without government benefits the agricultural sector just might fail on a massive scale. Thus, the original recipients of federal program benefits owed their position not to lobbying outlays but to a national agricultural crisis.

Suppose now that it is widely perceived that, once established, the program will continue into the indefinite future. Intergenerational land sale then will occur somewhere in that future. Before they die, many first-generation farmers will sell their land to members of the second generation. Now consider the following question: How much will the original generation of farmers receive for their land, given that it is known that government payments in the future will be made to whoever owns the land? In the most extreme case, with perfect land markets, perfect foresight, and no uncertainty about the program's continuance, it can be shown that all of the benefits from the program, in the first and in all future periods, must accrue to the original generation of farmers. The original landowners, knowing the size of future payments, can obtain a price for their land that is sufficiently high to extract all of the program benefits. Any benefits that were left for the second generation would be bid away until the returns to owning land and the returns to other economic activities are equalized. Thus, competition in the land market will bid away any extranormal returns to land ownership, which means that landowners in the second and subsequent generations are no better off as a result of the government program than nonfarmers. In this extreme case, all of the current and future program benefits are capitalized into the price of land. Farmers in all ensuing generations, then, pay for the benefits they will receive throughout their careers in the form of increased land prices.

Let us turn now to the question of why the political forces at work in the situation we have described make it difficult or impossible to put an end to the program. To a very large degree after the first generation, government payments are used to pay farm mortgages. Should those government payments be taken away, many people would be unable to afford their mortgage payments, and would go bankrupt. Because land values would drop, rural banks would be forced to take large losses on their now-"bad" loans, and many of them would go out of business as well. It is not difficult to imagine the resultant serious financial crisis in rural areas, and how such a crisis might upset large parts of the wider economy. Therefore, it might be a common sentiment among almost all political pressure groups to avoid such a crisis. Then even if none lobbied, the common sentiment might be sensed and responded to by politicians. Even to the extent that a rural financial crisis did not affect the financial well-being of nonfarmers, it might be that altruism would reinforce the common sentiment to avoid the rural financial crisis.

The data appear to show that rent-seeking expenditures fall very far short of the program benefits received by farmers and landowners. We have outlined five possible explanations for this fact. The first four would lead one to conclude that farmers and landowners are, and have been for many decades, the beneficiaries of federal largesse that comes at almost no cost to themselves. The arguments related to farm fundamentalism, to the especially effective political capital possessed by farmers, to the power of voting institutions to minimize the need for lobbying activity—all suggest that agricultural programs that deliver tens of billions to farmers annually have no good economic justification.

If any one of these is the primary explanation for the low observed lobbying expenditures in agriculture, then a deep conundrum remains. Why is the case of agriculture special in the sense that program benefits come at such a low cost? Furthermore, the literature on political economy provides little help in understanding this question. We know of no models and no experimental evidence that can explain the facts at hand: a generous and continuing program that attracts almost no active lobbying opposition and that requires so little political expenditure by beneficiaries.

If, on the other hand, the capitalization explanation is correct, a very different view of the situation comes into focus. Current farmers, in this version of events, are not the recipients of enormous levels of benefits. Rather, they paid the price of their benefits when they purchased their artificially expensive land. The increasing prevalence of cash rental arrangements in farming complicates the story somewhat, but that is a discussion we will leave aside for now.

Our interest is in the size of rent-seeking expenditures and in the level of rent dissipation. What are the implications of our results for Tullock's dissipation idea? His point was that, under certain conditions, any rent created by a government policy can be expected to attract wasteful rent-seeking expenditures equal to the rent itself. In short, government-sponsored rent is waste. This is exactly true whenever dissipation (the ratio of lobbying expenditures to program benefits) is equal to 1, and waste exceeds the rent if dissipation is greater than 1.

The effect of capitalization of benefits into land values, though, changes everything. It suggests that the rents are not what they appear to be at all. In this version of our story, we believe that the rent is not total program payments in a given year but rather the incremental change in program payments due to lobbying. Thus, because benefits are smaller than they appear to be, dissipation is higher than it would be if rent-seeking expenditures were compared to total payments. On one hand, then, we have argued that the correct measure of dissipation brings us closer to Tullock's observation that a rent will attract wasteful rent seeking of an equal amount. On the other hand, one could say that agricultural rents, and therefore welfare losses due to agricultural policy, are small. If capitalization is fairly complete, perhaps there is some truth to the view that the billions of dollars of payments doled out under the various agricultural programs since the Great Depression have done nothing more than fulfill an obligation to the farmers of that original generation. According to this view, the benefits paid to farmers during the early years of the programs are sunk costs, and since that time farmers have enjoyed very little real benefits.

Even if this were true, though, it must still be said that artificially high land prices represent a significant distortion in the economy. In a sense, then, even in our simple and relatively extreme account, while actual rents are small the welfare losses due to agricultural programs may be substantial. This issue is deserving of future research.

EXTENSIONS

This chapter is not very technical. It contains an idea regarding the role of capitalization in determining how lobbying should be viewed. We have attempted to show why this is an important consideration for understanding both the ultimate fate of agricultural programs and the evaluation of their welfare effects.

The idea of capitalization of program benefits into land prices and the resulting inertia embodied in agricultural policy bears a striking resemblance to a relatively recent literature on the persistence of policy. Coate and Morris (1999), for example, develop a model that they claim explains why policies tend to be persistent. Their model is built up from a solid conceptual foundation and relies upon a game-theoretic framework to show how a long-lived policy can be an equilibrium outcome. It is, however, sufficiently abstract that it appears to shed little practical light on the specific details of agricultural policy and the lobbying activities surrounding that policy.

In future work we intend to develop a formal version of our simple model that will permit a similar investigation of the causes of persistence in U.S. agricultural policy. What is the political justification for the continued existence of a set of policies whose beneficiaries are now better off than the average citizen? To be sure, this is not unusual, but still the question seems to be important. If the benefits conferred upon recent generations of American landowner/farmers are significantly diluted due to elevated land prices, then perhaps this provides one more argument in favor of eliminating them. Yet at the same time, if the benefits are significantly diluted then, for precisely this reason, current farmers would be seriously hurt if the programs were eliminated.

The attempt in the 1996 farm bill to wean the sector from production-based benefits has largely been foiled. Generous deficiency payments for the crops that had received them have been traded for expanded disaster-relief and insurance benefits. A coherent model of the persistence of agricultural policy, of the role of capitalization, and of the effects of rent-seeking behavior on both could provide a useful addition to our understanding of farm programs.

CONCLUSIONS

As is often true, things in agricultural policy may not be as they seem. Farmers appear to receive large benefits from federal programs. We have argued that if these benefits are capitalized into land prices to a large degree, from the farmers' perspective the benefits may be more illusory than real. Likewise, rent dissipation in agriculture appears to be very small. We have argued that this may be incorrect as well, depending on the appropriate definition of rents. It is possible that, in a dynamic world with an intergenerational land market, farmers who were not active at the time the program was initiated benefit very little from the policy. Because the supply of land is inelastic, farmers in the first generation are able to extract future rents when they sell their land.

This has important implications for the political economy of agriculture. In general, the literature has not provided a satisfactory method for measuring rent-seeking expenditures and neither is rent itself easy to identify. Agricultural policy provides an excellent example of both difficulties. The struggle to measure the value of rents due to farm programs must address the question of how land values capture those rents, possibly requiring the continuation of a set of policies that are unable to deliver benefits commensurate to their cost.

NOTES

1. The following appeared in a story in the *St. Petersburg Times* (May 15, 1994): "Big Sugar for decades has thrived by playing the inside of the political process with an acumen even its critics admire. 'No one's beat them to date,' said Tom Hammer, Washington lobbyist for the Sweetener Users Association. 'When the battle is engaged, Big Sugar will bring to it what it always brings: a fierce single-mindedness, and a good deal of money.'"

REFERENCES

Anderson, S. P., J. K. Goeree, and C. A. Holt. "Rent Seeking with Bounded Rationality: an Analysis of the All-Pay Auction." Unpublished manuscript. Charlottesville: Department of Economics, University of Virginia, 1996.

Aoki, M. 2000. "Towards a Comparative Institutional Analysis." Unpublished manuscript. Stanford University Department of Economics. (Forthcoming from MIT Press.)

Baye, M. R., D. Kovenock, and C. G. De Vries. "The Solution to the Tullock Rent-Seeking Game when R > 2: Mixed-Strategy Equilibria and Mean Dissipation Rates." *Public Choice* 81(1994): 363–80.

Becker, G. S. "A Theory of Competition Among Pressure Groups for Political Influence." *Quarterly Journal of Economics* 58(1983): 371–400.

————. "Public Policies, Pressure Groups, and Deadweight Costs." *Journal o. Public Economics* 28(1985): 329–47.

Bullock, D. S. "The Countercyclicity of Government Transfers: A Political Pressure Group Approach." *Review of Agricultural Economics* 16(1994): 93–102.

Bullock, D. S., and E. E. Rutström. "The Size of the Prize: Testing Rent-Dissipation when Transfer Quantity Is Endogenous." Unpublished manuscript. Urbana: University of Illinois Department of Agricultural and Consumer Economics, 2000.

Coate, S., and S. Morris. "Policy Persistence." *American Economic Review* 89(1999): 1327–36.

Cochrane, W. W., and M. E. Ryan. *American Farm Policy.* Minneapolis: University of Minnesota Press, 1981.

Coggins, J. S. "Rent Dissipation and the Social Cost of Price Policy." *Economics and Politics* 7(1995): 147–66.

Davis, D. D., and R. J. Reilly. "Do Too Many Cooks Always Spoil the Stew?: An Experimental Analysis of Rent-Seeking and the Role of a Strategic Buyer." *Public Choice* 95(1998): 89–115.

Ellingsen, T. "Strategic Buyers and the Social Cost of Monopoly. *American Economic Review* 81(1991): 648–57.

Fabella, R. V. "The Social Cost of Rent-Seeking under Countervailing Opposition to Distortionary Transfers." *Journal of Public Economics* 57(1995): 235–47.

Friedman, J. W. *Game Theory with Applications to Economics.* Oxford: Oxford University Press, 1990.

Gardner, B. L. "Causes of U.S. Farm Commodity Programs." *Journal of Political Economy* 95(1987): 290–310.

Harberger, A. C. "Monopoly and Resource Allocation." *American Economic Review* 44(1954): 77–87.

———. "The Measurement of Waste." *American Economic Review* 54(1964): 58–76.

Hotelling, H. "The General Welfare in Relation to Problems of Taxation and of Railway and Utility Rates." *Econometrica* 6(1938): 242–69.

Krueger, A. O. "The Political Economy of the Rent-Seeking Society." *American Economic Review* 64(1974): 291–303.

Millner, E. L., and M. D. Prat. "An Experimental Investigation of Efficient Rent-Seeking." *Public Choice* 62(1989): 139–51.

Olson, M. *The Logic of Collective Action: Public Goods and the Theory of Groups.* Cambridge: Harvard University Press, 1965.

Schotter, A., and K. Weigelt. "Asymmetric Tournaments, Equal Opportunity Laws, and Affirmative Action: Some Experimental Results." *Quarterly Journal of Economics* 107(1992): 511–39.

Shogren, J. F., and K. H. Baik. "Reexamining Efficient Rent-Seeking in Laboratory Markets." *Public Choice* 69(1991): 69–79.

Stigler, G. J. "The Theory of Economic Regulation." *Bell Journal of Economics and Management Science* 2(1971): 3–21.

Tullock, G. "The Welfare Costs of Tariffs, Monopolies, and Theft. *Western Economic Journal* 5(1967): 224—232.

———. "Efficient Rent-Seeking." In J. M. Buchanan, R. D. Tollison, and G. Tullock, eds., *Toward a Theory of the Rent-Seeking Society.* College Station: Texas A&M University Press, 1980.

———. "Long-Run Equilibrium and Total Expenditures in Rent-Seeking: a Comment." *Public Choice* 43(1984): 95–97.

———. "Back to the Bog." *Public Choice* 46(1985): 259–63.

———. "Another Part of the Swamp." *Public Choice* 54 (1987): 83–84.

———. "Editorial Comment." *Public Choice* 62(1989): 153–54.

Tweeten, L. *Farm Policy Analysis.* Boulder, CO: Westview Press, 1989.

———. *Foundations of Farm Policy.* Lincoln, NE: University of Nebraska Press, 1979.

U.S. Department of Agriculture, Economic Research Service. *Agriculture Income and Finance*, AIS-72, September 1999.

9

Coalitions and Competitiveness: Why has the Sugar Program Been Resilient?

Charles B. Moss and Andrew Schmitz

The FAIR Act of 1996 appears to conform to the economic paradigm that traditional commodity programs, such as the sugar program, impose economic costs on society. However, some of these programs have survived despite several attempts to eliminate them. Many factors combine to explain this resilience. This study examines the role of coalitions and vertical integration in the policy debate over the U.S. sugar program. This debate can be characterized by the differing interests of the American Sugar Alliance and the Coalition for Sugar Reform. The effectiveness of the American Sugar Alliance is enhanced by vertical integration in the U.S. sugar industry. Coalition theory and new institutional economics are used to describe the pressure that each organization exerts on the policy process and the possibility of changes based on the coalition's structure.

INTRODUCTION

Professor Luther Tweeten, one of the leading scholars of U.S. agricultural policy, has been a proponent of free-market economics, arguing that government should play a limited role in agriculture (Tweeten 1970, 1989, 1992). He has testified many times in Washington, D.C. to essentially remove government from the broad area of agricultural income policy. His forthcoming chapter in *Agricultural Trade, Globalization, and the Environment* (Tweeten 2001) is very clear on his position as to why governments should reduce their involvement in agriculture. Tweeten has argued in many forums that decoupling farm programs is essentially impossible. He argues that most farm programs have an impact on agricultural production and hence distort domestic and international trade.

Despite arguments by Tweeten and others, however, U.S. agricultural policy is still very much alive. In fact, U.S. farm subsidies are at historic highs in spite of the fact that in the Freedom to Farm Act of 1996 (FAIR), the federal government promised to reduce its role in agriculture. In 2000, a subset of U.S. farm subsidies totaled $23.3 billion (production flexibility payments, conservation reserve programs, loan deficiency, and emergency assistance). The total is much higher when one includes crop insurance and water subsidies.

One of the most controversial farm programs is U.S. sugar policy. That sugar policy generates significant benefits for producers but imposes costs on U.S. consumers and users (GAO). Moves to reduce these costs through policy reform, however, have been unsuccessful (Knutson, Penn, and Flinchbaugh 1998):

> The 1996 farm bill was a test of the power of sugar. Decoupling support from production was applied to crops. The support program was phased out in dairy. Peanut quotas became transferable. But sugar lost nothing. In reality, however, the big gainers once again were HFCS [High Fructose Corn Syrup], ADM [Archer Daniels Midland] (a major HFCS processor), and corn producers. (p. 94)

Tweeten (1989, 1992) as well as others have been very critical of the U.S. sugar program. Many studies (see Schmitz and Christian 1993) show the costs and benefits of the U.S. sugar program. Producers clearly gain while consumers and industrial users lose. Many of the studies estimate that the net social (or deadweight) cost of the sugar program exceeds $1 billion annually. More recent studies find that the cost of the sugar program has fallen substantially. For example, the General Accounting Office study (GAO 2000) estimated the cost of the program to be $532 million for 1998 when factoring in the U.S. benefits received by providing preferential treatment for sugar exporters to the United States.

The question then becomes, why have the sugar interests in the United States been able to maintain the program? To undertake program reform, it is necessary to understand the many forces that shape and support the program. Dealing with agricultural policy in general is tempting; however, we think that a case study concentrating specifically on why the sugar program has been in place for many years provides guiding principles as to why U.S. farmers have been very successful at receiving government transfers.

The political clout of the industry became evident once again in the 2000 harvest period when the sugar lobby was successful in obtaining payment-in-kind (PIK) to increase the returns for sugar producers by dealing with the oversupply of sugar. The use of PIK implies a direct government outlay for sugar, which is in violation of the Dole amendment requiring the sugar program to be zero cost. However, mounting Commodity Credit Corporation (CCC) stocks of sugar under nonrecourse loan programs also may imply significant cost to the government.

It is our hypothesis that the high degree of vertical and horizontal integration of the sugar industry, along with the existence of a strong coalition, the American Sugar Alliance (ASA, which includes high fructose corn syrup [HFCS] producers), provides the rent-seeking vehicle to protect U.S. sugar producers from the production and trade subsidies of other countries. In addition, U.S. sugar interests have been successful at log-rolling for political clout by joining forces with other agricultural interests, including peanut and tobacco growers. We explore how the theory of public choice can explain why strong political support exists for the sugar program. Included in our analysis is an assessment of the role of the ASA and the Coalition for Sugar Reform (CSR). The ASA is a strongly knit coalition that supports the sugar program under FAIR. The CSR, officially formed in 1997, is composed of a group of sugar users and environmental groups that oppose sugar policy.

COALITIONS AND THE U.S. SUGAR PROGRAM

One explanation for the resilience of the sugar program over time is the presence of cohesive sugar beet-, sugarcane-, corn-, and HFCS-producer coalitions. These producers have invested large sums of money to garner the support of politicians to extend the sugar program. The highly integrated nature of the sugar industryemthus, its wide political baseemmakes political support of the program relatively easy to garner. Sometimes producers buy political support from both major political parties at the same time. An example in the state of Florida is the Fanjul family (the owners of Flo-Sun), one of whose members is a major contributor to the Democratic party; another is a major contributor to the Republican party. The ASA has been more successful in their attempts to gain this support than has the CSR.

The American Sugar Alliance

The relative success of the ASA in sustaining the sugar program in the face of well-organized (and potentially well-funded) opposition is an important dimension of how well sweetener (sugar and HFCS) markets function in the United States. According to ASA's Web page, (http://www.sugaralliance.org)

> The American Sugar Alliance is a national coalition of cane, beet, and corn farmers, processors, suppliers, workers, and others dedicated to preserving a strong domestic sweetener industry. The sweetener industry is made up of more than one million beet, cane, and corn farmers who produce sugar and corn for sweeteners, as well as thousands of other Americans who work in sweetener production and processing and in other businesses providing goods and services to the industry. The American Sugar Alliance works to assure that farmers and workers in the U.S. sweetener industry survive in a world of heavily subsidized sugar. Only

through a united effort can these dedicated Americans continue to offer a plentiful and secure domestic supply of sweetener at a reasonable price. The American Sugar Alliance provides the backing and support to meet this national need.

Membership in the American Sugar Alliance ranges from the sweetener producers themselves — including farmers all across the country, and sugar and corn processors and refiners — to the implement dealers, local banks, community businesses of all types, fertilizer distributors, factory workers producing heavy equipment needed in sugar and corn sweetener production, and countless others in services and supply.

Among the major players in ASA are the American Sugarbeet Growers Association, American Sugar Cane League, Archer Daniels Midland Company, Florida Sugar Cane League, Gay and Robinson, Inc., Hawaii, National Corn Growers Association, Rio Grande Valley Sugar Growers, Inc., Sugar Cane Growers Cooperative of Florida, and United States Beet Sugar Association. In addition to these groups, sponsors at the ASA 2000 meetings included the Amalgamated Sugar Company, LLC, American Crystal Sugar, ED and F Man Sugar, Inc., Florida Crystals Corporation, Imperial Holly Corporation, Minn-Dak Farmers Cooperative, and Southern Minnesota Beet Sugar Cooperative.

Historically, the ASA, made up of sugar and HFCS interests, has been an effective force in generating support for U.S. sugar policy. The HFCS-producing group supports sugar import quotas, the key component of U.S. sugar policy, because of the sugar-price umbrella provided by the quotas. In addition, corn producers support the sugar program because corn is used in the production of HFCS. Some contend that the domestic sugar program has given sweeteners, such as HFCS, the impetus to develop. Corn and HFCS producers (for example, Archer Daniels Midland) have joined the ASA, which gives additional geographical and political support to the coalitions that favor the maintenance of the U.S. sugar policy. The ASA and other coalitions that support the sugar program have been extremely successful having their united voice heard in Congress (Schmitz 1995).

Despite their general alliance in support of domestic sugar policy, there may well appear to be policy conflicts among the diverse membership in the ASA. For example, even between the two large integrated sugar producers in Florida (U.S. Sugar Corporation and Flo-Sun, both involved in the growing, milling, refining, and marketing of sugar in the United States) a policy conflict may exist. Flo-Sun, a multinational firm, also produces sugar in the Dominican Republic and receives preferential export quota treatment under the U.S. sugar policy. However, such is not the case for U.S. Sugar Corporation, whose sugar production operations are confined to the U.S. The U.S. Sugar Corporation, for example, supports highly restrictive quotas, whereas Flo-Sun supports more liberalized trade because of its operations within the United States and abroad. There are many other examples of conflicting objectives. One of these, discussed in detail later, are the expanded exports of HFCS in the face of tight sugar quotas.

Coalition for Sugar Reform (CSR)

The coalition that wants U.S. sugar policy reformed is less well developed and is less well understood than is the coalition in support of the sugar program. The CSR was organized in 1997, partly because of the lack of sugar policy reform under the FAIR Act of 1996. This coalition includes American Bakers, Chocolate Manufactures, Independent Bakers Association, and the National Soft Drink Association. In addition, it includes the United States Cane Sugar Refiners' Association, the Everglades Trust, and the National Audubon Society. According to the coalition, the sugar policy costs U.S. consumers $1.4 billion dollars annually in the form of higher food costs. However, the ASA disagrees with these findings, arguing that the decreased sugar prices that would result from policy reform would not be passed to consumers (fig. 9.1). They argue that the benefits would be captured instead by industrial users (such as candy manufacturers) in the form of higher profits (Roney 1999).

The CSR has many members that also belong to an organization often called the Sugar Users Group. This group, which sponsors the annual International Sweetener Colloquium, is highly vocal. It strongly opposes U.S. sugar policy. Supporters of the 2001 Colloquium included the New York Board of Trade, Kraft Foods, Philips Morris, Hershey Foods Corporation, National Soft Drink Association, and M and M/Mars.

OPPOSITION TO TRADITIONAL COST/BENEFIT ANALYSIS

Perhaps one reason that the U.S. sugar program remains in place is the disagreement over the impact of the program. Although academic economists generally contend that U.S. sugar policy imposes net costs to society, many others disagree. For example, Jack Roney (1999), a leading economist for the ASA, contends that many studies are biased against the sugar industry in that they do not account for the fact that a free-trade sugar price would be much higher than the distorted U.S. border price that is used in the cost-benefit calculations. (The CSR appears to support the use of distorted world prices in cost/benefit calculations. Thus, the stand taken by CSR appears to be less of a call for free markets and more of a plea to take advantage of the distorted world-market price for sugar.) However, studies accounting for the cost of the U.S. sugar program based on free trade market prices give results ranging from zero (Koo 1999) to significant net positive cost (Tweeten et al. 1997).

Supporters of U.S. sugar policy rest many of their arguments on the fact that the world sugar market is highly distorted and the so-called world price for sugar (New York futures prices) is far below free-market levels. The world-market price

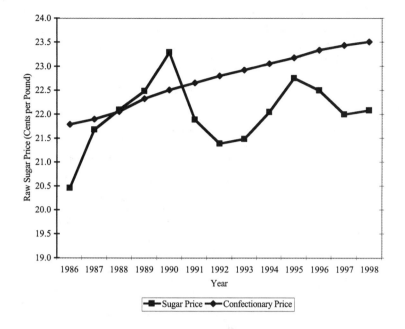

Figure 9.1. Comparative prices of sugar and confectionery.

is often referred to as a dumping price. Josling states that sugar is one of the most protected agricultural markets in the world. Thus, consumer gains might be obtained by the elimination of these protectionist measures. Sugar protection, however, is quite resilient worldwide. Both regional trade agreements (RTAs) and multilateral agreements, such as the Uruguay Round of the WTO, have left these policies more or less intact. The result is a residual world market for sugar that is extremely inelastic, causing significant price volatility. In addition, the world price tends to be lower than maintained prices in major markets such as the United States and the European Union (Borrell 1999; Koo 1999; Schmitz and Vercammen 1990).

In addition to ignoring the free-trade price debate, past studies have failed to address several other major issues. Little is known, for example, regarding the extent to which a drop in sugar prices from removing U.S. sugar import quotas would be passed on to consumers. Much of the analysis assumes 100 percent passthrough. However, if this is not the case, industrial users could gain from the removal of U.S. sugar supports. It appears that the potential gains to industrial users would be significant; why else would they invest large sums of money fighting to have the sugar program removed? The potential gains to industrial users are clearly illustrated in the GAO, which included a model in which not all the benefits from a drop in sugar prices are passed to consumers. Food manufacturers could be gainers or losers from the removal of the sugar program, depending on the degree of passthrough of lower raw sugar prices to consumers (table 9.1).

Table 9.1. User Groups and the Benefits of Eliminating the Sugar Program, 1996 and 1998

	1996		1998	
	Partial Pass-through	Full Pass-through	Partial Pass-through	Full Pass-through
	Million 1999 dollars			
Distribution of Benefits				
Final Consumers	$587	$1,434	$769	$1,960
Food Manufacturers	715	(60)	999	(85)
Sugarcane Refiners	95	97	61	63
Total	$1,397	$1,471	$1,829	$1,938

Source: GAO/RCED-00-126 Sugar Program (2000, p. 23).

COALITIONS AND VERTICAL MARKET STRUCTURES

Coalitions and Agricultural Policy

CSR and the ASA are coalitions. In this case, the purpose of each coalition is to capture economic rents through the policy process. In more detailed form, the study of coalitions involves cooperative behavior within a noncooperative situation. The simplest framework for this is the search for a partner in the three-person, zero-sum game (von Nuemann and Morgenstern 1944). In this scenario, coalitions are formed by two of the players, who choose each other to extract rent from the third player. The basic structure of the three-person, zero-sum game can be translated into a simple voting model in which coalitions are formed based on voting behavior. We imagine a democracy of three individuals: sugar farmers, processors, and consumers. Coalitions are then formed when two of these three players choose to override the vote of the third. For example, the farmers and sugar processors may choose to form a coalition to extract rents from the consumers by enacting a tariff and/or quota (as is done by the ASA). Alternatively, processors could form a coalition with consumers to increase sugar imports, which would cost farmers rent (which describes the goals of the CSR). Two other possibilities exist: (1) farmers could form a coalition with consumers to extract rents from the processors, or (2) each player could play a noncooperative game.

The relationship between the theory of coalitions and rent-seeking activities is undefined in several aspects. First, it is unlikely that rent-seeking behavior could be characterized as the simple act of looking for a partner. Exceptions include the voting model or majority-rule framework. In the three-person game, two votes form a majority, so choosing a partner sets the stage for rent-seeking activities. The second exception, which follows from the first, is over the control of the political process. It is possible that a single player could win the game by submitting a bid. If the consumer is the odd one out in the coalition structure, the size of his or her payment could be sufficient to win the game.

The concept of economic coalitions can be explained within the constant-sum game presented by Roth (1980). Specifically, we assume three possible allocations of rent:

$$V(\{1,2\}) = \left\{ \tfrac{1}{2}, \tfrac{1}{2}, 0 \right\}$$
$$V(\{1,3\}) = \left\{ \tfrac{1}{4}, 0, \tfrac{3}{4} \right\} \tag{1}$$
$$V(\{2,3\}) = \left\{ 0, \tfrac{1}{4}, \tfrac{3}{4} \right\}$$

In each case, $V(\{i,j\})$ denotes the allocation of economic rents that result from the cooperation between agent i and j. Specifically, if agent 1 and 2 cooperate in a game, agent 1 earns one-half of total rents and agent 2 earns one-half. The game is a constant-sum game in that the total return to society remains unchanged. Under this scenario, a stable coalition is formed without side payments only in the case of $V(\{1,2\})$. Note that either agent 1 or 2 is worse off by cooperating with agent 3. Hart and Kurz (1983) expand this concept to develop a general case for the endogenous formation of coalitions. Their analysis further develops the case of stability and the value of coalitions.

The economic process that generates the game is important. This study presumes two types of coalitions: (1) vertical coalitions, in which firms are at different levels of the marketing channel and guide organizations to lobby for policies that benefit each member of the coalition; and (2) horizontal coalitions, in which the cooperative players are at the same level of the marketing channel.

Abstracting slightly from the constant-sum game, figure 9.2 presents the economic rents resulting from two possible configurations of the sugar industry. The first panel of figure 9.2 presents the current sugar market, ignoring for the moment sugar imported under quota, whereas the second panel depicts the economic rents under free trade at a fixed world price that is less than the current domestic price. These two scenarios give rise to two potential coalitions:

$$V(\{F,R\}) = \{\pi_F^1, \pi_R^1, \pi_C^1\}$$
$$V(\{R,C\}) = \{\pi_F^2, \pi_R^2, \pi_C^2\} \tag{2}$$

Under the first scenario, farmers F form a coalition with the sugar processors and refiners R; under the second scenario, candy manufacturers C form a coalition with the sugar processors and refiners. It is clear from the graph that $\pi_F^1 > \pi_F^2$, or that farmers are definitely worse off if they must sell their sugar at the current world-market price. Similarly, it is clear that $\pi_C^2 > \pi_C^1$, or that the candy manufacturers are better off purchasing sugar at the current world-market price. The coalition formation then depends on whether $\pi_R^1 > \pi_R^2$, in which case the stable coalition is $V(\{F,R\})$, or whether $\pi_R^2 > \pi_R^1$, in which case the stable coalition is $V(\{R,C\})$.

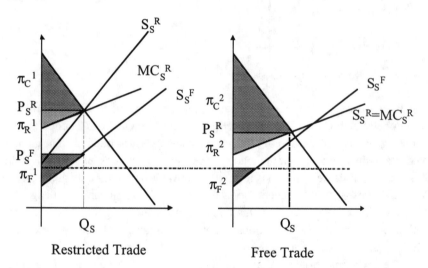

Figure 9.2. Comparison of economic rents under restricted trade and free trade.

The stability of the coalition structure developed in equation (1) is dependent upon the lack of other choices. However, if we allow the players of the game to generate a new set of payoffs, we can envision an infinite number of choices based on side payments in the current game. Specifically, we can envision a fourth allocation of the constant-sum game as:

$$V^*(\{1,3\}) = \left\{ \frac{1}{4} - p_1^2 + p_3^1, p_1^2 + p_3^2, \frac{3}{4} - p_3^1 - p_3^2 \right\} \tag{3}$$

where p_i^j is a side-payment from agent i to agent j. This game is based on the inferior coalition between player one and player three. However, it allows player three to make a side payment to either player one $(p_3^{\,1})$ or player two $(p_3^{\,2})$. Player one can make a similar payment to player two. This structure allows for an expansion of the original three allocations to an infinite number of allocations based on a single outcome of the original game. An interesting variant of this game (and possibly the long-run equilibrium) is the scenario in which $p_3^{\,1}=1/12$, $p_3^{\,2}=1/3$, and $p_1^{\,2}=0$. This set of side payments yields an equal distribution of rents. It could be argued that the FAIR Act of 1996 represents a game in which domestic farmers are offered a side payment by government (and, implicitly, by consumers). Expanding this scenario to the sugar example, we can construct an alternative allocation,

$$V^*\big(\{R,C\}\big)=\left\{\pi_F^2 + p_R^F + p_C^F, \pi_R^2 - p_R^F, \pi_C^2 - p_C^F\right\} \qquad (4)$$

in which $p_R^{\,F}$ is a side payment from refiner to farmer and $p_C^{\,F}$ is a side payment from consumer to farmer. This side-payment scheme would be similar to the shift to direct payment commodity programs under the FAIR Act of 1996 and, hence, would probably involve the federal government as the instrument for making the side payment. In addition, it is important to note the relative-risk effect on the value of the returns in the coalition structure. The amount of the side payment required to change the coalition structure declines if future rents within the coalition are perceived as relatively more risky. Thus, an increased likelihood of success in the WTO negotiations may reduce the side payments required by farmers.

Given the full model of coalitions with side payments, a more complete set of coalitions in the domestic sugar industry can be developed. In this development, we focus on the vertical integration of production and the role of HFCS in the coalition. The new theory of the firm, as developed in the following section, can be used to describe the integration between sugarcane farmers and sugar mills, but the integration of these organizations into sugar refineries is not supported. Arguments favoring vertical integration could then rely on coalition theory. The purchase of sugar refineries could be viewed as a side payment to the refinery. In return for the side payment, the sugar refinery becomes a member of the ASA and not of the CSR. Other arguments exist, such as the economies of scope and the potential for market power in the market for refined sugar, but we do not discuss these issues here.

Rent-seeking Activities by Integrated Firms
Many policy discussions lack detail on vertical markets, which include producers and users. Processors of commodities play a major role in policy formulation whether they are cooperatives or privately owned. The ASA has become a rent-

seeking instrument for sugarcane and sugarbeet producers and processors, and the CSR has become the instrument for sugar users to lower sugar prices through the removal of the sugar program. From the perspective of the sugar producer, this rent-seeking behavior is enhanced by the ownership structure of sugar production, processing, and refining in the United States. In sugar, there is a large degree of vertical integration directly through producer ownership and indirectly through cooperatives in the domestic sugar industry. Also, extensive integration exists between producers and processors through the use of contracts.

The Theory of the Integrated Firm

It is our hypothesis that any change in U.S. sugar policy is conditioned, in part, upon vertical and horizontal linkages between producers and processors of sugar and corn within the domestic industry. Horizontal linkages between sugarbeet, sugarcane, and corn growers support the sugar program. Support also comes through a high degree of vertical integration within the sugar production, processing, and marketing sectors. Two sugar entities in Florida are integrated from the farm to refinery level (table 9.2). Thus, these sugar farmers maximize rents from both production and sugar refining. This is not the case, however, with Louisiana producers in which the degree of integration between sugar producers and raw-sugar-mill owners is less. Thus, if more imports were allowed into the U.S., Louisiana producers could not capitalize from the added refining.

Why do different firms choose to produce raw materials for sale to other firms which, in turn, transform these raw materials into the finished product? One central question raised by Coase (1937) is "why do two firms exist and not one or three?" Another question we raise in the context of the sugar industry is, "Why do some firms vertically integrate from the production to the marketing of the final product, while other firms simply contract with private processors?" Vertical integration in an industry depends on transaction costs; on the ability of one firm to extract profits through noncooperative behavior; and on the economies of scale. Coase (1937), Williamson (1985), and Grossman and Hart (1986) focus on transaction costs. For example, Grossman and Hart examine the relationship between contracting and integration. They compare the economics of integration through ownership to contracting by examining the completeness of contracts. If a complete contract can be specified, contracting may yield lower transaction costs than does direct ownership. Situations not conducive to complete contracting, however, favor ownership as a mechanism to reduce transaction costs.

The framework provided by the traditional theory of the firm appears to be well suited to the explanation of the ownership of sugarcane mills in Florida. The theory, however, does less well when explaining recent moves by Florida growers

Table 9.2. Sugarcane Sugar-Refining Companies (1898–1998)

Company	Refinery Location	Capacity[a]	Year Built
California & Hawaiian Sugar Co.[b] Crockett		3,400	1898
830 Loring Avenue			
Crockett, California 94525			
[510 787-2121]			
Florida Crystals Refinery[b]	South Bay	925	1978
PO Box 86			
South Bay, Florida 33493			
[407 996-9072]			
Imperial Sugar Company[c]	Clewiston, FL	850	1889
PO Box 9	Gramercy, LA	2,150	
Sugar Land, Texas 77487	Port Wentworth, GA	3,100	
[281 491-9181]	Sugar Land	1,950	
Refined Sugars, Inc.[bd]	Yonkers	2,000	1938
One Federal Street			
Yonkers, New York 10702			
[914 963-2400]			
Tate & Lyle North American Sugars Inc.[b]	Baltimore	3,000	1922
1114 Avenue of the Americas	Brooklyn	2,000	1865
24th Floor	Chalmette, LA	3,000	1908
New York, New York 10036			
[201 896-6066]			
U.S. Sugar Corporation	Clewiston	1,800	1998
PO Drawer 1207			
Clewiston, Florida 33440			
[941 983-8121]			

Total Capacity (tons of raw sugar per day) 24,175 Percentage owned by producers 33.6%

[a] 24-hour melting capacity, short tons, raw value
[b] Member of United States Cane Sugar Refiners' Association
[c] Owned by Flo-Sun
[d] Purchased by the Flo-Sun and others in the late 1990s

to integrate the sweetener industry from the production level to the refinery level. Explaining this type of increased integration may be easier within either an economies-of-scale framework or a strategic-planning mechanism that deals with changes in trade policy.

Rents in a Vertical Contracting Chain

Regardless of the reasons for integration, viewing the integrated firm as a marketing channel allows for the decomposition of the economic rents into processes within the firm. In figure 9.3 we depict the economic rents gleaned from sugarcane production and the numerous upstream market activities of the integrated sugarcane firm. S^F is the traditional supply of sugarcane at the farm level. Adding the marginal cost of milling sugarcane into raw sugar yields the supply function of raw sugar, S^M. Next, the supply function of refined sugar, SR, is derived by adding to the above the marginal cost allocated to the refining of raw sugar into refined sugar. Finally, the retail supply function, S^T, is derived by adding the marginal cost of marketing to the supply of refined sugar. The intersection of the marketing-supply curve with the retail-demand curve yields the total quantity of sugar produced by sugarcane. Following through the marketing channel, R^F is the economic rent generated at the farm level, RM is the economic rent generated by milling, R^R is the economic rent generated by refining the raw sugar to refined sugar, and R^T is the economic rent generated by marketing.

The allocation of rents for the integrated sugarcane production complex can be contrasted with the rents generated in the integrated sugarbeet production complex (fig. 9.4). In the sugarbeet production complex, the economic rents to milling and refining are captured in one process.

U.S. refineries can also import raw sugar for processing and sale in the domestic and international markets. Under the sugar program, raw sugar can be imported both for domestic use and for re-export as refined sugar (most of the imported raw sugar is for domestic use). Expanding the original sugarcane channel presented in figure 9.3, the ability to import sugar changes the rents to the integrated firm (fig. 9.5). Under market integration, the processors can import sugarcane under the quota to refine and sell it in the domestic market. In the first panel of figure 9.5, we depict the production and milling activities as similar to the first two steps of sugar production that we illustrated in figure 9.3. We allow the firm to import a fixed quantity of raw sugar for refining and marketing in the United States. Given that the domestic price of sugar exceeds the world price of sugar, the supply of imported sugar under the quota can be modeled as perfectly inelastic. The supply of raw sugar in the United States is then the horizontal summation of the raw-sugar supply from sugarcane production and the amount imported under quota S'M. This supply curve lies to the right of the original supply curve of SM. Adding the marginal cost of refining to the marginal cost of marketing yields a supply curve at the consumer

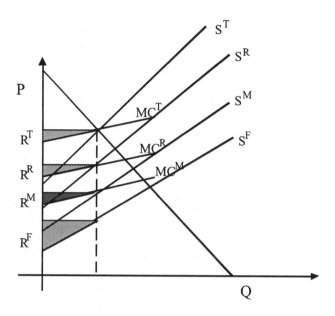

Figure 9.3. Decomposition of rents in the integrated sugarcane firm.

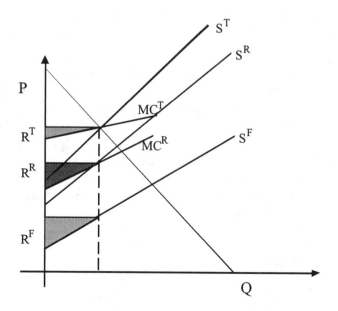

Figure 9.4. Decomposition of rents in the integrated sugar beet firm.

level of S^{T+R} before the import quota and S'^{T+R} after the import quota. This shift in supply causes domestic consumption of sugar to increase from Q to Q'. Associated with this increased quantity is an increase in the economic rent of refining and marketing. As depicted in the first panel, the gain in economic rent to refining and marketing when sugarcane production falls from Q^F to Q'^F must be weighed against the reduction in economic rent to the production of sugarcane.[1]

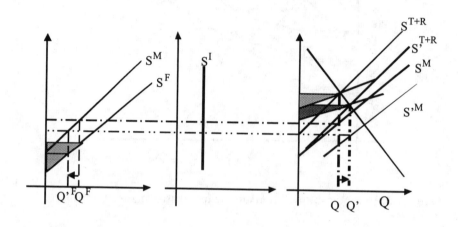

Figure 9.5. Economic rents to integrated sugarcane producers through refining imported raw sugar.

Figure 9.5 implies that integrated sugarcane producers in Florida may exact addition rents through the processing of imported raw sugar. By contrast, the sugarbeet refiners in figure 9.4 do not have a similar opportunity. Although their process may be modified to add value to imported raw sugar, geographical location is clearly a disadvantage to processing imported raw sugar. Also, it is important to recognize that sugarbeet producers are not completely separate from sugarcane producers, nor are sugarbeet refiners completely separate from sugarcane refiners. Certain multinationals, who are major players in the U.S. market, are involved in both sugarbeet and sugarcane production, processing, and marketing. In addition, several U.S. sugar refiners have joint marketing agreements with sugarbeet processors, which add to the complexity of coalition formation and of support for U.S. sugar policy. This coalition arrangement is highlighted in figure 9.6.

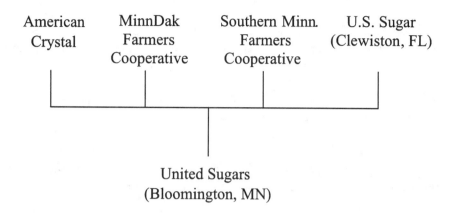

Figure 9.6. Sugar marketing coalition.

The integrated nature of sugar production in the United States has several implications for maintaining domestic sugar policy. The physical attributes of sugarcane (and, to a lesser extent, sugarbeets) closely link the sugarcane harvest and the milling of sugarcane into raw sugar. This linkage leads to significant vertical coordination through either contracts or ownership. In addition, some producers are vertically integrated from production through to the final marketing of sugar. This integration complicates the nature of rent-seeking activities within policy analysis. Fourth, the coalition structure supporting domestic sugar policy may be weakening under current trade agreements and agricultural prices. Last, we have not explored the competitive nature of the U.S. sweetener industry. The existence of ASA may well maintain competition within the sugar industry that might not be present if this coalition did not exist.

Mexico and U.S. Disputes Over Sugar and HFCS

Mexico's involvement in the sugar industry has been significant in the past through government ownership of the sugar mills and government regulation via the *ejido* system. In 1999, Mexico exported 25,000 tons of sugar to the United States under NAFTA. This is about 1.6 percent of imports under the U.S. sugar quota. More important, under a U.S.-Mexico side agreement, Mexico could export to the U.S. an additional 250,000 tons of sugar annually if Mexico maintained net-exporter status for at least two years. As of March 2001, the United States has not honored this agreement even though Mexico is a net sugar exporter. Complicating matters is the push by the HFCS industry to export HFCS to Mexico. Lynn Jensen, the president of the National Corn Growers Association, states that "…U.S. corn refiners have a growing market in Mexico, largely in the soft drink arena, and Mexican sugar producers are not anxious to give up that part of their market to corn sweeteners" (quoted in Maixner 2000, p. 3). Added imports by Mexico would replace sugar use in Mexico and hence add to Mexico's net sugar export position. However, Mexico has brought several lawsuits against the United States, attempting to block imports of HFCS from the United States. This competition between domestic sugar and foreign HFCS in the Mexican sweetner industry raises an interesting potential conflict between the U.S. sugar industry and its traditional HFCS manufacturer allies.Significant growth has occurred in the production and demand for HFCS. As a search for alternative sweetener markets abroad continues, this growth may weaken the incentives for HFCS producers to support the ASA. For example, the U.S. producers of HFCS want freer trade in sugar with Mexico in order to expand their exports of HFCS, provided that freer trade with Mexico in sugar does not result in a significant drop in U.S. sugar prices. If a significant decline in the price of domestic sugar is a result of freer trade with Mexico, the price of HFCS will fall. Thus, the HFCS industry has to weigh both the costs and benefits from freer trade in sugar with Mexico.

Using research by Williams and Bessler (1997) and by Moss and Schmitz (1999, 2001), it appears that within a fairly wide price range, the HFCS market has become separated from the sugar market. Thus, sugar prices could fall without having an impact on the domestic market for HFCS. As a result, the HFCS industry may support the sugar quota, but at a higher level than other groups in the ASA.

Log-Rolling and Sugar Programs
The FAIR Act of 1996 made significant changes in the support programs for agriculture in the United States. It included the elimination of deficiency payments and acreage set-asides. Although sugar emerged relatively unscathed, the changes in other agricultural programs had a significant effect on domestic sugar producers in the late 1990s.

Devadoss and Kropf (1996) predict that the reductions in trade distortions negotiated in the Uruguay Round of the WTO would move the sugar market toward an equilibrium more in line with the world market.[2]

Their model projects that sugarcane production in the United States would fall to 830,000 acres in 2000, while sugarbeet production would decline to 1,332,000 acres. Combined, these declines would lead to a reduction in U.S. domestic sugar production of 1.89 percent (that is, 6.559 mmt). Aggregate consumption would increase to 7.995 mmt and imports would increase to 1.952 mmt. These expectations have not been met. Sugarbeet planting increased to 1,560,000 acres in 1999 and sugarcane acreage increased to 989,200 acres (table 9.3). As a result the domestic supply of U.S. sugar increased to 8.257 mmt. (For example, in Louisiana alone, production increased from 1.05 million tons in 1996/97 to 1.68 million tons in 1999/00.) This rapid increase in production has undermined the basic foundation of the U.S. sugar program.[3]

Table 9.3. Sugarbeet and Sugarcane Plantings, 1,000 acres

Sugarbeet Plantings 1996 and 1999

	1996 Plantings	1999 Plantings	Average Log Change
Great Lakes			
Michigan	153.0	194.0	0.0791
Ohio	4.9	1.8	-0.3338
Total	157.9	195.8	0.0717
Upper Midwest			
Minnesota	441.0	480.0	0.0282
North Dakota	226.6	251.6	0.0349
Total	667.6	731.6	0.0305
Great Plains			
Colorado	54.8	72.1	0.0915
Montana	57.7	61.8	0.0229
Nebraska	55.8	72.7	0.0882
Wyoming	58.0	58.0	0.0000
Total	241.5	264.6	0.0304
Far West			
California	84.0	110.0	0.0899
Idaho	187.0	211.0	0.0402
Oregon	17.4	20.1	0.0481
Washington	13.0	27.5	0.2497
Total	301.4	368.6	0.0671

Sugarcane Plantings 1996 and 1999

	1996 Plantings	1999 Plantings	Average Log Change
Florida	438.0	460.0	0.0163
Hawaii	46.0	37.3	-0.0699
Louisiana	370.0	465.0	0.0762
Texas	34.9	31.0	-0.0395

The United States now finds itself in a surplus situation, and policymakers are grappling to find new solutions to this problem, including possible implementation of a PIK program.

In the debate leading up to the anticipated farm bill in 2002, several commodities are experiencing financial difficulties. Declining agricultural prices add an additional coalition argument in the guise of log-rolling (Stratmann 1992). Table 9.4 presents the producer subsidy equivalents (PSEs) for seven commodities in the United States (OECD 2000). These figures show the overall decline in agricultural subsidization in the first year of the FAIR Act. The table indicates that the PSEs fell for every commodity between 1986-88 and 1997. The largest reductions were for maize (corn) and rice, falling 63.2 and 80.8 percent, respectively. The smallest decline was for sugar, falling only 24.1 percent. Clearly, to shift support from other agricultural programs for the U.S. sugar program would have been politically difficult in 1997. PSEs for other commodities have increased as their prices have declined. The PSE for wheat increased 92.9 percent in 1998, and in 1999 approached the 1986-88 level. In addition, the PSEs for other crops such as soybeans grew at a phenomenal rate (although soybeans received little direct support under the 1985 farm bill, such support was included in the loan deficiency payments under the FAIR act). Thus, in 2001, log-rolling again may support the U.S. sugar program as other commodities attempt to garner support for a new farm bill. Given the high commodity prices in 1996, the log-rolling appeared to be less than in previous farm bills, yet the sugar program was able to survive. The reason for this survival is partly, as argued by Schmitz and Christian (1993), that sugar support is strong in the absence of log-rolling.

Overproduction and PIK
The United States supports its internal sugar price through import quotas. Domestic sugar policy is implemented through a tariff rate quota (TRQ) consistent with the Uruguay Round of the WTO. That policy maintains domestic sugar prices above the world market price by the high "second" tariff on imports above the minimum access level. In addition, the United States has a loan rate for sugar. However, the operation of the TRQ traditionally guaranteed that the loan rate was never truly effective before the year 2000. In 2000, domestic production reached a level that appeared to make forfeitures eminent.

U.S. policymakers face an excess supply problem (fig. 9.7). Given domestic supply S_d and an import quota q_2-q_1, the internal price is supported at P. However, given technological change and supply shifts that are due to changing relative commodity prices favoring sugar production, production shifts to S_d' . If import quotas remain unchanged, the internal price falls to P_0 in the absence of government storage. To maintain prices at , the government has to essentially store the amount imported (q_2-q_1). This was the situation the government and industry faced in 2000.

Table 9.4. Producer Subsidy Equivalents (PSEs) for the United States (%)

Commodity	Level PSE in Each Year(s)					Growth in PSE		
	1986-1988	1997-1999	1997	1998	1999	1986-88/97	1997/1998	1998/1999
Wheat	49	37	25	39	46	-49.0	56.0	17.9
Maize	38	24	14	27	30	-63.2	92.9	11.1
Other Grains	40	34	23	40	40	-42.5	73.9	0.0
Rice	52	17	10	15	26	-80.8	50.0	73.3
Oilseeds	8	15	4	15	25	-50.0	275.0	66.7
Sugar (Refined Equivalent)	58	56	44	56	68	-24.1	27.3	21.4
Milk	60	54	45	61	57	-25.0	35.6	-6.6

Source: OECD Agricultural Policies in OECD Countries: Monitoring and Evaluation 2000.

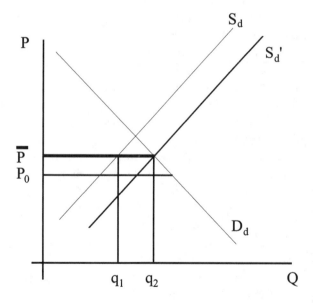

Figure 9.7. Excess supply of sugar.

Government responded with a PIK program, idling sugarbeet production. The government ended up holding more than 700,000 tons of sugar as of March 1, 2001. Future discussions on U.S. sugar programs will have to deal with this supply issue. Some have proposed supply controls in addition to import controls (Fernandez 2001). Regardless, the PIK program had far-reaching implications at least for processors of sugarbeets. Many now face bankruptcy or reorganization. In the spring of 2001, several were trying to sell their facilities to farmer cooperatives. The possible resulting structure when cast in the earlier framework of vertical markets has important implications for the design and implementation of future U.S. sugar policy.

Conclusions

One of the least understood elements of agricultural policy is how the structure of market channels affects the design and implementation of policy. We have emphasized both horizontal linkages and vertical market structures in our examination of the U.S. sugar policy. The increasing and high degree of vertical market integration in sugar will likely play as important a role in formulating sugar policy as do horizontal linkages among producers.

In view of the generally low commodity prices in U.S. agriculture, log-rolling will play a part in determining the new agricultural program. The form that the U.S. sugar program will take is not clear. It may well contain elements of supply control (Fernandez 2001), especially in view of excess production problems facing the U.S. industry in 2000. Many other proposals have been suggested, including deficiency payment schemes (Orden 2001). Regardless, our hypothesis is that given the discussion above on the strength of coalitions that support U.S. sugar policy, U.S. producers will not be left to operate at the world-distorted market price, which is well below the U.S. producer price.

The ultimate beneficiaries from the reduction of government intervention in agricultural markets would be consumers. Following the free market paradigm, increased imports from lower-cost foreign producers would imply lower food prices for consumers in the United States. One limitation of this study is that we do not explicitly model consumer coalitions. However, the policy debate appears to be devoid of an organized consumer voice. One possibility to explain the lack of consumer interest is that the gain to consumers from the reduction in sugar prices represents a very small share of consumer's food expenditure. This significance is further reduced by the relatively small percentage of their income that consumers spend on food in the United States. Thus, consumer's may not perceive agricultural programs to have a significant effect on their individual budgets, whereas sugar producers and industrial users have significant monetary amounts at stake with regard to changes in the sugar program.

NOTES

1. Apart from the complexities added by vertically integrated firms, geographic concentration, and domestic sugar policy, the economics of the sugar industry are also complicated by the role of multinationals. Multinationals exist in the sugar industry at several levels. On the production side, some sugarcane producers operate

in several countries. Hence, the potential for change in economic rents, which is present because of the imports presented in figure 9.4, are complicated by firm-level considerations. Specifically, a multinational sugarcane producer may choose to locate his firm among other countries, which includes the United States, as long as he can export his sugar into the United States under U.S. import-quota rules. Under such rules, the United States is allowed to grant preferential treatment to certain sugar exporters. We hypothesize that domestic sugar producers may well support different levels of protection than do multinational producers. These considerations, along with the possibility of importing for re-export, add a further complication to the integrated firm example presented in figure 9.5.

2. Although the stated purpose of the Uruguay Round of the WTO was the elimination of trade distortions, the first set of reforms put a tariff upon the current quotas and provided minimum access provisions. The accord then called for the eventual phase-out of all quotas through time. For the period covered by Devadoss and Kropf (1996), however, the Uruguay Round of the WTO can only be described as a move toward freer trade.

3. The baseline projections for crop 1999/2000 prices in 1997 put corn at $2.55 per bushel, wheat at $3.95 per bushel, rice at $10.00 per cwt, and soybeans at $5.90 per bushel. The February 2000 USDA baseline has corn at $1.80 per bushel, wheat at $2.50 per bushel, rice at $5.75 per cwt, and soybeans at $4.90 per bushel. In general, 1999/2000 agricultural prices for these commodities were 38 percent lower than the 1997 projections. The data in Table 9.4 presents some insights into the sugar production response from 1996 to 2000. All states, except Ohio, exhibit increased sugarbeet acreage between 1996 and 2000. When taken together, Florida and Louisiana account for 93.5 percent of U.S. domestic sugarcane production, both in increased plantings between the 1996/1997 crop year and the 1999/2000 crop year.

REFERENCES

Borrell, B. "Sugar: The Taste Test of Trade Liberalization." Paper presented at the World Bank Conference on Agriculture and the New Trade Agenda in the WTO 2000 Negotiations, Geneva, Switzerland, October 1–2, 1999.

Coase, R. H. "The Nature of the Firm." *Economica* 4 (1937): 386–405.

Devadoss, S. and J. Kropf. "Impacts of Trade Liberalization under the Uruguay Round on the World Sugar Market." *Agricultural Economics* 15 (November 1996): 83–96.

Fernandez, L. J. "Sugar Policy Needs of Florida Cane Growers and Processors." Speech at the Agricultural Outlook Forum 2001, Alexandria, Virginia, Feb 23, 2001.

GAO. *Sugar Program: Supporting Sugar Prices Has Increased Users' Costs While Benefiting Producers.* RCED–00$en126, Washington, DC June 9, 2000.

Grossman, S. and O. Hart. "The Costs and Benefits of Ownership: A Theory of Vertical and Lateral Integration." *Journal of Political Economy* 94 (August 1986): 691–719.

Hart, S. and M. Kurz. "Endogenous Formation of Coalitions." *Econometrica* 51 (1983): 1047–64.

Josling, T. "The Place of Sugar in Regional and Multilateral Trade Negotiations." Paper presented at Sweetener Markets in the 21st Century, Miami, Florida, November 15, 1999.

Knutson, R. D., J. B. Penn, and B. L. Flinchbaugh. *Agricultural and Food Policy, Fourth Edition.* New Jersey: Prentice Hall, 1998.

Koo, W. "The U.S. Cane and Beet-Sugar Industries Under Alternative Trade Liberalization Policy Options." Paper presented at Sweetener Markets in the 21st Century, Miami, Florida, November 15, 1999.

Maixner, E. "Corn Growers, Refiners Appeal for Mexican Sugar Dispute Settlement." *Feedstuffs* 72 (34)(April 14 2000): 3.

Moss, C. B. and A. Schmitz. "The Changing Agenda for Agribusiness: Sweetener Alliances in the 21st Century." Paper presented at Sweetener Markets in the 21st Century, Miami, Florida, November 15, 1999.

Moss C. B. and A. Schmitz. "Price Behavior in the U.S. Sweetener Market: A Cointegration Approach." *Applied Economics,* Revision March 2001.

OECD. *Agricultural Policies in OECD Countries: Monitoring and Evaluation 2000.* Paris. Organization for Economic Cooperation and Development, 2000.

Orden, D. "Farm Policy Reform in the United States." Forthcoming in Moss et al., eds. Agricultural Trade, Globalization, and the Environment. Norwell, MA: Kluwer Academic Publishers, 2001.

Roney, J. "U.S. Sugar Policy: Domestic and Trade Policy Challenges." Paper presented at Sweetener Markets in the 21st Century, Miami, Florida, November 15, 1999.

Roth, A. E. "Values for Games Without Side Payments: Some Difficulties with Current Concepts." *Econometrica* 48 (1980): 457–77.

Schmitz, A. "Sugar: The Free-Trade Myth and the Reality of European Subsidies." International working paper IW95–20, Gainesville, Florida, December 1995.

Schmitz, A. and D. Christian. "The Economics of Politics of U.S. Sugar Policy." *The Economics of Politics and World Sugar Policies,* Stephen V. Marks and Keith E. Maskus, eds. 49–78. Ann Arbor: University of Michigan Press, 1993.

Schmitz, A. and J. Vercammen. "Trade Liberalization in the World Sugar Market: Playing on a Level Field?" Berkeley: Department of Agricultural and Resource Economics, University of California, November 1990.

Stratmann, T. "The Effects of Logrolling on Congressional Voting." *American Economic Review* 85 (December 1992): 1162–76.

Tweeten, L. *Foundations of Farm Policy*. Lincoln: University of Nebraska Press, 1970.

———*Farm Policy Analysis*. Boulder, CO: Westview Press, 1989.

———*Agricultural Trade: Principles and Policies*. Boulder, CO: Westview Press, 1992.

———Trade, Uncertainty, and Farm Programs." Forthcoming in C. Moss, et al., eds. *Agricultural Trade, Globalization, and the Environment*. Norwell, MA: Kluwer Academic Publishers, 2001.

Tweeten, L., J. Sharples, and L. Evers-Smith. "Impact of CFTA/NAFTA on U.S. and Canadian Agriculture." Working paper 97–3. St. Paul, MN: IATRC, 1997.

Von Nuemann J. and O. Morgenstern. *Theory of Games and Economics Behavior*. New Jersey: Princeton University Press, 1944.

Williams, O. and D. Bessler. "Co-Integration: Implications for Market Efficiency of HFCS and Refined Sugar Markets." *Applied Economics* 29 (1997): 225–32.

Williamson, O. E. *The Economic Institutions of Capitalism*. New York: The Free Press, 1985.

10

Farmland is Not Just for Farming Any More:
The Policy Trends

Lawrence W. Libby

Farmland policy in the United States seeks to protect the various nonfood amenity services of land valued by nonfarm people. A trend toward market-based approaches protects private property rights and lets consumers demonstrate willingness to pay. But pressure also exists for a regulatory structure that emphasizes the responsibilities of land ownership. Future policy will strike a delicate balance of rights between landowners and consumers as farming adapts to the realities of the twenty-first century.

INTRODUCTION

Farmland is important to people for many reasons, including but certainly not limited to the corn or alfalfa that happens to grow there. Land is valued by people for various natural resource services and amenities. People seek to reduce or avoid off-farm pollution and other damages from farmland. This chapter deals with land amenities and services, not with pollution. All states and many local governments have adjusted land market rules or signals in the interest of encouraging or assuring provision of amenity services. Land other than farmland may provide those services — open land that is not part of a farm, wetlands, and other ecologically significant areas. Emphasis here, though, is on farmland. The purpose of this chapter is to review the history and rationale for farmland policy, identify the most prominent themes for future policy, and examine the role for markets in that policy.

EVOLUTION OF U.S. FARMLAND POLICY

Policy interest in the rate and pattern of farmland conversion has evolved over the past sixty years. Actual policy remains a highly diverse mix of public inducements and restrictions for many different purposes.

The Scarcity Argument

Bunce (1998) has described a strong "resourcist" motivation in the initial emergence of land policy. A general neo-Malthusian concern about adequacy of the land base to meet food needs carried the movement through the early 1980s and remains important today, even in the face of statistical and market evidence of food abundance (see Tweeten 1998). The general sentiment is that long-term food security is indeed something to pay attention to, not as a clear crisis but as a matter of obligation to, or option for, future generations. Many people have the intuitive sense that farmland is too important a national asset to be thoughtlessly and irreversibly squandered on poorly considered development patterns. The resourcist perspective comprises a risk-averse attitude about the future food supply, a desire to take actions today that reflect caution with the nation's productive land base to avoid the high adjustment costs of underestimating future food needs.

That posture obviously has a cost, measured as the net value of foregone economic growth on land protected for farming. In economic terms, what is the appropriate discount rate for current actions to protect farmland expected to pay off only in the very long run (presumably more than seventy-five years)? Weitzman (1999, 29) proposed that the discount rate on actions to avoid potential environmental and resource depletion catastrophes (such as global warming) declines as project time increases, with a low-normal rate for the first twenty-five years (three to four percent) and decreasing to zero percent beyond three hundred years. Given the inherent uncertainties about deep future technologies and food demands, and the relative consequences of being wrong on the short or long side, discounting still makes sense, but at a low rate.

The "precautionary principle" is well established in European environmental policy, essentially suggesting that when potential negative consequences of a proposed action are severe, society should err on the side of caution. Much debate centers on the particulars of a proposed action and degree of caution desired, but the principle and its policy applications are compelling (Foster, Vecchia, and Repacholi 2000). Potentially irreversible depletion of an essential resource may be justification for establishing a "safe minimum standard" decision rule that accepts some reduction in measured net benefit in the interest of long-term sustainability, outside the discounting procedures of conventional benefit-cost analysis (Farmer and Randall). These notions are relevant to the farmland case.

Ecological and Amenity Services

A more "ecocentric" posture on farmland protection has been part of policy debates since the 1980s, including the various ecological and amenity services associated with actively farmed land (Bunce 1998). Substantial literature addresses the nature of those services and their implicit economic value. Kline and Wichelns (1998), for example, determined in their survey that protecting groundwater, wildlife, and natural areas was considered the most important reason for protecting farmland, with food supply and controlling growth seen as less important. Other amenity services include the scenic open space character of rural landscapes and a certain heritage or cultural value associated with farming. Various agrarian attitudes also underlie farmland policy evolution, from basic recognition that food production is an essential activity to more fundamentalist emphasis on the farming lifestyle as a superior relationship among people, nature, and community.

Federal Policy
The history of federal farmland policy goes back to the New Deal era as the nation responded to fifty years of land abuse culminating in the Dust Bowl. In 1934, the National Resources Board asserted that "heedless and unplanned land exploitation should give way to federal land policies geared to the national welfare" (Lehman 1995, 18–19). The purpose was to collectively acknowledge the comparative advantage of certain lands for crop production. States, counties, and land-grant colleges rebelled at this national initiative, resulting in the formation of county land-use committees of farmers throughout the country. County land-use plans were to be coordinated at the state level for a national network of farmland plans that would guide national policy. The system never matured, falling victim to interagency squabbles and a general lack of technical competence in the Bureau of Agricultural Economics that had been given leadership.

The only other serious national policy proposals came as part of the general environmental movement of the late 1960s and 1970s. Senator Henry Jackson of Washington proposed a national system of state comprehensive plans, recognizing the primacy of states in land-use policy for private lands. The Nixon administration had its own bill and the two were merged in the Land-Use Policy and Planning Assistance Act of 1973. The rhetoric of the time referred to "the cost of sprawl" (Real Estate Research Corporation), "the price of planlessness...in ugly sprawl, blight, wasted resources" (*The New York Times* 1972, A24), and the importance of national guidelines for local land-use planning. The bill looked to be a sure thing but faded away in 1974, a victim of Watergate and more general suspicion of land-use policy from Washington rather than from the states or county court house. The Chamber of Commerce, National Association of Homebuilders, National Association of Realtors, and the American Farm Bureau Federation were prominent in their opposition. Representative Jeffords of Vermont picked up the torch in 1977 with the National Agricultural Land Policy Act proposal that federal agencies be required to formally acknowledge and analyze the effect of various federal development subsidies on the nation's farmland supply. Once again, fears of federal tampering in state and local programs and the suggestion that this was the first step in national land-use planning led to the proposal's demise. The most positive outcome of nearly four years of debate on the Jeffords bill and its Senate counterparts was the formation of the National Agricultural Lands Study (NALS), co-chaired by Bob Gray, Jeffords' former chief aide (Lehman 1995).

The NALS report of 1981, released in the last four days of the Carter administration, was the last major federal initiative in farmland protection (except for federal dollars distributed for state and local purchase of development rights under the 1996 farm bill). NALS foundered on bickering over alleged misuse of land-use statistics and deep philosophical differences between the study co-directors (see Lehman 1995, 133–156; Libby and Stewart 1999). Also in 1981, the Farmland Protection Policy Act was passed as part of the farm bill. This action required federal agencies to minimize the impacts of their programs on the nation's farmland, retaining at least some of what Jeffords had sought four years earlier (American Farmland Trust 1997).

State Policy

States are the center of policy action, as they in essence have been since the 1960s. The politics of private property and legal history of delegation of powers favor action in states, ·with further delegation to local governments. Logical, but little political, support exists for a guiding national structure. The range of state policy instruments is addressed in the following sections.

Taxes

Maryland was the first state to enact special property tax treatment for farms with its deferred tax program in 1956. As of mid-2000, all fifty states provide some sort of property tax incentive to encourage farmers to remain in farming. Only Michigan lacks a differential land tax system in which farmland tax is based on productive value rather than full market value.[1] Michigan, New York, and Wisconsin use a circuit breaker approach, allowing a state income tax credit for a portion of an eligible farmer's property tax bill (see AFT 1997, 155–56).

Zoning

Land-use zoning has been applied to farmland retention efforts since the 1970s, when California, Pennsylvania, and Washington enacted enabling statutes for counties. Local units in twenty-one other states now have agricultural protection zoning in place. Agricultural zones are required for all counties under Oregon's growth-management program, and Hawaii has established an agricultural zone in its statewide ordinance. Some zoning ordinances limit land use in an agricultural zone to farming and related uses only (exclusive agricultural zoning), whereas others rely on large minimum lot sizes to discourage residential development.[2] Because agricultural zoning seeks to discourage nonfarm development in farming areas, it usually reduces market value of the zoned land, implicitly shifting the cost of growth management onto current rural landowners. That society realizes a benefit from such regulations may in fact provide little comfort to the landowner.

Purchase

Purchase programs enable the public seeking the amenity services of farmland to pay the owner for the land-use discretion necessary to retain those noncommodity services. Under these programs, the farmer relinquishes forever the right to sell for development or to convert his land to a nonfarm use. The government or nonprofit land trust acquiring the rights must ensure that the land is used appropriately under the new deed and essentially "retires" the development rights.[3] Suffolk County, New York was the first to undertake such a program, in 1974. Maryland, Massachusetts, and New Hampshire followed in the late 1970s, and several more states enacted programs following release of the National Agricultural Land Study in 1981. As of July 2000, nineteen states possess specific authority for purchase of development rights (or conservation easements) on farmland (AFT 1997, 84–85; Thompson). Virginia, West Virginia, and Ohio added their programs in 2000.

FUTURE POLICY TRENDS

Public actions to alter the farmland market are designed to affect the balance between the land-use opportunities of owners and those of non-owners. The result is reflected in the flow of land services. As discussed previously, the full range of policy instruments is represented in the current farmland policy setting. Land-use regulation implies that the landowner has the basic responsibility to manage his or her land in ways that provide certain valued services to the broader public, consistent with some income from the farm, and must avoid taking actions that unacceptably compromise those services. Acquisition, at the other end of the spectrum, implies that any change in the mix of farmland services in the interest of the broader public requires compensation. That is, the farmer has the right to manage the farm to generate income over a reasonable planning horizon; any additional services desired of that private land must be paid for through purchase of selected rights. The purchaser may be the individual desiring that particular service or a government entity acting on behalf of many non-owners seeking the services of privately owned farmland.

Two strong threads lead to future farmland policy. First, we are in an era of privatization in all areas of resource policy. The basic notion is that "command and control" instruments are insensitive to differences in land, landowner, and community attributes from place to place and are inconsistently administered. Further, privatization enables consumers of land services to demonstrate willingness to pay. The other major thread in future farmland policy is the growing sense that land ownership carries responsibilities as well as rights. The idea is that private rights of ownership do not include the right to pollute the waterways, deplete the natural productivity of land, or eliminate certain land services that are nonrival in consumption or are nonexcludable, valued by the broader public but not readily marketed. Rights of the owner are not absolute, limited by the rights of non-owners. These two threads are at the extremes of the owner/non-owner rights spectrum, yet both will be prominent in future farmland policy.

Privatization

Policy actions in this category include efforts to price various currently unmarketed farmland services and create a mechanism by which consumers can express willingness to pay. This approach reduces tendencies for non-owners of farmland to "free ride" and generally makes coherent the demand for land services. In the "polluter pays" spirit of dealing with land *dis*-amenities, some land amenity programs are "producer gets" in that either a public or private entity pays farmers to provide selected nonexclusive and nonrival services *or* to forego actions that would compromise those services. Others are "beneficiary pays," meaning that services can be withheld from noncontributors (Hanley et al.) and people express effective demand for landscape amenities.

"Producer-Gets" Programs

More state and local programs will provide public funds to buy land use rights on behalf of the general public. As noted, nineteen states had such programs as of July 2000; others will follow. Services purchased in this manner are generally as follows: the nonexcludable open-space amenities; natural resource services such as groundwater recharge, biodiversity, and certain hydrologic functions; and the general interest in managed growth.

Taxpayers have supported these efforts by approving dedicated funding sources, usually through the sale of bonds rather than higher taxes. Maryland, with the nation's oldest purchase program, has spent more than $200 million for farmland development rights during the past twenty-three years with funds from an agricultural land-conversion tax and transfer fees. Pennsylvania's Environmental Stewardship Act of 1999 allocates $20 million a year for five years for purchase of agricultural conservation easements (Bowers, January 2000). In 1998, New Jersey voters supported a $1 billion bond issue, half for farmland preservation over the next decade (Bowers, March 2000). Federal funds for state match were part of the 1996 farm bill and will likely be expanded in future farm bills. Linking farmland protection with other open-space interests has likely increased voter appeal of such programs in several states.[4] All such programs involve permanent transfer of development rights from farmer to a public entity, considered by many farmers to be too great a sacrifice. An Ohio proposal would lease development rights for a renewable thirty-year period (Libby 1999).

A huge information need in state purchase-of-development-rights (PDR) programs concerns the value of those services or attributes that people say they want but are not priced in the market. Policymakers and interest groups alike need better evidence of what people *would* pay for these land services if given the chance. At issue is not only how much the services are really worth but also what is a fair price to the landowner. This information would help sharpen a rationale for easement purchase programs, building the case for a defined public interest, and would help individual groups better define their own policy priorities.

Several studies have estimated willingness to pay to retain farmland services in the presence of development pressures. Bergstrom et al. (1985) surveyed residents of Greenville County, South Carolina, regarding what they would pay to avoid developing an area of farmland; Beasley et al. (1986) used bidding games to estimate annual willingness to pay to avoid development of a specific area of Alaska, and Halstead (1984) used a similar approach in Massachusetts. Breffle et al. (1998) estimated an average willingness to pay at $192 per local resident to protect a neighborhood open area near Boulder, Colorado, using a contingent valuation survey. Other studies are reviewed by Racevskis et al. (2000). All are credible estimates of hypothetical willingness to pay for the amenity attributes of specific areas, though they are subject to the usual bias confronting hypothetical rather than actual payments. Rather than this general "benefits received" rationale, OECD countries employ an opportunity-cost procedure, paying farmers the cost they incur, including income foregone, in providing the conservation amenity (Hanley et al. 1998).

People also value the various ecosystem services of farmland (such as flood mitigation, composting, and new biological medicines). That value includes the direct utility of those services and the broader "heritage value" of assuring ecosystem opportunities for future generations, or just the inherent existence value of those systems (Toman 1995). Costanza et al. (1997) made a heroic and quite useful attempt to estimate the value of all ecosystem services in terms of damage avoided or the cost of acquiring those services from an alternative source.

Revealed preference analyses infer how much consumers value certain ecosystem or amenity services of land from the actions they take. Hedonic pricing models help clarify the increment of value associated with particular amenity services of farmland. Demand for amenity value also can be estimated from travel costs: people will bear substantial cost and inconvenience to visit a land resource setting, presumably an indication that the benefit received is at least as much as the cost to get to it (see Smith 1993).

All such methods of estimating revealed preference, contingent value, or stated preference values associated with the nonmarket services of farmland have their limitations. Boxall et al. (1996) argue that choice experiments, including conjoint analysis of multiattribute products or services, are preferable to contingent valuation techniques because the former permit the consumer to choose between attributes of the target landscape. Repeated sets of choices permit statistically reliable conclusions about *which* landscape attributes are valued and by how much. Contingent valuation studies, on the other hand, seldom offer the consumer a chance to consider substitutes for the attributes in question. Land services associated with an endangered species habitat or the innate value of species diversity are particularly difficult to value and their estimates are less reliable (Hodge 2000). But contingent valuation has the advantage of being more transparent to consumers and therefore is more easily administered (Scarpa 2000).

Private alternatives to public purchase of farmland amenity rights are functioning in the United Kingdom. Conservation, Amenity, and Recreation Trusts (CART) enable those who want these farmland services to express and implement that demand through private conservation organizations. A particular CART may concentrate on such farmland services as bird habitat or scenic landscape. Although some of these services are nonrival but excludable, most are pure public goods provided from private donations. Conservation organizations such as the Nature Conservancy or Ducks Unlimited apply membership fees and donations in a similar way in the United States. These organizations may purchase full-fee simple rights to land or just the specific rights needed to implement their land use preferences. Voluntary contributions for desired land-use patterns would seem to be the middle ground between producer-gets and beneficiary-pays programs more private than general support of public PDR programs, but more public than user charges for exclusive land services.

Monetary incentives that encourage farmer actions to reduce water pollution and protect fragile ecosystems are prominent in the 1996 Federal Agricultural Improvement and Reform Act (Ogg 1999) and in general give landowners the chance to find the least expensive way to meet an environmental standard (Batie and Ervin 1999). Incentive-based instruments are clearly central to future environmental policy affecting agriculture. Early discussion of conservation provisions for the expected 2002 farm bill has focused on incentive instruments for changing farmer behavior in the interest of water quality, protecting wetlands, and taking the most erosive or environmentally sensitive lands out of production. "Green payments" are proposed to compensate farmers for the various services provided. A proposed amendment to the 1985 Food Security Act would establish the Conservation Security Program through which farmers and ranchers would undertake defined nutrient management and other sound conservation practices in return for rental payments from USDA. Level of payment would depend on level of conservation employed (S-1426, Senator Tom Harkin, July 22, 1999). Incentive payments to secure certain environmental services or reduction in environmental damage from conventional farming practices are a "new generation program which reflects the needs of a global economy" and a GATT-legal payment to farmers (Batie 1998).

Beneficiary-Pays Programs

Transfer of development rights (TDR) is more clearly a "beneficiary pays" approach to privatizing provision of farmland amenity services. The landowner with zoning-granted permission to develop must first purchase an additional right from a landowner in a zoning district that does not permit development (AFT 1997, 121–144). Government creates the market for development rights and establishes the number of development rights to be exchanged in the "market for development" and the zoning structure within which it must operate. The rights being acquired are rival and exclusive, voluntarily conveyed from one owner to another at a price that reflects their market value. Montgomery County, Maryland, has by far the most successful TDR program with a significant amount of farmland permanently protected while new housing is being built in other areas of the county. The program was created following a significant "downzoning" that reduced development potential on large areas of farmland in the county. TDR made the zoning change more acceptable to the farmers involved.

Other user-purchased services of farmland include hunting and fishing rights, on-farm recreation opportunities, wildlife viewing, and gathering wild mushrooms (Merlo 1996; P. Bromley 1990). Homebuyers will pay a premium to be located near open farmland. Again, those desiring the specific point source amenity services are able to express effective demand and purchase them without direct government interference. Advertising and labeling can help market those farmland amenities, as in promoting the specific features of a farm "bed and breakfast" opportunity, or the scenic virtues of a residential community near farmland.[5] Eco-labeling can enable consumers to acquire good farmland stewardship along with the food product. California Clean Growers, for example, grant their seal of approval only when growers follow certain stewardship practices while protecting wildlife, scenic countryside, and other land services. No monitoring of the precise service at its source takes place, but consumers can express their general support for such services through their buying decisions (van Ravenswaay and Blend 1999).

Mandatory Action by Landowner

Farmer and former Assistant Secretary of Agriculture Jim Moseley recently observed about voluntary and compensatory approaches to soil conservation, "About 20 percent of the farmers use good conservation practices because it's the right thing to do. They will protect the land with or without payment. Another 75 percent respond to targeted incentive payments. But 5 percent of the farmers will oppose *everything* a government asks them to do. So successful land protection requires *both*—incentives and disincentives" (Moseley 1999). "Often misunderstood, or deliberately downplayed, is the need for a pre-existing regulatory framework that provides the incentive to reduce environmental problems" (Heimlich and Claassen 1998, 101). All management choices by farmers are made in the context of the institutional structure that defines a market. Any market is a collection of rules that establishes conditions for transactions among buyers and sellers. Changes to these conditions come as policy-driven adjustments to rights and obligations of market participants. Land-use zoning, conservation compliance, water-quality standards, and other regulatory mechanisms establish the boundaries of acceptable action by farmers managing the many services of farmland. Although past policy has relied heavily on voluntary, incentive-based techniques for agriculture, future U.S. farmland protection programs will include this important regulatory element—privatization of amenity services *within* a redefined market structure.

Limits on regulatory or mandatory approaches to achieving certain land-use patterns are a function of both law and political culture. Acceptable limits on public regulation of private owners differ among nations and among states within the U.S. Regulatory approaches imply that landowners have the general obligation to provide a mix of services that meets at least the minimal expectations of the broader public. Put another way, non-owners have rights to a certain bundle of land services and rights of the owner are limited by those non-owner rights. Land-use zoning is undertaken to "protect the health, safety, and general welfare" of the public. Measures of those general standards reflect prevailing social attitudes and are established in legislative language and court tests of specific regulations.

Agricultural Zoning
Agricultural zoning to direct development away from good farming areas and encourage a land market that supports farming will be an increasingly important part of the farmland policy mix.[6] Legal challenges of specific ordinances have generally focused on procedural and substantive due process (that is, how the ordinance relates to a valid public interest and assurance that appropriate procedures were followed) and "the taking issue." Much has been written about regulatory taking with little clear consensus. The legal criterion for "taking" is that land-use regulations are so restrictive on the owner that they remove virtually all income-earning potential from the land, thus "taking" the land for public use without compensation. Public limitation of landowner discretion is a deeply held ideological concern, both by those supporting and those opposing zoning. Some argue that regulation of private farmland is an unfair distribution of the cost of protecting amenity values, contrary to the basic freedoms under which the United States was founded and (worst of all) a distortion that makes land markets inefficient (Mills 1989).

Others emphasize the responsibilities of land ownership that accompany rights. No owner's rights are absolute. Property is a social construct, created for various practical purposes (see D. Bromley, 1993). The persistent scholar can find precedence for nearly any position on the matter (see Yandle 1995; Cordes 1999). In his thoughtful critique of agricultural zoning and various voluntary methods, environmental attorney Mark Cordes (1999) concludes that "Effective farmland preservation programs will have to restrict a landowner's ability to convert (farmland to nonfarm use) by relying on techniques that place decision authority elsewhere, most notably the government. The most common and least expensive way this can be done is . . . agricultural zoning" (p. 1047). He acknowledges the need to blend the public purpose of farmland protection with other public goals, such as affordable housing.

Supreme Court testing of agricultural zoning statutes has generally held them to be acceptable restriction of private landowners, and not regulatory taking. The significant Penn Central case in New York City found that although city zoning to protect the heritage value of Grand Central Station would alter Penn Central's "investment-backed expectations" in their plan to demolish the station and build a fifty-five story skyscraper, it did not constitute a taking, because adequate economic opportunities remained with the owner (403 US 104, 1977). The more recent *Lucas v. South Carolina Coastal Council* (505 US 1003, 1992) found that all economically beneficial and productive use of land was regulated away, and the coastal ordinance was declared an unconstitutional taking.[7] Cordes states that in general agricultural zoning is more like Penn Central than Lucas, and farming remains an economically viable use. An agricultural preference district would permit the business adjustments that a farmer must make to remain viable. If a particular ordinance did remove *all* economic options from a currently nonviable farm, then it could presumably constitute a regulatory taking. The facts of any case are unique, though precedence seems to support agricultural zoning.

Mere loss of economic potential is not adequate basis for invalidating a zoning ordinance. Courts have asserted that *some* level of land-use restriction should be anticipated by all landowners in today's society. Some background risk is associated with any ownership of property. All people are regulated to some degree, and a certain reciprocity exists among land owners. Regulation will expand opportunities for some while limiting them for others, and most people must come out ahead once in a while. Even "downzoning" that reduces allowed residential density on farmland would probably not constitute a taking if the land remains a viable farm. Downzoning is *politically* difficult but usually legally valid. Any ordinance must be internally logical, well designed, and consistent with a comprehensive plan or a state court may find it invalid.

Zoning more often hinges on political rather than legal acceptability. Political acceptability of agricultural zoning goes beyond legal questions to matters of equity—whose rights should be altered to achieve public purpose? Conventional wisdom holds that farmers and other landowners will always oppose zoning. In fact, however, many farmers realize that consistent and reasoned zoning can *expand* their opportunities, not just reduce them. Zoning can reduce farmer/neighbor conflicts, relieve rural congestion, and in some cases increase rural land values

(Heneberry and Barrows 1990). Recent analysis by Esseks et al. (1998) indicated farmers' general willingness to share the cost of regulations to achieve "smart" growth.[8] A national survey by Esseks et al. (1998) of 1,729 owners of at least five acres of farmland showed that 58 percent preferred regulation, 16 percent preferred incentives, and 13 percent said leave it to private settlement in court or the marketplace in considering how to protect farmland from residential development.

Approximately 75 percent of respondents felt that any compensation to landowners for zoning-induced loss of land value should net out any government "givings" in the form of a newly paved road next to a farm. Among members of eighty-one county advisory councils for the Ohio Farm Bureau, 51 percent supported stronger county or township rural zoning (Ohio Farm Bureau 1998). A large number of county farmland preservation committees in Ohio identified better zoning as a key strategy for future farmland protection efforts.[9] Thus, ample evidence shows that the more mandatory approaches to providing the amenity services of farmland have support among those who are likely to be most directly affected by the regulations. Assurance of future land-use patterns and reduced uncertainty are important to landowners. Zoning *that is well designed and consistently administered* (essential qualifiers) has a prominent place in future farmland policy.

Total Maximum Daily Load

Another set of potentially mandatory limits on farmer management choices concerns implementation of "total maximum daily load" (TMDL) provisions of the U.S. Clean Water Act of 1972. The idea is that a certain ambient water-quality level should be sustained in all water bodies by tracing pollutants back to their source and imposing controls necessary to ensure that the daily load of a particular pollutant for that stream segment is not exceeded. Point sources ("end of pipe") have been reasonably well controlled, and nonpoint sources (including agriculture) are now the center of attention. In fact, agricultural runoff is the single largest contributor to remaining water-quality problems of U.S. rivers and lakes. The U.S. Environmental Protection Agency (USEPA 2000) has asserted that 59 percent of excessive loading in analyzed stream segments comes from agriculture. TMDLs have always been a part of the Clean Water Act but were largely ignored until USEPA issued a renewed call for action in 1996. Draft rules for implementing TMDLs and other provisions were issued in August 1999 and finalized in July 2000.

Agriculture has largely avoided mandatory water quality standards in the past, but the socio-political pressure for cleaner water and the general sense that point sources are already regulated will likely place greater demands on farms. A joint statement by USDA and USEPA (May 1, 2000) renewed support for voluntary and incentive-based approaches to reducing farm-generated water pollution. Farmers will be given credit against TMDL for voluntary reduction in runoff through the use of conservation practices. Various techniques for trading "pollution rights" among point and nonpoint source polluters are being studied and tested. States will have considerable discretion in identifying specific nonpoint contributors to water-quality problems and developing best-management practices for forestry point sources and livestock operations. The requirement for a USEPA permit for livestock operations of fewer than one thousand animal units and forestry operations was dropped from the final rule, though new USEPA proposals out for comment in early 2001 bring the possibility back.

Thus, agriculture is still handled gently but, if improvement is not forthcoming, more mandatory measures will surely follow. Large farms will likely be at the top of the enforcement list, but any farm permitting pollution from livestock or crop operations will be under scrutiny. In fact, large farms with the better land and other resources may be more able to reduce runoff than smaller farms operating with limited resources. Greater attention to agricultural nonpoint pollution could thus be an additional influence on structural trends toward large farms.

The inherent difficulty of determining responsibility among nonpoint sources will remain, but the pressure is on. Uncertainty of source alone will not stop the TMDL strategy, though it will likely become the key point of contention in the implementation phase (Boyd 2000).

Conservation Compliance

Mandatory features of conservation policy were first introduced into farm legislation with the 1985 Food Security Act, updated in the 1990 farm bill and modified, though retained, in the 1996 Federal Agriculture Improvement and Reform Act. The *conservation compliance* feature is only quasi-mandatory because farmers not participating in federal farm benefit programs need not comply. The concept is that only by providing certain environmental and amenity services of farmland could the farmer retain eligibility for such payments as environmental quality incentives, production flexibility payments, disaster payments, Conservation Reserve Program, crop insurance, storage loans—virtually any USDA payment. This mandatory approach is not popular with farm groups or legislators but is strongly supported by natural resource elements of the coalition for farm legislation. Further, the intuitive logic of requiring certain farmland behavior as a condition for USDA payments seems reasonable to many people. Although these mandatory features will be strongly debated with the next farm bill, popular sentiment seems to expect sound land stewardship from the nation's farmers as a condition for continuing the generous farm programs.

Tweeten and Zulauf (1997) propose an "environmental compliance program" that could be voluntary with green payments, or mandatory. The ECP would go beyond current compliance policy to include all aspects of a farm that affect the environment. They argue that farmers should be expected to eliminate (or reduce) negative externalities because they are being asked only to stop doing "bad" things ("taking" from others) but should be compensated for providing positive environmental amenities enjoyed by the larger public. If Moseley and Napier are correct, voluntary measures alone will not work; mandatory environmental compliance could follow the current mode of linking to eligibility for other USDA programs or follow the more straightforward enforcement procedures of USEPA.

CONCLUDING OBSERVATIONS

Markets Redefined
Markets will continue to play the central role in future farmland allocation decisions. Preferences of buyer and seller are paramount in determining the products or services from farmland. As with all markets, land markets consist of sets of rules that structure relationships among buyers and sellers and verify results of transactions. Some farmland services purchased involve direct personal experience on the land itself, as habitat for game or for other recreation opportunity. But other services are nonexclusive and nonrival in character. For these, market rules may be altered to provide opportunities for people who do not own land and cannot purchase desired services from those who do. Buyer and seller still prevail, but with a slightly different set of options to choose from. *It is not a question of market versus nonmarket, but a more complete understanding of what markets really are* (see Bromley, D., 1997). This conclusion is consistent with the Tweeten and Zulauf (1997) "new paradigm for agricultural policy" in which commodity markets replace price supports and other direct interventions.

Combinations of Privatized and Mandatory
The two major threads of future farmland policy, privatization and mandatory limits on landowners, are not discrete alternatives but rather parts of an evolving strategy. Privatization of natural resource and environmental policy is clearly the banner, but the regulatory context within which privatization occurs will change as well. Taxpayers and voters will increasingly demand sound land stewardship and adherence to general standards of growth management. People will expect the same from farmers as from other businesses and land users. Regulations will be a part of the policy setting, and plenty of evidence shows that farmers understand and accept the broader social responsibilities of ownership.

The Twenty-First Century Farmer
American farms and farmers are the envy of the world, central to the general quality of life enjoyed throughout the nation. New farm technologies have released people from the land and freed millions of acres for other uses. As the absolute quantity of

farmland diminishes, its scarcity creates substantial value, *only some of which is reflected in current land, commodity, or service markets.* People care about farms, even those who have little idea what a real farm is. They have images of openness, freshness, health, peacefulness, and beauty—unrealistic, perhaps, but genuine. That perception is shared by consumers at all income levels. Farmers are among the most highly respected members of twenty-first century society (perhaps there is a scarcity value there, as well). Farmers have many allies, for many different reasons, and should expect continued support when the land they manage continues to create a social value. That pool of support is broad but shallow and can dissipate quickly if the trust is abused. Mandatory measures should provide the safety net as we move beyond Moseley's 20 percent who accept the many responsibilities of land ownership.

Farmers have a real stake in land policy options that leave the primary discretion with the business manager. Consumers who want those other nonfood services also have a stake in the continued viability of the farm business. An obvious need exists for farmers and consumers to work more closely together in designing and implementing policies that serve their mutual interests.

NOTES

1. Debate on the appropriate rebate procedure to apply when a participating farmer decides to sell for nonfarm development sidetracked the Michigan use value proposal in mid-2000. Farm interests prefer a rollback payment based on the difference between farm and taxable market value, whereas other land-use groups insist that rollback be a certain percentage of market value at time of conversion. The rollback is to be used to purchase development rights to Michigan farmland. The Farm Bureau approach would generate much less revenue for that purpose, adding to the political debate (Bowers, June 2000).

2. Ohio is listed as a state with agricultural protection zoning, but enabling language for county and township ordinances does not include the "general welfare" purpose for local ordinances. Thus, unless there is a clear public health or safety rationale, Ohio counties and townships have felt a lack of authority for open land or agricultural zones (Meck and Pearlman 2000). Even though twenty-nine of Ohio's eighty-eight counties report having some local "agricultural preference zones," 87 percent of those zoned townships permit minimum lot sizes of less than three acres, hardly a barrier to residential development in agricultural zones (Stamm 1999).

3. Most state programs have "escape clauses" that enable the farmer to buy back development rights under extreme unforeseen circumstances in which farming is no longer a reasonable option for those lands. Massachusetts requires specific legislative action to reverse an easement sale. Maryland and Pennsylvania require a twenty-five-year waiting period; New Jersey has no escape provisions at all

(Daniels and Bowers 1997).

4. In Ohio, development interests critical of farmland preservation in a time of food abundance and the apparent lack of farmland shortage attempted to have "farmland" removed from the list of eligible lands in the bond initiative proposed by the Governor. The legislation ultimately passed with farmland included and was voted on in a November 2000 referendum.

5. Tryon Farm near LaPorte, Indiana offers "... a rare combination of new simple houses, environmentally preserved natural settings and the traditional farming cycle of planting and harvest. Approximately three quarters of this farm will always be preserved as rolling farmland, meadows, woods, and ponds." Kane County, Illinois, calls attention to the historic and architecturally significant farmhouses and barns as key to the county heritage, making Kane County a pleasant place to live and work (Kane County Development Department 1991). Thus, farmland amenities are specifically highlighted in seeking a pattern of private spending decisions that will benefit the county economy. The farmland amenities are thus privatized.

6. A frequently used definition of agricultural preference zoning is those ordinances with minimum residential lot size of at least twenty acres that support agricultural activities and significantly restrict nonfarm uses. By that definition, Ohio and perhaps others of the twenty-two states listed by AFT (1997)have little if any real agricultural preference zoning.

7. In this case, David Lucas paid nearly $1 million for two developable coastal lots. The Coastal Council subsequently passed a preservation law that prohibited development. The U.S. Supreme Court found that to be a regulatory taking, but added that were the ordinance simply prohibiting a common law nuisance, it would not be a taking (Cordes 1999, 1053–1059; see also Rinehart and Pompe).

8. Much of the "cost" of land use regulation is actually value generated by government spending for roads, sewer and water, tax incentives for home ownership, and other "givings" of public policy. Thus taxpayers have created a large part of the private land value for which they may be asked to once again compensate the landowner in a "taking" case (Runge et al. 2000).

9. As of August 1, 2000, thirty-five of sixty-one eligible Ohio counties have submitted farmland protection plans, partially funded by grants from the Ohio Department of Development. A summary of plan elements was prepared by Jill Clark of the Ohio office of The American Farmland Trust.

REFERENCES
403 U.S. 104, 1977.

505 U.S. 1003, 1992.

American Farmland Trust (AFT). *Saving American Farmland: What Works.* Northampton, MA: The American Farmland Trust, 1997.

Batie, S. "Green Payments as Foreshadowed by EQIP." Staff Paper 99–45. East Lansing: Department of Agricultural Economics, Michigan State University, July 1998.

Batie, S., and D. Ervin. "Flexible Incentives for Environmental Management in Agriculture: A Typology." In *Flexible Incentives for the Adoption of Environmental Technologies in Agriculture.* F. Casey, A. Schmitz, S. Swinton, and D. Zilberman, eds., Norwell, MA: Kluwer Academic Publishers, 55–78, 1999.

Beasley, S., W. Workman, and N. Williams. "Estimating Amenity Values of Urban Fringe Farmland: A Contingent Valuation Approach." *Growth and Change* 17 (1986): 70–78.

Bergstrom, J., B. Dillman, and J. Stoll. "Public Environmental Amenity Benefits of Private Land." *Southern Journal of Agricultural Economics.* 17 (July 1985): 139–149.

Bowers, D. "Michigan Farm Bureau Joins Homebuilders: Legislation Fails." *Farmland Preservation Report* 10 (June 2000): 8.

Bowers, D. "Pennsylvania's Current Year Funding is Highest in Nation." *Farmland Preservation Report* 10 (January 2000): 3.

Bowers, D. "Nation's Farmland Programs Vary Widely in Funding, Politics." *Farmland Preservation Report* 10 (March 2000): 5.

Boyd, J. "Unleashing the Clean Water Act: The Promise and Challenge of the TMDL Approach to Water Quality." *Resources.* 139 (Spring 2000): 7–10. Washington, DC: Resources for the Future.

Boxall, P., W. Adamowicz, J. Swait, M. Williams, and J. Louviere. "A Comparison of Stated Preference Methods for Environmental Valuation." *Ecological Economics* 18 (1996): 243–253.

Breffel, W., E. Morey, and T. Lodder. "Using Contingent Valuation to Estimate a Neighborhood's Willingness to Pay to Preserve Undeveloped Land." *Urban Studies* 35 (1998): 715–727.

Bromley, D. "Regulatory Takings: Coherent Concept or Logical Contradiction?" *Vermont Law Review* 17 (1993): 653–665.

Bromley, D. "Rethinking Markets." *American Journal of Agricultural Economics* 79 (1997): 1383–1393.

Bromley, P. "Wildlife Opportunities: Species Having Management and Income Potential for Landowners in the East." In *Income Opportunities for the Private Landowner Through Management of Natural Resources and Recreational Access*, W. Grafton and A. Ferrise, eds. Morgantown, WV: West Virginia Cooperative Extension Service, 1990.

Bunce, M. "Thirty Years of Farmland Preservation in North America: Discourses and Ideologies of a Movement." *Journal of Regional Studies* 14 (1998): 233–247.

Constanza, R., R. d'Arge, R. deGroot, S. Farber, M. Grasso, B. Hannon, K. Linsburg, S. Noeem, R. O'Neill, J. Paruelo, R. Raskin, P. Sutton, and M. vandenBelt. "The Value of the World's Ecosystem Services and Natural Capital." *Nature* 387(1997): 253–260.

Cordes, M. "Takings, Fairness, and Farmland Preservation." *Ohio State Law Journal* 60 (1999): 1033–1084.

Daniels, T. and D. Bowers. *Holding Our Ground: Protecting America's Farms and Farmland.* Washington, DC: Island Press, 1997.

Esseks, D., S. Kraft, and L. McSpadden. "Owners' Attitudes Towards Regulation of Agricultural Land." CAE Working Paper 98–3. DeKalb, IL: Center for Agriculture in the Environment, May 1998.

Farmer, M. and A. Randall. "The Rationality of a Safe Minimum Standard." *Land Economics* 74 (August 1998): 287–302.

Foster, K., P. Vecchia, and M. Repacholi. "Science and the Precautionary Principle." *Science* 288 (May 12, 2000): 979–981.

Hanley, N., H. Kirkpatrick, I. Simpson, and D. Oglethorpe. "Principles for the Provision of Public Goods from Agriculture: Modeling Moorland Conservation in Scotland." *Land Economics* 74 (February 1998): 102–113.

Halstead, J. "Measuring the Non-Market Value of Massachusetts Agricultural Land." *Journal of the Northeast Agricultural Economics Council* 13 (1984): 12–19.

Heimlich, R. and R. Claassen. "Agricultural Conservation Policy at a Crossroads." *Agricultural and Resource Economics Review* 27:1 (April 1998): 95–107.

Heneberry, D. and R. Barrows. "Capitalization of Exclusive Agricultural Zoning into Farmland Prices." *Land Economics* 66 (1990): 249–258.

Hodge, I. "Current Policy Instruments: Rationale, Strengths, and Weaknesses." Unpublished paper prepared for a joint OECD-USDA workshop. Washington, DC: U.S. Department of Agriculture, June 2000.

Kane County Development Department. "Built for Farming: A Guide to the Historical Rural Architecture of Kane County." Geneva, IL: Kane County Development Department, May 1991.

Kline, J. and D. Wichelns. "Measuring Heterogeneous Preferences for Preserving Farmland and Open Space." *Ecological Economics* 26 (1998): 211–224.

Kline, J. and D. Wichelns. "Public Preferences Regarding the Goals of Farmland Preservation Programs." *Land Economics* 72: 538–549.

Lehman, T. *Public Values, Private Lands.* Chapel Hill: The University of North Carolina Press, 1995.

Libby, L. "Improving the Farmland Policy Options for Ohio Local Government." Unpublished staff paper. Columbus: Department of Agricultural, Environmental, and Development Economics, Ohio State University, June 1999.

Libby, L. "In Pursuit of the Commons: Toward a Farmland Protection Strategy for the Midwest." CAE/WP97-2. DeKalb, IL: American Farmland Trust, 1997.

Libby, L. and P. Stewart. "The Economics of Farmland Conversion." In *Under the Blade: The Conversion of Agricultural Landscapes*, R. Olson and T. Lyson, eds. Boulder, CO: Westview Press, 1999.

Meck, S. and K. Pearlman. *Ohio Planning and Zoning Law.* Cleveland, OH: West Group, 2000.

Merlo, M. "Commoditisation of Rural Amenities in Italy." *Amenities for Rural Development: Policy Examples.* Paris, France: Organization for Economic Cooperation and Development, 85–95, 1996.

Mills, D. "Is Zoning a Negative Sum Game?" *Land Economics* 65 (February 1989): 1.

Moseley, J. Unpublished remarks at The Nature Conservancy conference, Purdue University, West Lafayette, IN, October 28, 1999.

Napier, T. "Regulatory Approaches for Soil and Water Conservation." In *Agricultural Policy and the Environment*, L. Swanson and F. Clearfield, eds. Ankeny, IA: Soil and Water Conservation Society, Chapter 15, 1994.

The New York Times. June 26, 1972: A24.

Ogg, C. "Evolution of EPA Programs and Policies that Impact Agriculture." In *Flexible Incentives for the Adoption of Environmental Technologies in Agriculture,* F. Casey, A. Schmitz, S. Swinton, and D. Zilberman, eds. Norwell, MA: Kluwer Academic Publishers, (1999): 27–42.

Ohio Farm Bureau Federation. "Farmland Preservation: Defining Our Community." Report of Advisory Council Discussions. Columbus: Ohio Farm Bureau Federation, January 1998.

Racevskis, L., M. Ahearn, A. Alberini, J. Bergstrom, K. Boyle, L. Libby, R. Paterson, and M. Welsh. "Improved Information in Support of a National Strategy for Open Land Policies: A Review of Literature and Report on Research in Progress." Unpublished paper. Columbus, OH: Department of Agricultural, Environmental, and Development Economics, 2000.

Real Estate Research Corporation. *The Costs of Sprawl.* Washington, DC: Government Printing Office, 1974.

Rinehart, J. and J. Pompe. "The Lucas Case and the Conflict Over Property Rights." In *Land Rights: The 1990s Property Rights Rebellion,* B. Yandle, ed. Lanham, MD: Roman and Littlefield Publishers, 1995.

Runge, S., T. Duclos, J. Adams, B. Goodwin, J. Martin, R. Squires, and A. Ingerson. "Public Sector Contributions to Private Land Value: Looking at the Ledger." In *Property and Values: Alternatives to Public and Private Ownership*, C. Geisler and G. Daneker, eds. Washington, DC: Island Press, 2000.

Scarpa, R. "Contingent Valuation Versus Choice Experiments: Estimating the Benefits of Environmentally Sensitive Areas in Scotland: Comment." *Journal of Agricultural Economics* 51 (January 2000): 122–128.

Smith, K. "Non-Market Valuation of Environmental Resources: An Interpretive Appraisal." *Land Economics* 69 (February 1993): 1–26.

Stamm, J. "The Ohio Zoning and Land Use Survey." Unpublished report. Columbus: Ohio State University Extension, September 1999.

Thompson, E. "Raising the Bar: Farmland Protection Confronts 21st Century Sprawl." Unpublished presentation at American Farmland Trust PACE Conference, East Windsor, NJ, April 10, 2000.

Toman, M. "Ecosystem Valuation: An Overview of Issues and Uncertainties." Discussion Paper 94–43-Rev;3ed. Washington, DC: Resources for the Future, Inc. March 1995.

Tweeten, L. "Competing for Scarce Land: Food Security and Farmland Preservation." Occasional Paper, ESO# 2385, Columbus, OH: Department of Agricultural, Environmental, and Development Economics, 1998.

Tweeten, L. and C. Zulauf. "Public Policy for Agriculture after Commodity Programs." *The Review of Agricultural Economics* 19 (Fall/Winter 1997): 263–280.

U.S. Environmental Protection Agency. "Final Rules on Clean Water Act Implementation." Washington, DC: *U.S. Federal Register* 65:135 (July 13, 2000): 48585–43745.

van Ravenswaay, E. and J. Blend. "Using Ecolabeling to Encourage the Adoption of Innovative Environmental Technologies in Agriculture." In *Flexible Incentives for the Adoption of Environmental Technologies in Agriculture,* F. Casey, A. Schmitz, S. Swinton, and D. Zilberman, eds. Norwell, MA: Kluwer Academic Publishers, (1999): 119–138.

Weitzman, M. "Just Keep Discounting, But ...." In *Discounting and Intergenerational Equity*, P. Portney and J. Weyant, eds. Washington, DC: Resources for the Future, 1999.

Yandle, B., ed. *Land Rights: The 1990s Property Rights Rebellion.* Lanham, MD: Rowman and Littlefield, 1995.

11

Kuznets Curves for Environmental Degradation and Resource Depletion

Aref A. Hervani and Luther Tweeten

This chapter tests the hypothesis that the relationship between environmental variables and income is an inverted U-shape. Notable new contributions of this study are (1) to estimate impacts of income on the environment through population growth, and (2) to include a wider range of environmental (including natural resources) variables than in previous studies. We estimate environmental Kuznets curves using cross-sectional data by country for several greenhouse gases, organic water pollution, and some natural resources. Our results mostly support the hypothesis of an inverted U-shaped relationship between income on the one hand and environment degradation and resource depletion on the other.

INTRODUCTION

Environmental quality and food security may be compromised if greenhouse gas emissions (GHG) and natural resource- (such as petroleum) depletion continue to mount as income rises. On the other hand, rising income under economic growth can slow resource depletion and environmental degradation for reasons discussed later. For example, greater productivity in agriculture from investments in science and education made feasible by economic growth enables cropping of fewer and environmentally safer acres while freeing more land to be used for grass, trees, recreation, and biodiversity-protecting resources. An environmental dilemma is that environmental degradation and natural-resource depletion caused by rising per-capita income on a global scale may severely degrade the environment before the turnaround to positive benefits of higher income can occur. The demands placed on the environment and natural resources in early economic growth stages can be of lesser concern if further economic progress protects the environment and natural resources.

The objective of this chapter is to test the hypothesis that the income elasticity of environmental degradation and resource depletion turns from positive at lower income levels to negative at higher income levels. In testing this hypothesis, we evaluate the impact of per-capita income growth on (1) per-capita environmental degradation and natural-resource depletion and (2) population growth. Rising environmental degradation and resource depletion *per capita* could be offset by less population growth under rising income, causing the total income elasticity to turn from positive to negative at some income level.

BACKGROUND
Several recent studies of the relationship between economic growth and environmental degradation have found the inverted U-shaped relationship of the Kuznets curve (Grossman and Krueger 1995; Ruddle and Manshard 1981; Selden and Song 1994; Tucker 1995; Holtz-Eakin and Selden 1995; Stern and Common 1996; Lopez 1994; Suri and Chapman 1998; and Agras and Chapman 1999). The inverted U-shaped curve suggests that countries at the early stages of economic growth exploit the environment. Low income per capita tends to be attended by high discount rates on present versus future consumption. A high discount rate discourages spending on environmental improvement capital as high population growth places pressure on the environment. Low-income countries cannot afford efficient waste control or disposal technologies.

As their income rises, consumers spend a smaller share of their income on goods and a larger share of their income on services whose production and disposal are less detrimental to the environment. Education under economic progress can promote awareness of environmental hazards, which in turn can generate effective policy responses. As countries become more affluent, they invest more in science and technology that can reduce pressures on the environment. Higher income allows nations to invest more in environmental regulation, control, and enforcement. Higher income also may bring lower birth rates and population growth, further reducing pressure on the environment. Thus, environmental degradation may reach a turning point as countries reach higher income levels per capita.

Antle and Heidebrink (1995) tested Vernon Ruttan's (1971, 707, 708) hypothesis that "in relatively high-income countries the income elasticity of demand for commodities and services related to sustenance is low and declines as income continues to rise, while the income elasticity of demand for more effective disposal of residuals and for environmental amenities is high and continues to rise. This is in sharp contrast to the situation in poor countries where the income elasticity of demand is high for sustenance and low for environmental amenities." Antle and Heidebrink (1995, 620) found that the two classes of environmental amenities they examined, parks and forests, were negatively related to income in countries with less than $1,200–2,000 income per capita. Income elasticities were greater than unity at higher incomes. Stokey (1998) looks at the shape of the cost function for regulation of the environment as a primary factor driving the Kuznets curve.

Studies by Eckholm (1976), Ruddle and Manshard (1981), and World Bank (1984) concluded that increasing population—land ratios energize processes of extensification and intensification. In the process of extensification, the demand for fuel, building materials, and land for crops and livestock grows as population increases from low levels, forcing people onto new land. As population increases from low levels and new lands are unavailable, demographics pressure for intensification as farmers apply more inputs in cultivating for higher yields.

Soil degradation is exacerbated by a combination of low income and high population density, fostering intensification. After analyzing data from various countries, Grepperud (1996), Kuru (1986), Srinivasan (1985), and Whitlow (1987) concluded that increasing population is associated with degradation of land. Soil degradation increased in the United States for decades as population and cultivation expanded. But rising national income afforded investments in science and technology that in turn enabled the nation to reduce sheet and rill (water) erosion rates on cropland from nine tons per acre in 1938 to three tons per acre by the late 1990s (see Tweeten and Amponsah, 1998, 49). Wind erosion of soil also decreased.

Cropper and Griffiths (1994) look at the interaction of population growth and environmental quality through an examination of the effect of population pressures on deforestation in sixty-four developing countries. They estimated this relationship using pooled cross-sectional and time-series data for each continent and, to capture the effects, they used the rural population density and the rate of population growth in the equation. Their results indicated that the levels of income at which rates of deforestation peak are $4,760 for Africa and $5,420 for Latin America. Chaudhuri and Pfaff (1998) found an inverted U-shaped curve between air quality and environment.

Gale and Mendez (1996) concluded that pollution rises with the capital abundance of a country and falls with land abundance. Whitlow (1987) examined erosion and land tenure patterns of Zimbabwe; Kuru (1986) examined land degradation in Ethiopia; and Grepperud (1996) examined the population pressure hypothesis for the Ethiopian Highlands; their research found that higher population pressure caused higher livestock stocking rates and intensive cultivation, which in turn caused degradation of land. High population growth coupled with poverty in the tropics is causing species extinction rates to rise in "hot spots" such as Madagascar and the Amazon Basin (World Resources Institute 1994, 147–162).

Turning from rural areas to cities, urban air quality, water quality, and sanitation have improved in many countries with increased income (World Resources Institute 1994, 163). Income growth supporting science and industry in urbanized developed countries gives rise to fiber optics and numerous other technologically improved inputs that substitute for and hence help conserve natural resources such as copper.

An early study by Van Bergeijk (1991) showed no consistent relationship between economic growth (GDP and trade) and emissions of nitrous, carbon, and sulfur oxides for OECD countries. Later studies by Shafik (1994) and by Grossman and Krueger (1995) suggest an inverted U-shaped relationship between pollution and GDP levels. Seldon and Song (1994) found the inverted U-shaped relationship between pollution and economic development (GDP per capita) from cross-national panel data on emissions of four important air pollutants: suspended particulates, sulfur dioxide, oxides of nitrogen, and carbon monoxide. This result, implying that emissions will decrease in time with economic growth, is strengthened if lower per capita emissions are reinforced by the trend to zero or negative population growth induced by higher income per capita.

Holtz-Eakin and Selden (1994), using panel data, found a diminishing marginal propensity to emit carbon dioxide as GDP per capita rises. Hilton and Levinson (1998) investigated the relationship between automotive lead emissions and national income for forty-eight countries over twenty years. They concluded that lead emissions exhibit the inverted U shape of the environmental Kuznets curve and that this turning point occurs at $7,000 per capita. Shafik and Bandyopadhyay (1992) analyzed the relationship between particulate and sulfur dioxide emissions and GDP per capita and found an inverted U-shaped curve. Holtz-Eakin and Selden (1995) found an inverted U-shaped curve between carbon emission and GDP per capita

Two principal environmental problems that may not improve per capita with higher income per capita are greenhouse gas (GHG) accumulation and natural resource depletion. GHG accumulation contributes to global warming, and depletion of natural resources such as fossil fuels and rock phosphate that lack economic substitutes can threaten food supplies and living standards.

Most affluent industrialized nations have fertility rates below replacement rates and seem destined for zero population growth. This study tests empirically whether the reduction in natural resource depletion and GHG emissions from less population induced by income growth offsets any per capita growth in these variables from higher per capita income.

METHOD OF ANALYSIS

The conceptual framework for this study rests on numerous previous studies, including many cited in the previous section that conclude that environmental degradation and natural resource depletion are nonlinear functions of population level and density, and of income (Grossman and Krueger 1995; Selden and Song 1994; World Resources Institute 1994; Tucker 1995; Holtz-Eakin and Selden 1995; Suri and Chapman 1998; and Agras and Chapman 1999).

Because we desire to separate the income elasticities for per capita environmental efforts from population effects, and because reactions to incremental income differ between environmental and population equations, two specifications are required. Using cross-sectional data, single-equation demand models are estimated per capita for energy and components (oil, gas), greenhouse gases (CO_2, NO_2, SO_2, methane), heavy suspended particulates, phosphate fertilizer consumption, and organic water pollution as a function of income per capita and population densities. A base model (I) and variants (II) are presented, the latter to reduce multicollinearity and test sensitivity of results to alternative specifications.

The coefficients from the equations provide income elasticities for various environmental and resource variables. These elasticities are compared with the income elasticities (including direct and indirect effects) of population growth to determine whether income growth is a net positive or negative influence on environmental variables.

Environmental Equations
Environmental equations were expressed in the following form:

$$X_i = f (GDP, GDP^2, POPD, POPD^2, GDP * POPD, \varepsilon) \qquad (1)$$

where

X_i = environmental variables per capita;
GDP = gross domestic product per capita;
POPD = population density;
ε = error term.

Variable sources, units, and definitions are found in table 11.1.

Empirical models include an interaction term (GDP * POPD) to allow income growth to have a different effect on X_i in the presence of high population density, or conversely, for population density to have a different effect on X_i in the presence of high income growth. The squared terms allow a nonlinear relationship between environmental variables and explanatory variables. The model allows a statistical test of the Kuznets relationship and can be solved for a turning point, if any, where the per capita impact of incremental income on the environment turns from positive to negative. The coefficients on the linear variables are hypothesized to be positive, on the squared terms negative, and on the interaction term indeterminate. The latter is indeterminate because, although income growth per capita is expected to be more damaging to the environment in the presence of high population density, that income growth may afford the means to protect the environment if high population density creates pressures to alleviate environmental problems.

Table 11.1. Variable Descriptions, Units of Measurement, and Sources

Variables	Definition	Sources
CO_2	Carbon dioxide emissions from the burning of fossil fuels and the manufacture of cement. (annual metric tons of CO_2 emissions per capita; average of period 1990-1995)	International Energy Agency (1997) CO_2 Emissions From Fuel Combustion, 1972-1995
NO_2	Nitrogen dioxide emissions. (annual micrograms per capita; average for period 1990-1995)	The World Bank (1998) World Development Indicators, 1998
SO_2	Sulfur dioxide emissions. (annual micrograms per capita; average for period 1990-1995)	The World Bank (1998) World Development Indicators, 1998
Methane	Methane emission from solid waste sources (annual metric tons per capita; average for period 1990-1995)	The World Bank (1996) The World Resources, 1996-1997
SPM	Total suspended particulate emission from combustion that are in air. (annual micrograms per cubic meter of air, average for period 1990-1995)	The World Bank (1998) World Development Indicators, 1998
Worg	Organic water pollution emission measured by biochemical oxygen demand (kilograms per capita in 1993)	The World Bank (1998) World Development Indicators, 1998
Energy	Total final consumption of energy by all sectors in the economy (annual metric tons of oil equivalent (Mtoe) per capita; annual average for period 1990-1995	International Energy Agency (1997) Energy Statistics and Balances of NON-OECD Countries, 1994-1995
Oil	Total oil product consumption in the country (metric tons of oil equivalent (Mtoe) per capita; annual average per period 1990-1995)	International Energy Agency (1997) Energy Statistics and Balances of NON-OECD Countries, 1994-1995
Fertz	Phosphate fertilizer consumption. (metric tons of plant nutrients per capita; annual average for period 1990-1995)	FAO (1996) FAO Yearbook , FAO Statistics Series No. 126 Vol. 44. 1994.

POPG	Population growth (average annual growth rate for the period 1990-1995)	United Nations Development Programme (UNDP), 1996 Human Development Report 1996
GDP	Gross domestic product (GDP) (Laspeyres Index weighted by 1985 international prices in U.S. dollars per capita, 1990	Published for the World Bank, The John Hopkins University Press, Baltimore and London, World Tables 1996
POPD	Population density is the rural population divided by the total land area (people per square kilometer, 1990)	United Nations Development Programme (UNDP), 1996 Human Development Report 1996
URBAN	Urban population (percentage of population living in urban areas, 1990)	United Nations Development Programme (UNDP), 1996 Human Development Report 1996
ILTC	Adult illiteracy rate (adults aged 15 and above who cannot read and write as percentage of total population, 1990)	United Nations Development Programme (UNDP), 1996 Human Development Report 1996
FLBR	Women in labor force (percentage of the total labor force, 1990)	United Nations Development Programme (UNDP), 1996 Human Development Report 1996
FEMED	Females with education (percent of total population, 1990)	United Nations Development Programme (UNDP), 1996 Human Development Report 1996
CALC	Daily calorie supply per capita (calorie equivalent of the net food supplies per day per capita, 1990)	United Nations Development Programme (UNDP), 1996 Human Development Report 1996
IMR	Infant mortality rate (number of deaths of infants under one year of age during the indicated year per 1,000 live births in the same year, 1990	United Nations Development Programme (UNDP), 1996 Human Development Report 1996

Following Hotelling (1931), prices belong in equation 1 to measure rationing pressure as natural resource reserves are exploited. However, we do not include price for several reasons. One is that we measure environmental emissions or resource use, not reserves. Second, price data are too inaccurate to measure rationing constraints over time. Price data facing consumers of petroleum, for example, were available for only a few developed countries. So, like authors of other studies cited herein (Grossman and Krueger 1995; Selden and Song 1994), we did not include price. In addition, a study undertaken by de Bruyn et al. (1998) tested the inverted U-shaped Kuznets curve for three pollutants [CO_2, NO_2 SO_2] with energy prices included in the model for four individual countries and found energy prices to be insignificant in most cases.

Simple correlation coefficients among independent variables shown in table 11.2 indicate that some of the variables are highly correlated. Multicollinearity between and among GDP, GDP^2, POPD, $POPD^2$, and the interaction term (GDP * POPD) could destabilize individual regression coefficients, making estimates unreliable. Therefore, along with the original Model I containing all independent variables from (1), each model was re-estimated omitting highly correlated variables. Whether a more completely specified Model I provides more reliable coefficient estimates than a less completely specified Model II but with less multicollinearity is a judgment call.

Table 11.2. Simple Correlation Coefficient Matrix

	GDP	GDP^2	POPD	$POPD^2$	GDP*POPD
GDP	1.00	0.96	0.18	0.12	0.30
GDP^2		1.00	0.13	0.06	0.25
POPD			1.00	0.96	0.95
$POPD^2$				1.00	0.88
GDP*POPD					1.00

Source: See Table 11.1 for data.

In most instances, data for the dependent variable were an average for the 1990–1995 period to include information from recent years and to reduce error with mean data. The period average is used as in previous studies (Selden and Song 1994) to allow longer term average impact of the explanatory variables to be expressed without employing a lagged dependent variable or other complex structure introducing autocorrelated error structures. The independent variable in most instances was for 1990 because longer series were not available or reliable. The choice of years also helps to make the independent variable prior to and hence causal to the dependent variable.

Data sources range from low-income countries to high-income countries. Countries with missing data were omitted from the study. Numbers of countries, which necessarily differed among equations depending on availability of data, are reported with results.

Population Growth Equation

The equation for population growth (POPG) is specified differently than for the environmental variables X_i. Population growth is expressed as a function of per-capita income (GDP), infant mortality rate (IMR), illiteracy rate (ILTC), percentage of females in the labor force (FLBR), female education levels (FEMED), urbanization (URBAN), and food calories as a proportion of recommended food requirements per capita (CALC) (for conceptual framework, see Mankin et al. 1990):

$$POPG = f\,(GDP, GDP^2, ILTC, FLBR, IMR, URBAN, FEMED, CALC, \mu)$$

$$(2)$$

where the error term is μ. Variable units, sources, and definitions are given in table 11.1. The dependent variable is the average annual population growth rate for years 1990–1995, whereas all explanatory (right side or causal) variables are for 1990 unless otherwise indicated.

Income is hypothesized to have both direct and indirect impacts on population growth. The general model specified in equation (2) expresses a possible direct curvilinear relationship between population growth (dependent) and income (independent). Income also has an indirect impact on POPG by influencing the infant mortality rate (IMR), illiteracy rate (ILTC), and other independent variables. The indirect effects of income on population growth are derived through estimation of single equations where each explanatory variable Y_j in (2) is regressed on income per capita as in equation (3)

$$Y_j = f\,(GDP, GDP^2, \xi)$$

$$(3)$$

where ξ is the error term and other variables are defined as before. An upper estimate of the total effect of income per capita on population growth is measured by the summation of the direct and the indirect effects after inserting (3) into (2).

Equations (2) and (3) were estimated by two-stage least squares (2SLS) to account for endogeniety. Results were similar to results for single equations presented in this chapter, hence results for 2SLS are not presented but are available from the author.

Income Elasticity Equation

As indicated previously, a general hypothesis tested in this study is that, as income rises, slower population growth offsets some of the rising natural-resource depletion and GHG emissions per capita. Elasticities differ by income level, hence the hypothesis is tested at various levels of income per capita.

The total level X_i of an environmental variable or natural resource i is the per capita level X_i times population P or $\overline{X}_i = X_i$ P. The elasticity $E_G \overline{X}_i$ of $\overline{X}_i$ with respect to income per capita G is the sum of the elasticity $E_G X_i$ of the environmental or natural resource variable per capita X_i with respect to income per capita G plus the elasticity E_G P of population with respect to income per capita G, or

$$E_G \overline{X}_i = E_G X_i + E_G P. \tag{4}$$

It follows that if rising per capita income increases pollution per capita X_i, then overall pollution will decline with economic growth only if the elasticity of population with respect to income is sufficiently negative to offset $E_G X_i$.

Because the population growth rate dP/P is more sensitive than population to economic growth, dP/P rather than P is regressed on income per capita G. The elasticity of total population growth with respect to G or E_G P is calculated as

$$E_G P \cong G \ (d \ POPG \ / \ d \ G) \tag{5}$$

where the latter derivative is from equations (2) and (3), and G is GDP.

EMPIRICAL RESULTS

Results are presented first for greenhouse gas emission equations; then for natural resource equations; and finally for the population growth equation. Equations were estimated with cross-sectional data by ordinary least squares.

Heteroskedasticity is common in cross-sectional data and this study is no exception. The data in each equation were given Newey and West's (1987) heteroskedasticity test, which also tests for autocorrelated residuals. Because heteroskedasticity was found in nearly all equations, each of the reported empirical equations (including population-growth equations described later) was adjusted with Newey and West's correction for heteroskedasticity.

Turning points for environmental degradation are derived for Model I (full specification) or Model II (omitted variables) where applicable. Turning points in Model I are affected by population densities due to the interaction term. The population density levels are assumed to be the average for the sample of 121 countries because of no systematic relationship between GDP per capita and POPD.

Greenhouse Gas Emission Equation Estimation Results: CO_2, SO_2, NO_2, Methane as Independent Variables

CO_2 Equation

Two empirical equations for CO_2 were estimated and are reported in table 11.3 (see Gale and Mendez 1996; Grossman and Krueger 1995; Selden and Song 1995; Tucker 1995; and Agras and Chapman 1999). Model I in the table, specified previously as equation (1), is estimated with data from 120 countries (see table 11.1 for data definitions, sources, and other information).

The independent variables account for 62 percent of the variation in CO_2 emissions per capita. The Durbin-Watson statistic of 2.1 does not indicate a problem with autocorrelated residuals.

In Model I, the coefficient of POPD was insignificant. Multicollinearity could influence the result. Consequently, the interaction term GDP * POPD was removed from Model I, to form Model II, which indicates that CO_2 emissions per capita initially increase with greater population density but decrease when population density reaches high levels.

The finding of a U-shaped relationship between environmental quality and economic growth accords with previous studies (Gale and Mendez 1996; Grossman and Krueger 1995; Selden and Song 1994). Our estimated Model II indicates a turning point for income at $19,863 per capita, hence a high-income country. Seldon and Song (1995) estimated a turning point at approximately $16,000 per capita; Holtz-Eakin and Selden (1995) found a turning point at $35,428; and Agras and Chapman (1999) calculated a turning point of $13,630. Holtz-Eakin and Selden and Tucker (1995) estimated a much higher turning point. Ten countries in our study had incomes above $20,000 per capita.

NO_2 Equation

The equation was estimated using NO_2 emissions per capita as the dependent variable. Coefficients in Model I (table 11.3) are consistent with expectations and indicate that NO_2 per capita rises and later falls as ever-higher income per capita is achieved. Model II, with the GDP * POPD interaction term and the POPD variable dropped to reduce multicollinearity, shows a positive and statistically significant coefficient on the linear term for GDP. The coefficient of GDP squared had a negative and statistically significant sign. The results indicate a possible turning point for NO_2 emissions per cubic meter of air ranging from $15,727 per capita with Model I to $17,748 per capita with Model II. Seldon and Song (1994) estimated a turning point at approximately $11,000 income per capita, and List and Gallet (1999) estimated a turning point at $9,000 per capita GDP.

SO_2 Equation

The SO_2 equations shown in table 11.3 feature total emissions per capita as the dependent variable. Income per capita showed a positive impact and GDP^2 showed a negative but statistically insignificant impact in Model I. Again in deference to multicollinearity problems, variables were dropped to form Model II. The resulting significant positive coefficient on GDP and significant negative coefficient on GDP^2 indicate that rising national income per capita increases sulfur dioxide emissions per cubic meter to $15,853 of income per capita. With higher income, total emissions decline even without a decrease in population. The results in table 11.3 for SO_2 lie above the turning points of $4,053 estimated by Grossman and Krueger (1995); $8,000 by Seldon and Song (1994); $6,654 by Agras (1995); and around $3,670 by Shafik and Bandyopadhyay (1992). However, our turning point for SO_2 lies between the turning points of $12,500 estimated by Kaufman et al. (1998) and $21,000 estimated by List and Gallet (1999).

Methane Equation

The methane (from solid waste) equation is instructive in revealing the relationship between income per capita on the one hand and greenhouse gas emissions and solid waste accumulation per capita on the other. Model I for methane emissions from solid waste reported in table 11.3 indicates that rising income per capita significantly increases emissions but at a decreasing rate.

To reduce multicollinearity, the interaction term and the squared term on population were dropped to estimate Model II. Signs and significance of coefficients of income variables remain unchanged. The turning point for Model II occurs at an income level of $24,094 per capita. Use of the methane from waste to replace natural gas can diminish environmental damage. It is important to note that the variable used in this study does not include methane from sources other than solid waste.

Table 11.3. Greenhouse Gases: Carbon/ Nitrogen/ Sulfur Dioxides /Methane (From Solid Waste) Emission Equation Estimates. Dependent Variables 1990–1995: Independent Variables 1990

Variables	Dependent Variable: CO_2		Dependent Variable: NO_2		Dependent Variable: SO_2		Dependent Variable: Methane	
	Model I	Model II	Model I	Model II	Model I	Model II	Model I	Model II
Intercept	979.83*	178.09	21.481*	10.969*	29.491*	22.539*	2.6287**	2.1711**
	(503.4)	(397.4)	(8.257)	(5.411)	(11.583)	(10.104)	(0.4564)	(0.3383)
GDP	1.1502**	1.1918**	0.00386**	0.00436**	0.00559*	0.00602*	0.00139**	0.00133**
	(0.2016)	(0.2043)	(0.0013)	(0.0013)	(0.0034)	(0.0029)	(0.0002)	(0.0002)
GDP^2	-3.06E-5**	-2.94E-05**	-1.23E-07*	-1.23E-07*	-2.43E-07	-1.90E-07*	-2.36E-08*	-2.76E-08*
	(7.45E-06)	(7.85E-06)	(5.29E-08)	(4.70E-08)	(1.60E-07)	(9.32E-08)	(1.18E-08)	(1.11E-08)
POPD	-1.364	8.8231**	-0.1208*		-0.0745		-0.0073	
	(2.721)	(3.565)	(0.0540)		(0.1042)		(0.0057)	
$POPD^2$	-0.0008**	-0.0013*	1.20E-04*		2.63E-05		9.54E-06	
	(0.0002)	(0.0005)	(5.98E-05)		(0.00013)		(1.00E-05)	
GDP*POPD	0.0005**		3.94E-06		1.12E-05		-9.26E-07	
	(0.0001)		(3.38E-6)		(1.19E-5)		(1.12E-6)	
No. of Obs.	120	120	63	63	63	63	104	104
Adjusted R^2	0.6224	0.5894	0.1716	0.1766	0.1119	0.1071	0.5425	0.5258
Durbin-Watson	2.103	1.792	1.972	1.851	2.132	2.075	1.870	1.865
Turning Point for GDP (U.S. $/ per cap.)	19,845	19,863	15,727	17,748	11,531	15,853	29,449	24,094

** Significant at 1% or below * Significant at 10% or below Numbers in parenthesis represent standard errors (SE)

Suspended Particulate Matter (SPM)

SPM is not a greenhouse gas but is often emitted with GHGs and therefore is included in this section. The independent variables in Model I accounted for 19 percent (adjusted R^2) of the variation in the dependent variable, SPM per cubic meter (table 11.4). The linear GDP coefficient was insignificant and the sign was suspect. The interaction term and $POPD^2$ were dropped to reduce multicollinearity and to focus on the impact of income per capita on particulates. Model II yields a positive sign for the linear GDP coefficient and a negative and statistically significant coefficient for the squared term. Results provide only weak support for an inverted U relationship between income and particulates, with a possible turning point at $1,743 income per capita in Model II. Shafik and Bandyopadhyay (1992) found a turning point at $3,280 per capita GDP. Grossman and Krueger's (1995) calculated turning point for suspended particulates at just under $5,000 of per capita GDP and Seldon and Song's (1994) turning point of over $8,000 provide more realistic

Table 11.4. Particulate Equation Estimates: Dependent Variable: Particulates (SPM) 1990–1995 Average; Independent Variables 1990.

Variables	Dependent Variable: Particulates (SPM) (Coefficients)	
	Model I	Model II
Intercept	0.03919	0.04815
	(0.0356)	(0.0396)
GDP	-2.32E-05	1.29E-05
	(4.05E-05)	(2.63E-05)
GDP ^{2}	-1.15E-08*	-3.71E-09*
	(5.64E-09)	(2.19E-09)
POPD	0.00154	9.60E-04*
	(0.0010)	(3.40E-04)
POPD ^{2}	-3.18E-06	
	(4.17E-06)	
GDP*POPD	3.71E-07	
	(3.02E-07)	
No. of observations	39	39
Adjusted R ^{2}	0.189	0.178
Durbin-Watson	2.081	2.011
Turning Point for GDP (U.S. $/per capita)	none	1,743

**Significant at 1% or below
* Significant at 10% or below
Numbers in parenthesis represent standard errors (SE)

estimates. These results indicate that a turning point exists.

The SPM equation also was estimated using data expressed as per capita rather than as per cubic meter of air. The equations are not presented here because of poor statistical results. However, the results from this per capita model were consistent with results for total SPM per cubic meter in showing a reduction in total SPM when income per capita reaches a fairly low level even without considering the dampening impact of rising per capita income on population.

Natural-Resource Depletion

Natural-resource depletion is measured by phosphate fertilizer consumption, organic water pollution (a proxy for petroleum depletion through petroleum-based chemicals as well as a measure of resource degradation), and energy used (a proxy for fossil fuel reserve depletion). The conceptual equation is expressed in (1) and variables are defined in table 11.1.

Economic development increases consumption of some nonrenewable natural resources such as phosphate fertilizer and fossil fuels. Global reserves of petroleum seem adequate for only fifty to one hundred years at trend usage (Tweeten and Amponsah 1998). Energy is critical to agriculture, including fossil-fuel feedstock for nitrogen fertilizer production. Fossil fuels have no cheap, safe, and abundant alternatives.

Assuming triple the global reserves listed by the U.S. Geological Survey, phosphate fertilizer trend rates of consumption will deplete reserves in 233 years (Tweeten and Amponsah 1998, 51). And phosphate is of special concern because it is a basic building block of nature with no ready substitutes.

Phosphate Fertilizer

Two specifications of the phosphate fertilizer consumption per capita equation are reported in table 11.5. The coefficients on the linear and the squared GDP per capita variable display positive and negative signs, respectively. The population density coefficients display signs in Model I indicating that higher population densities first decrease and then increase phosphate fertilizer consumption per capita. The interaction term had a significant and negative sign, implying that the response of fertilizer use to higher income is less in the presence of high population density. Model II estimated by dropping the interaction term and POPD^2 does not improve on Model I—the adjusted R^2 declines.

The income turning point (above which increased GDP per capita reduces phosphate fertilizer use per capita) ranges from an income level of $18,145 with Model I to $15,040 with Model II. At higher incomes, investments in technology may produce improved farm plant production and fertilizer technology to reduce phosphate consumption per capita.

Table 11.5. Organic Water Pollution Equation Estimation: Dependent variable (Worg) 1993/ Independent Variables 1990–1993 Fertilizer Consumption Equation Estimation: Dependent variable (Fertz) 1990–1995 Average/ Independent Variables 1990

Variables	Dependent Variable: Phosphate Fertilizer (Fertz)		Dependent Variable: Organic Water (Worg)	
	Model I	Model II	Model I	Model II
Intercept	0.7872	0.3376	3.2908**	3.4175**
	(1.1010)	(0.9630)	(0.7856)	(0.6950)
GDP	0.0028**	0.0026**	7.10E-04**	7.10E-04**
	(0.0008)	(0.0008)	(1.50E-04)	(1.40E-04)
GDP^2	-7.84E-08*	-8.75E-08*	-1.60E-08**	-1.61E-08**
	(2.99E-08)	(3.34E-08)	(5.73E-09)	(5.63E-09)
POPD	-0.0074		0.0015	3.26E-04*
	(0.0146)		(0.0044)	(1.47E-04)
$POPD^2$	9.01E-06		-8.07E-08	
	(1.95E-05)		(1.71E-07)	
GDP*POPD	-2.92E-06**		-5.05E-08	
	(1.10E-06)		(2.53E-07)	
No. of observations	121	121	75	75
Adjusted R^2	0.3448	0.2757	0.3307	0.3489
Durbin-Watson	1.894	1.884	1.988	1.996
Turning Point for GDP (U.S. $/ per capita)	18,145	15,040	22,405	22,145

**Significant at 1% or below
* Significant at 10% or below
Numbers in parenthesis represent standard errors (SE)

Organic Water Pollution

Phosphate, nitrogen, and chemicals such as pesticides produced from fossil fuels play a role in water quality. Water quality is measured by Worg, the amount of oxygen consumed by microorganisms in water and measured in kilograms per capita (see table 11.1). At issue is whether organic water pollution per capita is a function of income per capita and population density. Substances that could cause water pollution increase with economic development; the empirical question is whether abatement measures offset so that pollution falls per capita at higher income levels. Availability of the dependent variable only for 1993 required a different choice of years for the data in table 11.5.

Model I (table 11.5) for Worg provides coefficient signs consistent with the hypothesis that organic substances in water increase at a decreasing rate with higher income and population density. The coefficient on the variable for interaction between income per capita and population density is negative and insignificant.

Model II reduces multicollinearity by excluding $POPD^2$ and the income-population density interaction variables. Model II coefficients on income are similar to those in Model I. The coefficients indicate that organic compounds in water do not turn down before income per capita reaches $22,405 (Model I) to $22,145 (Model II).

Energy Equations

Energy and oil equations provide insights into the impact of income on natural resource depletion. Equations also indicate potential contributions of fossil-fuel consumption to greenhouse gas emissions and hence global warming. Results for two models estimating energy use per capita are reported in the left columns in table 11.6. In Models I and II, income per capita variables (GDP and GDP^2) have significant coefficients with expected signs. They indicate that higher income leads to higher consumption of energy, but energy use per capita turns down above an income of approximately $22,404 (Model I) and $22,848 (Model II). Agras and Chapman (1999) found a turning point at $62,000 per capita GDP; Holtz-Eakin and Selden (1995) at $35,428 per capita; and Suri and Chapman (1998) at $55,000 per capita GDP. Thus, countries tend to lower their per capita consumption levels for energy only at relatively high income levels.

The estimated equations for petroleum oil consumption per capita in table 11.6 show significant and positive signs for income per capita and significant and negative signs for its squared term, implying that consumption of oil per capita peaks at approximately $19,379 (Model I) to $20,038 (Model II) of income per capita. Greater population density is associated with higher oil consumption per capita, but a turning point occurs at around 2,864 people per square mile, according to Model I for oil in table 11.6. The interaction term had a significant and positive sign, implying that high population density has a greater positive impact on oil consumption in the presence of high income per capita.

Data also were assembled for natural gas and coal consumption but were judged to be inadequate for measuring the impact of income. A major reason is that coal and natural gas are not as freely traded among countries as oil, partly due to transport problems. Hence usage among countries depends heavily on having local reserves of those fossil fuels.

Population Growth Equation

Table 11.7 reports empirical results for the population growth model specified earlier as equation (2) in the methodological section. Exogenous variables potentially influential in explaining population growth (POPG) include GDP, illiteracy rates (ILTC), female labor force participation (FLBR), infant mortality rate (IMR), urbanization (URBAN), female education (FEMED), and food supply

Table 11.6. Energy Demand Equation Estimation: Dependent Variable (Energy): 1990-1995/ Independent Variables 1990

Variables	Dependent Variable: Energy		Dependent Variable: Oil	
	Model I	Model II	Model I	Model II
Intercept	111.642	142.03	0.0792	-0.1730
	(169.67)	(121.09)	(0.0786)	(0.1427)
GDP	0.4750**	0.4478**	1.88E-04**	2.06E-04**
	(0.0975)	(0.0830)	(4.62E-05)	(5.06E-05)
GDP^2	-1.06E-05**	-9.80E-06**	-5.29E-09**	-5.16E-09**
	(3.40E-06)	(2.93E-06)	(1.59E-09)	(1.72E-09)
POPD	-0.0127		0.00026	0.0027**
	(0.8999)		(0.00032)	(0.00093)
POPD^2	-3.34E-05		-4.61E-07**	-4.04E-07**
	(9.66E-05)		(5.92E-08)	(1.35E-07)
GDP*POPD	-1.09E-05		1.62E-07**	
	(5.41E-05)		(2.94E-08)	
No. of Obs.	114	114	92	
Adjusted R^2	0.6643	0.6576	0.6252	92
Durbin-Watson	1.8591	2.172	1.824	0.5758
				1.6287
Turning Point for GDP (U.S.$/per capita)	22,404	22,848	19,379	20,038

** Significant at 1% or below
* Significant at 10% or below
Numbers in parenthesis represent standard errors (SE)

(CALC) as defined in Table 11.1.

Results from Model I indicate multicollinearity may be a problem. Thus some highly correlated explanatory variables were removed to focus on the impact of GDP in Models II and III (table 11.7). As is apparent from Model II, the income-per-capita variables have statistically insignificant coefficients. The positive coefficient for income per capita squared implies that a turning point occurs at some income level above which additional income has a positive rather than a negative impact on population growth. A square root term for GDP replacing the

Table 11.7. Population Growth Equation Estimation: Dependent Variable: (POPG): 1990-1995/ Independent Variables 1990

Variables	Dependent Variable: POPG		
	Model I	Model II	Model III
Intercept	2.4878	2.409665**	3.1335**
	(1.6673)	(0.40714)	(0.4661)
GDP	-6.00E-05	-1.28E-04	-4.56E-05*
	(8.37E-05)	(8.63E-05)	(2.10E-05)
GDP ^{2}	8.64E-010	5.18E-09	
	(3.16E-09)	(3.74E-09)	
ILTC	-0.0037	0.0137*	
	(0.0111)	(0.0056)	
FLBR	-0.0187	-0.0312*	-0.0260*
	(0.0122)	(0.0121)	(0.0147)
IMR	0.14296		
	(0.1213)		
URBAN	0.00522		
	(0.0086)		
FEMED	-0.0068		-0.0119*
	(0.0115)		(0.0047)
CALC	-0.0058		
	(0.0094)		
No. of Observations	110	110	110
Adjusted R ^{2}	0.1827	0.2318	0.1890
Durbin-Watson	2.129	1.840	2.019
Turning Point for GDP (U.S. $/ per capita)	9,883	7,994	none

** Significant at 1% or below
* Significant at 10% or below
Numbers in parenthesis represent standard errors (SE). Number of observations differ among equations because some countries lacked observations for variables in particular equations.

quadratic term did not improve the results.

In Model III (which does not include GDP^2), the coefficient of GDP per capita is negative and statistically significant. The t-value for the GDP of 2.6 in Model III compared with 1.5 in Model II provides some support for the hypothesis that income growth slows population growth over a wide range of incomes.

Other variables influenced by income could intensify the impact of GDP on population growth. The coefficient of female education rate (FEMED) has a significant and negative sign in Model III, implying that population growth tends to decrease when higher percentages of females are educated. Female labor force participation (FLBR) has a significant and negative sign, indicating that higher percentages of females in the labor force also tend to lower population growth. The coefficient of illiteracy rates ILTC had a significant and positive sign in Model II, implying that population growth was positively related to the rate of illiteracy and tends to decrease as a greater percentage of population is able to read and write. The effects of FEMED and FLBR variables in Model III, through which GDP operates indirectly to reduce population growth, are considered in subsequent analysis.

Summary of Results

This section presents the elasticities and tests the hypothesis that slower growth in population induced by income growth per capita offsets rising natural resource depletion and GHG emissions per capita, thus reducing overall environmental pressure under economic growth. Results are presented for an arbitrary geometric progression of income benchmarks. Those benchmarks are well within the range of data except for the top number. Only one country, Japan, had a GDP of more than $32,000 per capita in 1995. Income of $32,000 per capita is included to anticipate outcomes for affluent nations as incomes continue to grow and exceed that benchmark. Because the estimate applies to so few countries, it must be interpreted with caution.

Table 11.8 summarizes results from previous tables utilized to calculate the income elasticities for environmental variables and population growth equations. Income elasticities for direct, indirect, and total effects of population growth are reported in rows A–C; for greenhouse gases and particulates in rows D–H; and for resource depletion in rows I–L. To test hypotheses regarding the total income effect on environmental variables X_i, population growth elasticities (total effect) with respect to income are added to environmental variables X_i per capita with respect to income as shown earlier in equation (4). A positive value (+ in parenthesis) indicates that the incremental total effect of income growth on the environmental variable is unfavorable (X_i rises). A negative sign (– in parenthesis) in table 11.8 indicates income growth is favorable for the environment when effects of income growth on X_i and population are accounted for.

Population Growth Equation
The conceptual framework for calculating population growth with respect to income was presented in equations (3) and (5). Model III was selected from table 11.7 because it was the only equation with a coefficient on GDP significant at the 10 percent level or better. As is apparent from table 11.8 (rows A–C), the income elasticities calculated for direct, indirect, and total effects from empirically estimated population growth Model III in table 11.7 are negative and are of the greatest magnitudes (absolute value) among high-income countries.

Contributions to population elasticities with respect to income are surprisingly similar for the direct and indirect effects. Because results are from cross-sectional data, the elasticities tend to measure long-term effects of several years of adjustment to changing income.

For high-income countries considering direct and indirect effects, a 1 percent rise in income from a base of $16,000 per capita or greater reduces population by an estimated 1.2 percent in the long run. From a $32,000 income-per-capita base, a 1 percent increase in income reduces population by 2.5 percent from the combined direct effect and indirect effects.

Greenhouse Gas and Particulate Emission Equations
The most abundant GHG is CO_2, and its income elasticity per capita tends to fall but remains positive up to about $20,000 of GDP per capita. The income elasticities for environmental and resource variables X_i are computed by multiplying the coefficient from equations shown in earlier tables times the ratio of GDP per capita to X_i. To form the ratio, X_i is predicted from the respective equations with GDP set as shown at the top of table 11.8 and POPD set at the mean of the sample of 121 countries, 165 persons per square kilometer. The income elasticity becomes negative at higher income per capita. Combining the income elasticity with respect to population, the result (minus in parenthesis) indicates that incremental income growth reduces CO_2, NO_2, and SO_2 emissions in total in the long run above an income base of approximately $16,000 per capita. A slower growth in population induced by rising income leads to reduced methane emission levels at income level above approximately $20,000 per capita and to reduced suspended particulate matter (SPM) at much lower incomes.

SO_2 is a greenhouse gas and its elasticity is negative for higher income levels. Low SO_2 implies less acid rain. However, a negative elasticity does not necessarily imply less global warming because SO_2 tends to reduce rather than induce global warming.

Although negative income elasticities predominate at higher income levels for greenhouse gases studied herein, some possible complications could emerge. Production of one BTU of energy generates significantly more CO_2 with coal fuel than with either oil or natural gas fuel. Hence, as per capita income rises and consumers shift to coal from petroleum and natural gas as the latter reserves decline, GHG could accumulate.

Natural Resources Depletion

The income elasticities obtained for the energy demand equation (Row I) imply that a 1 percent rise in per capita income raises energy consumption per capita 2 percent or more for countries with income levels below $1,000 per capita. The elasticity declines as income rises and turns negative above incomes of $23,000 per capita. The turning point comes earlier if the income elasticity of population growth is considered.

The income elasticities obtained for oil consumption (Row J) give a similar turning point. The combined income elasticities for population and energy or oil reinforces the conclusion that additions to income per capita at high-income levels reduce overall consumption levels in the long run.

Income elasticities obtained for phosphate fertilizer consumption (Row K) imply that a rise in income raises phosphate-consumption levels in low-income and perhaps in middle-income countries. However, rising income eventually lowers phosphate fertilizer use per capita and also lowers population. Hence, rising income eventually lowers overall phosphate-fertilizer consumption.

Organic water pollution is both an indication of water-resource degradation and of natural-resource depletion, the latter because of fossil energy sources depleted to produce pesticides, fertilizers, and other compounds. The income elasticities derived for organic water pollution emission per capita (Row L) tend to fall with income. The combined income elasticities for organic water pollution and population imply that rising income above approximately $20,000 per capita brings lower total organic water pollution.

CONCLUSIONS

The environmental degradation and resource depletion variables in this study exhibit an inverted U-shaped curve, or Kuznets relationship, with economic development. The income elasticities of all environmental variables are positive at low-income levels, that is, resource degradation/depletion provides no basis for complacency regarding depletion/degradation of resources. One reason is that proper policy is essential for environment protection as income rises. That is, favorable environmental outcomes in part arise because of sound policies induced by higher per-capita income. Such policies include investment in education, science,

Table 11.8. Income Elasticities Derived from Estimated Equations

Income Levels per Capita	Low Income			Middle Income		High Income	
	$500	$1,000	$2,000	$4,000	$8,000	$16,000	$32,000
Population Growth Equation							
A Direct Effect	-0.0227	-0.0455	-0.0962	-0.1899	-0.3242	-0.7689	-1.3367
B Indirect Effect	-0.0363	-0.0329	-0.0558	-0.1316	-0.2197	-0.4454	-1.1277
C Total Effect	-0.0590	-0.0784	-0.1520	-0.3213	-0.5439	-1.2143	-2.4644
Greenhouse Gas and Particulate Equations							
D CO_2 equation (Model II) Combined Effect	0.2658 (+)	0.5183 (+)	0.9837 (+)	1.7553 (+)	2.6620 (+)	1.9298 (+)	-9.7178 (-)
E NO_2 equation (Model II) Combined Effect	0.0800 (+)	0.1547 (+)	0.2885 (+)	0.4928 (+)	0.6490 (+)	-0.0485 (-)	-10.244 (-)
F SO_2 equation (Model II) Combined Effect	0.1143 (+)	0.2211 (+)	0.4125 (+)	0.7058 (+)	0.9348 (+)	-0.0376 (-)	-7.7046 (-)

G Methane equation (Model II)	0.2059	0.4047	0.7809	1.4481	2.4411	3.0613	-1.1612
Combined Effect	(+)	(+)	(+)	(+)	(+)	(+)	(-)
H Suspended Particulate (Model II)	0.0216	0.0258	-0.0183	-0.3165	-1.7526	-7.9836	-33.8812
Combined Effect	(+)	(+)	(-)	(-)	(-)	(-)	(-)

Resource Depletion Equations

I Energy equation (Model II)	0.6025	1.1780	2.2485	4.0651	6.4047	5.9073	-15.794
Combined Effect	(+)	(+)	(+)	(+)	(+)	(+)	(-)
J Oil equation (Model II)	0.2764	0.5387	1.0206	1.8140	2.7188	1.8008	-10.945
Combined Effect	(+)	(+)	(+)	(+)	(+)	(+)	(-)
K Fertilizer Consumption (Model II)	0.7775	1.5008	2.7851	4.7037	5.9416	-1.9805	-59.416
Combined Effect	(+)	(+)	(+)	(+)	(+)	(-)	(-)
L Organic Water Pollution (Model II)	0.0892	0.1743	0.3321	0.5985	0.93381	0.8145	-5.5835
Combined Effect	(+)	(+)	(+)	(+)	(+)	(+)	(-)

technology, and conservation.

A second reason for concern is that income elasticities apply to the margin and veil high *average* environmental demand at high income levels per capita. The *environmental dilemma* consists of how and whether developing countries can pass through the stage of increasing environmental degradation and natural-resource depletion to higher per-capita income levels at which environmental preservation predominates—without irreversibly damaging the environment in the process.

The results of this study have additional limitations. Other data specifications and environmental variables could give different results. Attempts were unsuccessful to improve the analysis by replacing the population density variable with variables measuring urbanization and agriculture's share of GDP. Fortunately, results from some equations lacking strong statistical properties were broadly consistent with other studies.

ACKNOWLEDGMENTS
The comments of Carl Zulauf are greatly appreciated.

REFERENCES
Agras, J. "Environmental Development: An Econometric Analysis of Pollution, Growth, and Trade." Master Thesis. Ithaca: Department of Agricultural, Resource, and Managerial Economics, Cornell University, 1995.

Agras, J., and D. Chapman. "A Dynamic Approach to the Environmental Kuznets Curve Hypothesis", *Ecological Economics* 28 (1999): 267–277.

Antle, J., and G. Heidebrink. "Environment and Development: Theory and International Evidence." *Journal of Economic Development and Cultural Change* 44 (1995): 603–625.

Chaudhuri, S. and A. Pfaff. "Household Income, Fuel-Choice, and Indoor Air Quality: Microfoundations of an Environmental Kuznets Curve." Mimeo (1997), Columbia University.

Cropper, M., and C. Griffiths. "The Interaction of Population Growth and Environmental Quality." *The American Economic Review* 84(2) (1994): 250–254.

de Bruyn, S.M., J. Van den Bergh, and J.B. Opschoor. "Economic Growth and Emissions: Reconsidering the Empirical Basis of the Environmental Kuznets Curve." *Ecological Economics* 25 (1998): 161–175.

Eckholm, E.P. *Losing Ground.* New York: Norton Publishing, 1976.

Gale, L. R. and J.A. Mendez. "A Note on the Empirical Relationship between Trade, Growth, and the Environment." Working paper. Lafayette, LA: Department of Economics and Finance, University of Southwestern Louisiana, December 1996.

Grepperud, S. "Population Pressure and Land Degradation: The Case of Ethiopia."

Journal of Environmental Economics and Management 30 (1996): 18–33.

Grossman, G M. and A.B. Krueger. "Economic Growth and the Environment." *Quarterly Journal of Economics* 112 (1995): 353–377.

Hilton, F. G. H., and A. Levinson. "Factoring the Environmental Kuznets Curve: Evidence from Automotive Lead Emissions." *Journal of Environmental Economics and Management* 35 (1998): 126–141.

Holtz-Eakin, D., and T.M. Selden. "Stoking the Fires? CO_2 Emissions and Economic Growth." *Journal of Public Economics* 57 (1995): 85–101.

Kaufmann, R.K., B. Davidsdottir, S. Granham, and P. Pauly. "The Determinants of Atmospheric SO_2 Concentrations: Reconsidering the Environmental Kuznets Curve." *Ecological Economics* 25 (1998): 209–220.

Kelley, A. "The Population Debate: A Status Report and Revisionist Interpretation." Mimeo. Durham, NC: Center for Demographic Studies, Duke University, 1984.

Kuru, A. "Soil Erosion and Strategic Policy: the Case of Ethiopia." Publication of the Department of Environmental Conservation at the University of Helsinki, No. 7, 1986.

List, J.A., and C.A. Gallet. "The Environmental Kuznets Curve: Does One Size Fit All?" *Ecological Economics* 31 (1999): 409–423.

Mankin, G, D. Romer, and D. Weil. "A Contribution to the Empirics of Economic Growth." *Quarterly Journal of Economics* 106 (1990): 407–437.

Newey, W., and K. West. "A Simple Positive Semi-Definite, Heteroskedasticity and Autocorrelation Consistent Covariance Matrix." *Econometrica* 55(3) (May 1987): 703–708.

Ruddle, K., and W. Manshard. Renewable Natural Resources and the Environment: Pressing Problems in the Developing World. Dublin: Tycooly International Publishing, 1981.

Ruttan, V.W. "Technology and the Environment." *American Journal of Agricultural Economics* 53 (1971): 707–17.

Seldon, T.M., and D. Song. "Environmental Quality and Development: Is There a Kuznets Curve for Air Pollution Emissions?" *Journal of Environmental Economics and Management* 27 (1994): 147–162.

Seldon, T.M., and D. Song. "Neoclassical Growth, the J Curve for Abatement, and the Inverted U Curve for Pollution." *Journal of Environmental Economics and Management* 29 (1995): 162–168.

Shafik, Nemat. "Economic Development and Environmental Quality: An Econometric Analysis." *Oxford Economic Papers* 46 (1994): 757–773.

Shafik, N., and S. Bandyopadhyay. "Economic Growth and Environmental Quality: Time-Series and Cross-Country Evidence." Background Paper for *World*

Development Report 1992. Washington, DC: The World bank, 1992.

Srinivasan, T.N. "Population, Food, and Rural Development." Background paper prepared for the Working Group on Population Growth and Economic Development. Washington, DC: Committee on Population, National Research Council, 1985.

Stern, D., and M. Common., "Economic Growth and Environmental Degradation: The Environmental Kuznets Curve and Sustainable Development." *World Development* 24(7) (1996): 115–1160.

Stokey, N. "Are there limits to growth?" *International Economic Review* 39(1) (1998).

Suri, V., and D. Chapman. "Economic Growth, Trade, and Energy: Implications for the Environmental Kuznets Curve." *Ecological Economics* 25, (1998): 195–208.

Tucker, M. "Carbon Dioxide Emissions and Global GDP," *Ecological Economics* 15(3) (1995): 215–223.

Tweeten, L., and W. Amponsah. "Sustainability: The Role of Markets versus the Government." Ch. 4 in G. D'Sousa and T. Gebremedhin, eds., *Sustainability in Agricultural and Rural Development.* Brookfield, VT: Ashgate, 1998.

Van Bergeijk, P. "International Trade and the Environmental Challenge." *Journal*

12

Food Security, Trade, and Agricultural Commodity Policy

Daniel A. Sumner

National agricultural policies are often rationalized on "food security" grounds. This chapter evaluates such rationales. For example, I show that investments in local agricultural productivity may not significantly improve food security when trade can substitute for local production. Policymakers often claim that reliance on local food production enhances food security. An Index of National Food Security incorporates variable market prices but shows that for South Korea, blocking imports clearly lowers food security, although world rice prices are quite variable.

INTRODUCTION

No agricultural issue is more important than food security. Reflecting the importance of the topic, the literature on food security is long, broad, and deep.

Many disciplines make contributions in the area. Technical agricultural sciences focus on food production and the contribution of new agricultural knowledge and practices to more abundant food supplies. Nutritionists focus on the consumption side and how food intake relates to human well being and health. Others, including economists, emphasize diffusion of production or nutrition information, intra-family relationships, income distributions, and a myriad of factors that affect poverty and food prices. Papers included in El Obeid et al. (1999)provide perspectives on food security from a variety of disciplines.

Economists have made major contributions to understanding and perhaps even improving food security. Economic contributions have included analysis of poverty and income growth as well as food supply and food prices. Some of the work of economists takes a global perspective as reflected in Duncan (forthcoming) or Tweeten (1999); others take a local, household, or even individual perspective as reflected in much of the literature surveyed by Barrett. Time horizons also differ.

Some deal with long-term global progress on food production relative to human population; others consider periodic famine or local dislocations.

Food security may be defined as reliable access to a nutritionally adequate diet now and for the future. (I do not attempt to discuss distinctions between availability, access, and utilization. See Tweeten 1999.) The second part of this chapter compares technological innovation and policy reform as methods of attaining increased food security. The third section deals with the stochastic nature of the above definition of food security as represented by the term "reliable," and with the explicitly dynamic nature of the definition represented by the reference to access to food in the future. But first, I narrow the scope of this chapter by discussing the population for which I consider access to an adequate diet. The focus of this chapter is *national* commodity policy but, of course, there are other, perhaps more natural or important, aggregations to consider.

A global perspective on food security examines aggregate food production on the planet, productivity growth affecting the time-path of food supply, and the time-path of per-capita income growth, especially for the world's poor. Global population growth provides a denominator in these calculations. Global policy issues include institutional reform, macroeconomic and political stability, the level and variability of global food prices, and investments in human capital and in agricultural science and technology.

At the other end of the aggregation continuum is food security for households and even individuals within households. Again, the issues relate to access to food and the resources available to that individual or family. But, to understand individual or household food security, much of the focus must be on distribution. In particular, food security for a particular household requires not that the planet has adequate food but that the household in question has the resources to access an adequate diet given the prices (and other constraints) it faces. Policies to deal with food security of particular households may address the incomes of and effective food prices faced by a subset of the population that is particularly vulnerable.

This chapter deals with an intermediate aggregation. Because I address national agricultural policies, I focus on issues of food security defined for a single nation. I am motivated in part by the observation that national farm commodity and trade policies are often justified or at least rationalized with reference to national food security objectives, although I do not claim that the policies and issues discussed here are necessarily the most important considerations for food security. This chapter does show, however, that reforms of commodity policy can contribute to food security. It also shows how commodity policy reform may affect the food security consequences of other government policies. Finally, food security claims associated with some commodity polices may divert attention and resources from alternative policies that could make larger food-security contributions.

Food security is a stated objective of agricultural commodity policy in virtually all countries. Rich and poor, importer or exporter, governments seem to state food security of their population as an objective for agricultural policy. This has been true for several millennia as even a cursory look at ancient history or even fiction attests (Diamond 1997; Smith 1776; Lee and De Bary 1997; Roberts 1999).

The range of contemporary domestic and border policies tied to food security objectives is equally impressive. Subsidies for research, income redistribution, rural infrastructure, price floors, price ceilings, government stockholding, import barriers, and export subsidies are all listed as national food security policies. In the United States, for example, the 1985 farm bill (containing farm commodity programs, conservation programs, domestic food assistance programs, export subsidies, and a host of other loosely related policies and programs) was titled the "Food Security Act." In China, food security is often listed as the key objective of state trading in grains, a grain delivery quota system under which farmers receive less than market prices for part of their grain, and policies to encourage self-sufficiency for the nation and even for individual provinces. In South Korea, food security is listed as a primary reason that agriculture must be treated as a special case in World Trade Organization (WTO) negotiations. Food security is now discussed in the WTO under the currently popular rubric of the "multifunctionality" of agriculture.

AGRICULTURAL PRODUCTIVITY GROWTH AND POLICY REFORM
It is generally accepted that investments in agricultural science and similar contributions to agricultural productivity are fundamental to progress on global food security. By the same logic and evidence, the presumption is that national investments in agricultural productivity are necessary and important contributors to national or even local or regional food security. For example, at a national conference on commodity policy in 1999, the dean of agriculture at a major land-grant university argued that food security for the people of his state was a strong reason for increased research funding for his college. Economists may be inclined to dismiss such arguments as transparent sophistry and special pleading. But economists need to recognize that such statements influence people.

Economists have made considerable effort to assess the contribution of science to agriculture. This line of inquiry applies economic reasoning about long-run investments to costs and returns from agricultural research and extension and especially to public investments in research. Alston, Norton, and Pardey (1995); Huffman and Evenson (1993); and Alston and Pardey (1996), among others, document and represent progress in this area. The consensus is that returns are relatively high.

Investments in public policy reform have much in common with investments in agricultural science. For example, when Stigler (1976) poses and responds to the question, "Do Economists Matter?" he includes a rate of return calculation by

Ronald Coase that is quite favorable to the overall contribution of economic analysis leading to policy reform. A few papers have recently examined this question with respect to agricultural economics (see the session introduced by Smith and Pardey 1997.)

Before turning to an example, it may be useful to review some more general considerations. Few innovations create Pareto improvements—making some better off without making others worse off. This is clearly true for farm policy reform. But, even when agricultural research clearly raises world wealth and even improves the lot of the poor, some suffer losses. For example, most scientists and economists laud the contribution of the green revolution to food security for more than a billion people, but detractors argue that these innovations lowered welfare rather than improved it. They argue that certain new varieties were particularly well suited to farms that were already favored or that were owned by the relatively wealthy. In that situation, they argue, the poor suffer when a new seed improves the absolute and relative performance of these already favored farms. Distributional considerations complicate the assessments of returns for any innovation.

Agricultural policy reform that increases total social income almost always shifts income between groups. As with biological innovations, there are almost always losers. Those likely to suffer if policy were changed often argue (implicitly) that national welfare would be reduced with a reform because their well being is more important to national welfare than that of others. Nations often use agricultural policy explicitly to redistribute income. Thus, it may be reasonable to argue in some cases that the revealed social welfare function of society indicates who are the favored people. Of course, such an argument is based on a view that the existing political system is a legitimate reflection of social welfare. In that case, almost by definition, what is, is optimal. Policy change occurs only when the constraint set changes (or is thought to change), when the effective social welfare function itself becomes different because of some political change, or when new information becomes available.

Notwithstanding these concerns, in many real cases there is (almost) consensus. For example, few would argue that food price policy in Africa is optimal when it favors the urban elite to the detriment of the rural poor because of an insufficiently representative political system. In general, however, it remains problematic to incorporate the existing or potential political structures and outcomes in a social welfare function used in policy evaluation.

The natural approach for economists is to focus on aggregate national income as a welfare criterion even though no nation or social group maintains policies consistent with that social welfare function. The next step for economists is to argue that the potential for compensation implies that creating aggregate wealth, say through agricultural policy reform, provides welfare gains. Whether the society chooses to redistribute that wealth is a separable policy choice independent of the agricultural policy reform.

Issues similar to those involving income distribution are often raised in the context of environmental effects of agricultural innovations. The externality considerations that affect the calculation of returns to agricultural innovations also apply to policy reform. Environmental concerns about new varieties or pest-control measures are often more troublesome than the concerns raised with respect to policy reform. Indeed, policy reform is often seen to be environmentally benign.

EX ANTE EVALUATION OF PRODUCTIVITY VERSUS POLICY INNOVATION

Methods used for evaluating *ex post* contributions of investments in agricultural research or policy reform are also applicable, with appropriate adjustments, to *ex ante* projections of costs and benefits and, therefore, for allocation of resources. Such *ex ante* evaluation, conducted (explicitly or implicitly) before resources are allocated, must consider the cost of the effort, likelihood of achievement of results, likelihood of adoption of the results, schedule of adoption, and net payoff from the adoption for the adopters and for society generally. These issues and the difficulties they raise are very similar whether the *ex ante* allocation decisions are among plant genetics projects or among efforts that contribute to policy reform.

An example will help to illustrate issues in evaluating the payoff to policy reform compared to, say, biological research leading to agricultural productivity growth. To help fix ideas, I take a simple set of policy options and perform a simple analysis. Nonetheless, the example illustrates some real food security issues. First, we will consider equilibrium in a commodity market situation that has a common agricultural policy in place. I then consider the effects of introducing an agricultural productivity innovation stimulated by an investment in plant genetics and compare these with policy reform.

In figure 12.1, the demand for food is labeled Demand. The initial marginal cost (Supply) of domestic production is represented by the curve S_0, which starts at marginal cost P_{min}, and then, after a short horizontal segment, proceeds along the upward sloping portion of the curve. The potential import price facing this small country is P_w, but, in the initial situation, imports are banned by trade barriers justified on food security grounds.

Now, I evaluate the projected benefits of genetics research applied to this commodity market. Assume that the research would be adopted by producers. Crop yields would increase and cost of production would fall relative to the pre-research situation. The direct impact of plant breeding research is shown in figure 12.1 by a shift of the marginal cost function down and to the right. The new marginal cost curve is shown by S_1. For simplicity, I leave the minimum marginal cost at P_{min}, but the horizontal segment of the curve now extends a bit further to the right. The supply shift caused by genetics research has no effect on government policy, so imports are still not allowed. Market price now falls from P_0 to P_1 and consumers benefit from the price decline. Most economists would consider this a positive food security result. Of course, the innovations also benefit producers by lowering marginal costs and increasing equilibrium output, but the impact of the supply shift on net farm income may be positive or negative.

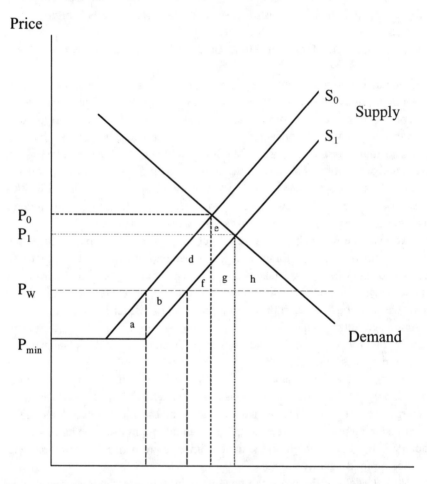

Figure 12.1. Returns to cost-reducing innovation relative to trade liberalizing policy reform.

How we view the social benefits from this research may depend on the total gain to society relative to the initial position, but our views may also be influenced by the distributional impacts. In figure 12.1, consumers gain from a lower market price and producers gain as long as the area between the two marginal cost curves is larger than the profits forgone from the lower price. Producer gains are more likely the higher the elasticity of demand (in absolute value) and the lower the elasticity of supply.

Figure 12.1 focuses on the distribution of annual income flows. Gross gain to the nation is area a+b+c+d+e. I have not discussed the costs of research or the timing of the benefits flow. With such information, one may calculate an aggregate net return and evaluate the payoff of this research relative to some target rate of return. Instead, imagine that the same resources could be devoted to policy reform that changes the trade policy associated with the commodity and, for simplicity, that it is timed in impact the same as the genetics project.

As with the genetics project, I assume that the policy reform effort would be successful and the implication would be a change in the policy. And, as with the plant genetics project, the policy change would affect both overall social income and the distribution of income. In this case, there is a large gain for consumers as price falls from P_0 to P_w. But producers lose from the lower price and the lower domestic quantity produced. Society as a whole gains the large triangle c+d+e+f+g+h.

If the costs of efforts to shift the technology or policy are the same, the policy change has a higher payoff in terms of the net present value of social income provided area f+g+h is greater than area a+b in figure 12.1.

It is instructive to consider how issues other than net present value of national income might affect the comparison of policy reform and genetics research. From figure 12.1 it is clear that consumers of food would prefer policy reform, whereas producers would lose from the policy reform. Which project is chosen might hinge mainly on the welfare or political weights applied to the change in income or wealth of the competing groups. Note also that the benefits of the policy change would be smaller if the plant-breeding research were undertaken so that the supply function is S_1. Likewise, the benefits of the genetics project would be much smaller in an open trade environment. Therefore, in this example, the social incentive for pursuing each innovation is smaller if the other is also achieved.

Now, I turn to the food and nutrition implications of research versus policy reform. This example shows that for a small country importer, investments in local agricultural productivity do little for access to food for the population. This is an extreme case but is broadly applicable. For example, it applies to almost all research conducted by individual U.S. states. Research that applies to local productivity of a crop grown widely elsewhere does not affect the market price and does not benefit consumers per se. Such research may have a social payoff by increasing producer profits and the value of land but generally will not improve food security. When access to outside supplies are limited, research will lower the price of food but, in that starting situation, opening the border is a much more effective tool.

Of course, innovations may improve the returns to owners of farming resources and thus add to the income of this group in society. In some cases, farm resource owners are themselves vulnerable to hunger and thus aid to them can be positive. But if, as in the case of trade barriers, the cost is raising the price of food to the whole population, the trade-off is almost surely negative for national food adequacy. This idea is explored in more detail later.

My point here is not that agricultural science is unimportant for global food supply or that productivity investments do not contribute to national or global welfare. Rather, my point is that as a matter of national food security policy, agricultural productivity growth is often overrated.

AN INDEX OF NATIONAL FOOD SECURITY

I now broaden notions of food security from looking simply at the price of a food commodity. This section defines and uses an index that links commodity policy, especially border measures, to common national objectives on food security. I incorporate random aspects of future food prices into this index. Only the barest outline is presented here. More detail and some background are provided elsewhere. (See in Sumner 2000; also, the website http://aic.ucdavis.edu/research/foodsecurity.html.) As background to this discussion, I refer the reader to Duncan (forthcoming) or Tweeten (1999), who provide assessments of the status of global food adequacy and patterns and prospects for food security. (See also FAO 1999 and data provided at www.fao.org.) Barrett (forthcoming) reviews food security definitions and relates nutritional adequacy to food security for vulnerable individuals. He also reviews the evidence on domestic and international food aid. Sumner and Tangermann (forthcoming) review the aspects of agricultural trade policy and the WTO that are used below.

To develop a workable index of food security, I begin with threshold food intake f_i^* above which a person, i, has adequate nutritional health. This means that the person has satisfied whatever standards are considered appropriate based on demographic, exercise, or medical conditions. Of course, different countries may define the threshold diet differently depending on the local situation.

To consider national implications of agricultural policy, it's necessary to aggregate food adequacy to the population of interest. I define the degree of population food adequacy as the share of the population consuming at least the minimum food required for their nutritional health. That is, for a given year t, the degree of national food adequacy, Fa_t, is measured as the probability that an individual from the population has $f_{it} > f_i^*$. The food intake distribution across the population is closely related to the income distribution.

Now I define a threshold of national food adequacy, Fa^*, which indicates the goal or policy objective of the nation. Any time this threshold is less than 100 percent, the nation acknowledges that it may not be possible to eliminate hunger or the possibility of inadequate diets for some share of the population over the relevant policy time horizon. But in every country, if the share of the population suffering food inadequacy is too large, as defined within that country, policymakers are willing to say that an issue of nationwide food inadequacy exists.

Now, I must add the stochastic nature of "security" to the notion of population food adequacy. I define the Index of National Food Security (INFS) as the probability that some given share of the population will be able to achieve adequate food intake in the future. Note that each part of my analysis—individual food adequacy, share of the population, and the stochastic nature of food supply and demand—has a role here. The stochastic future is introduced by defining the INFS as the probability that $Fa_t > Fa^*$ in some future time t. This index may be measured using the probability distribution of future events that affect food intake among those individuals most vulnerable to food shortfalls.

I claim that this index reflects in a reasonable way the concerns of agricultural and food policymakers who are responsible for agricultural trade policy. Other operational indexes that approach food security in this general way would share many of the same implications as INFS.

COMMODITY TRADE POLICY AND FOOD SECURITY
With this potentially measurable index of national food security (which focuses on access, not utilization), I may now investigate how market conditions and policies affect food security on a national level. The supply and demand for food are components of this investigation. Food demand is a function of the price of food and income and such variables as demographic characteristics and relative prices of other goods. I may consider food prices as mainly exogenous to the household or individual and as varying mainly over time and less across individuals. Income does vary widely across individuals and is endogenous. In many countries, income is mainly based on labor market earnings. However, to be applicable to poor rural populations, I may explicitly consider the income of farm households who sell food. For these households, the price of staple food crops affects both the cost of consumption and the family income.

The role of income in food adequacy may be introduced as follows. The functional form of the demand equation implies a specific relationship between the income distribution and the food intake distribution. For the point f*, which defines adequate food intake, holding price constant, there is an income I* which is the threshold income required for adequate food intake. The area of the income density function that is below I* maps into the area in the food intake distribution below f*. That is, the share of the population with less than adequate food intake is the share of the population with income below I*.

Next, consider the distribution of food price over time with income held constant. The distribution of price reflects the randomness inherent in agricultural markets. Unanticipated variation in price follows some probability distribution. Now, with the income distribution fixed, food intake relative to the threshold food-intake quantity, f*, is just a function of food price. The probability distribution of Fa is a transformation of the probability distribution of food price. The probability of a widespread national food shortfall is shown by the area under the price distribution to the right of P* or the area under the Fa distribution to the left of Fa*.

Policies that increase the income of the poor clearly raise IFNS. Such policies may be those that improve national average incomes or policies that shift income specifically to the poor who make up the left tail of the food intake distribution. Policies that make incomes less variable over time would also reduce the chance of widespread food shortage and thus raise INFS. Cross-national observations demonstrate clearly that the most important strategy for national food security relates to economic growth and widespread improvement in income. Thus, any food security policy must be evaluated against what it does to economic growth and particularly the opportunity for improving incomes of the poor.

Agricultural commodity policies affect incomes, but even more directly, these policies often affect the price of food. For example, although import barriers raise the price of food, many policy makers and even some economists claim that trade barriers emphasizing food self-sufficiency contribute to food security. Two possibilities could validate this claim.

First, if farmers comprise a large share of the poorest part of the national population and if these farmers derive a large share of their income from production of staple grains, then higher grain prices could improve national food security and raise the INFS. Note that this is not improved food security for others, but rather for the farmers themselves through an income effect. For subsistence farmers who consume all or almost all of their production at home, trade barriers cannot help, and for consumers who purchase food, trade barriers are positively harmful. But, in some poor countries, food security policy must pay close attention to the income of food producers.

Second, import barriers or other policies to enhance food self sufficiency may, in some cases, reduce the variability of food prices or, at least, reduce the perceived likelihood of a high price spike or other interruption in access to food. That is, the right tail of the food price distribution may be shortened. Thus, the left tail of the probability distribution of the share of the population with an adequate diet is shortened. In this case, INFS may rise although the mean price of food increases. This is a theoretical possibility. Its practical application depends on the facts of internal and global commodity price distributions.

As an example, consider the behavior of the INFS with reference to rice import policy in South Korea. (For earlier analysis of South Korean data, see Adelman and Berck.) Import barriers (essentially a ban on imports of table rice for human consumption) raise the price of rice for Korean consumers to about four times the average world price. In fact, the average price of rice in South Korea exceeds by a wide margin the highest level the world price of rice has reached in the past thirty years or more. Furthermore, the price of rice in Korea is potentially quite variable, given a variable climate and a relatively small rice-growing region. But Korean policymakers point to the potential of a dislocation in the world market that could cause an extreme and never previously observed import price spike facing South Korea if it were an importer. This is illustrated as a very long right tail on the implicit import price distribution. In terms of the INFS, this translates into a long left tail in the probability distribution of the share of the population with inadequate diet and a higher INFS than under autarky (no trade). Let us turn to data to examine the INFS for South Korea under autarky versus open trade.

In the following analysis, I calculate and compare the INFS for South Korea under two stylized rice policy regimes. The application of the INFS requires data or assumptions about a number of parameters of income and price distributions and how these are different under alternative policies. Here, the INFS calculations are for illustration only, and I make several simplifying assumptions.

The first policy considered is close to the current trade policy. Under the Uruguay Round Agreement on Agriculture of 1994, South Korea limits imports of rice to a small fixed quantity, which is not used for direct human consumption (Sumner and Lee 2000). To compute the INFS for South Korea under this policy, the internal mean price of rice is taken from FAO data and bears no connection to international prices. I use the internal price of rice in South Korea in 1995 as the mean of a log normal distribution (log price = 13.8532 per ton). To determine the prospective price variability, I take the rice yield history for Korea for the past twenty years and calculate a variance around the trend. This variance is used to provide supply shocks in a log linear supply and demand system with a price elasticity of supply of 0.25 and a price elasticity of demand of –0.25. With these data and assumptions, the variance of the change in log price is 0.037. This may be interpreted as a percentage variation in price.

To characterize the INFS, I use an individual threshold that defines a household as having adequate access to the staple grain if it spends less than 5 percent of its income on rice. This threshold may be readily adjusted to any criterion that policymakers or analysts prefer. I use data on the 1995 income distribution for South Korea and an income elasticity of demand for rice of 0.25 as the basis for simulations. For simplicity, I assume that all consumers face the same price and that rice producers who get a major share of their family income from rice production comprise a negligible share of the total population of the country. (Farmers comprise less than 10 percent of the South Korean population.)

With these data and assumptions, I am able to calculate an INFS for any threshold share of the population that spends less than the 5 percent of their income on rice. For example, with the share threshold, Fa*, set at 0.96, INFS is 0.9. Under the current autarky policy and with the data and assumptions outlined previously, there is a 90 percent probability that 96 percent of the population will spend less than 5 percent of their income on rice. When the Fa* is lowered slightly to 0.95, INFS increases to 0.95, which says that there is a 95 percent probability that 95 percent of the population will spend less than 5 percent of their income on rice.

Now I turn from autarky to a policy of open import markets for rice. For this analysis, I assume that South Korea is a small country and thus a price taker facing a variable world price of rice. I use the mean price of japonica rice on the world market converted to Korean won as the central tendency of the price distribution. (The mean world price is about one-fourth the internal Korean price.) The variance of the log price with these data is 0.060, about 62 percent above the variance calculated under autarky. I do not calculate an INFS under some probability that South Korea faces a rice embargo, but that could be easily done. The other parameters, relating to the income distribution and demand function, are the same as in the autarky case. Thus I ignore the effects of a more liberal trade policy on real incomes in South Korea.

Under these free trade assumptions, the probability is 99.9 percent that 99.8 percent of the population would spend less than 5 percent of their income on rice. In other words, with access to the world market, and using the past twenty years as a guide to price variability, essentially no chance exists that as much as 0.1 percent of South Koreans would spend more than 5 percent of their income on rice.

Therefore, even though world prices are substantially more variable than implied autarky prices for rice, the lower mean more than makes up for the higher variance in assuring food security for the poor in South Korea. With autarky, a significant probability exists that 4 or 5 percent of the population must spend more than 5 percent of their incomes on rice. If we take such an event as indicating national food security concerns, these concerns can be eliminated with an open border.

Two further comments are useful with respect to this example. First, the autarky policy banning imports would also require an export ban to keep food from flowing out of the country in the case of high world prices. As a policy to reduce high domestic prices caused by world price spikes, this insulation from exporting is key. Of course, this point does not apply when a country-specific embargo limits imports. Further, if the domestic industry is much smaller under open borders, as may be expected for Korean (or Japanese) rice, the potential for exports is relatively remote. Second, an import dislocation that interrupted access to rice would also affect agricultural inputs and thus affect domestic production as well as imports. Thus, staple grain self-sufficiency is ineffective even in the case that may seem most favorable for its support.

Finally then, under almost any food adequacy threshold and population share threshold, raising the average price of rice to four times the world price almost surely lowers the nutritional adequacy of the poor and lowers food security as measured by INFS or any other reasonable index.

CONCLUDING COMMENTS

This chapter examines two kinds of policies that are often supported by agricultural interests as contributing to national food security. I find that access to international markets can be a more effective contributor to food consumption than agricultural R&D. National food security arguments to support improvements in farm productivity are found to be weak. Indeed, with access to international markets, food security for a nation may hinge on agricultural R&D on a global basis rather than locally. With this approach, trade and R&D can work together to contribute to lower food prices and lower probabilities of high price spikes. This reasoning favors a system supporting agricultural productivity growth wherever it can be most effectively achieved, and not tied to the locale of consumers.

The Index of National Food Security used in this chapter recognizes that policymakers often express concerns about variable food prices. Even in nations with relatively low shares of income spent on staple foods, farm support policies are rationalized as reducing the prospects of food shortages despite the fact that national food production is much more variable than global food production. The INFS incorporates these considerations in a consistent way that shows the effects of parameters of the food supply distribution on national food security. Thus, we are able to evaluate through simulations how claims about food price variability translate into effects on food security of the poor.

Food security is a huge and important topic. This chapter provides only a few ideas that require more development to be operational. Nonetheless, I argue that emphasis on local food production for food security is often misguided and harmful to the world's poor.

REFERENCES

Adelman, I., and P. Berck. "Food Security Policy in a Stochastic World." *Journal of Development Economics* 34 (1991): 25–55.

Alston, J. M., G. W. Norton, and P. G. Pardey. *Science Under Scarcity: Principles and Practice for Agricultural Research Evaluation and Priority Setting.* Ithaca: Cornell University Press, 1995.

Alston, J. M., and P. G. Pardey. *Making Science Pay: The Economics of Agricultural R&D Policy.* Washington, DC: AEI Press, 1996.

Barrett, C.. "Food Security and Food Assistance Programs." In *Handbook of Agricultural Economics*, Bruce Gardner and Gordon Rausser, eds. Amsterdam: North Holland Press, forthcoming.

Diamond, J. *Guns, Germs, and Steel: The Fates of Human Societies.* New York: W.W. Norton and Company, 1997.

Duncan, J. "Food Security and the World Food Situation." In *Handbook of Agricultural Economics*, Bruce Gardner and Gordon Rausser, eds. Amsterdam: North Holland Press, forthcoming.

El Obeid, Amani E., S. R. Johnson, H. H. Jensen, and L. C. Smith, eds. *Food Security: New Solutions for the Twenty-first Century.* Ames: Iowa State University Press, 1999.

Food and Agriculture Organization, *The State of Food Insecurity in the World, 1999.* Rome: FAO, 1999.

Food and Agriculture Organization. Web site www.fao.org, August 2000.

Food and Agriculture Organization. Web site www.fao.org, "Global Information and Early Warning System." Periodic reports, 2000.

Huffman, W. E., and R. E. Evenson. *Science for Agriculture: A Long-Term Perspective.* Ames: Iowa State University Press, 1993.

Lee, P. H., and W. T. de Bary. *Source of the South Korean Tradition.* New York: Columbia University Press, 1997.

Roberts, J.A.G. *A Concise History of China.* Cambridge: Harvard University Press, 1999.

Smith, A. *An Inquiry into the Nature and Causes of the Wealth of Nations.* Oxford: Oxford University Press, 1976 (First published in 1776).

Smith, V. H., and P. G. Pardey. "Sizing Up Social Science Research" *American Journal of Agricultural Economics* 79, 5 (December, 1997): 1530–1533.

Stigler, G. "Do Economists Matter?" *The Economists as Preacher and Other Essays.* Chicago: University of Chicago Press (1982): 57–67. Reprinted from the *Southern Economics Journal,* January 1976.

Sumner, D. A. "Agricultural Trade Policy and Food Security." *Quarterly Journal of International Agriculture* 39, 4 (December 2000): 395–409.

Sumner, D. A., and H. Lee. "Assessing the Effects of the WTO Agreement on Rice Markets: What Can We Learn from the First Five Years?" *American Journal of Agricultural Economics* 82, 3 (August 2000): 709–717.

Sumner, D. A., and S. Tangermann. "International Trade Policy and Negotiations." In *Handbook of Agricultural Economics*, Bruce Gardner and Gordon Rausser, eds. Amsterdam: North Holland Press, forthcoming.

Tweeten, L. "The Economics of Global Food Security." *Review of Agricultural Economics* 21, 2 (Fall/Winter 1999): 473–478.

Tweeten, L., and D.G. McClelland, eds. *Promoting Third World Development and Food Security.* Westport, CT: Praeger, 1997.

13

Competing Paradigms in the OECD and Their Impact on the WTO Agricultural Talks

Tim Josling

Evidence exists of an emerging new paradigm in U.S. agriculture that emphasizes the ability of the sector to compete both with other sectors in the domestic economy and with overseas competitors. But a rival new paradigm has been gaining significance in Europe that stresses the public-good nature of agriculture and the inability of the sector to earn satisfactory incomes just by providing agricultural commodities. A third paradigm points to the nature of agriculture as increasingly one stage in an integrated supply chain that crosses borders in a global food system. The current WTO talks on agriculture will provide an opportunity for these paradigms to compete for the setting of international trade rules.

INTRODUCTION

The premise of this book is that a paradigm shift in agricultural economics is underway that is reflected in a new view of the nature and causes of "the farm problem." Essentially, this new view holds that U.S. agriculture has become a sector able to compete with others for resources without government price support, and able to compete on world markets without recourse to subsidies. Luther Tweeten is a major proponent of this view, and several of his writings describe the new paradigm and its domestic implications.

I focus on two aspects of this issue. First I cast the paradigm debate in the context of the OECD countries as a whole rather than just the United States. I argue that the discussion becomes more interesting if one asks whether a common paradigm exists in industrial countries and, if no common paradigm has existed in the past, has a "paradigm convergence" taken place in the last 15 years? Second, I look at the implications for trade policy discussions of having different paradigms "competing" in the international arena, specifically the WTO agricultural negotiations. Or, alternatively, if a convergence in paradigms is indeed under way, how does one negotiate among countries that are in a different phase of this paradigm shift?

The thorny question of the definition of a paradigm shift I leave to others.[1] For the purposes of this chapter, I assume that a policy paradigm is a common set of ideas that guide attitudes and justify policies. The paradigm provides a model that serves as a mental testing ground for policies. A paradigm shift is therefore a replacement by one such model with another, either suddenly or over time. Farm policies will tend to change as a result of this paradigm shift, subject to bureaucratic inertia or political entrenchment. Policies will therefore change at different speeds in different countries depending on specific conditions even if a common paradigm shift has occurred. Lagging policies are common in agriculture, because the ability to change policy instruments to respond to events is often limited by both political and institutional factors. Policy reform is the process of changing policies in response to a paradigm shift in the face of those forces that would defend the status quo.

No one would deny that farm policies in developed countries have changed markedly over the past two decades. The nature of these changes, however, seems still to be in doubt. Some have claimed that an irreversible policy reform has indeed occurred in response to a paradigm shift; others that the changes in these policies have been more along the lines of a natural evolution, reacting to structural, economic, and political forces but retaining their essential characteristics.

This debate is not just a matter of semantics and perspective. If a nonreversible change in policies has indeed taken place, then recognizing that change is crucially important for our understanding of the future relationship between the government and the agricultural sector. Moreover, the nature of the change has implications for trade policy as well as for domestic farm programs. In particular, the WTO process of trade liberalization is premised on a continuation of domestic policy reform. Some who argue for an irreversible shift also conclude that the relationship between domestic and trade policy has itself changed. Domestic policy change, at least in the OECD countries, has in effect been locked in by the WTO and the Uruguay Round Agreement on Agriculture. This effectively makes the case for a new "trade paradigm" for agriculture, in which the nature of world markets is no longer merely a residual of all the autonomous domestic policy decisions but is endogenous to the domestic policy process. This view rests on the notion that the Uruguay Round on the one hand constrained policy choice but on the other expanded the scope for such choice by providing new instrumentalities.

However, if the argument that a significant change has occurred in the underlying view of agriculture and of the relation between trade and domestic policy is found to be weak, then the prospects for further progress in the current agricultural trade negotiations are correspondingly dim. If all is "business as usual" in farm policies, then the implications for both domestic and trade policy as well as for the development of global agricultural markets are rather sobering. This chapter explores the question of whether a significant shift has occurred in agricultural policies in the OECD countries that can usefully be described as the result of a paradigmatic shift. But in the process, the issue is raised of whether the relationship between domestic and trade policies has changed, and hence whether the prospects for further trade policy reform have been thereby enhanced.

PARADIGM SHIFTS AND POLICY REFORM

The idea of a new paradigm for agricultural policy was explored explicitly by Coleman, Skogstad, and Atkinson (1997). They argue that a change has occurred from a state-assisted to a market-liberal paradigm. In the former paradigm, the state considers that the agricultural sector contributes importantly to realizing national policy objectives and that special support is required to enable that contribution to be maximized. In the latter paradigm, the sector is treated much like other sectors of the economy. Competitive markets should be the main source of income for commercial farmers, though public programs can still exist in areas where the market is inadequate. To Colman, Skogstad, and Atkinson, paradigm shifts can come slowly over time, rather than in a Pauline conversion of politicians to a new faith. They find some evidence of shifts of this nature but recognize that the process was by no means complete.

Tweeten has also argued strongly for the emergence of a new paradigm. His "new economic paradigm" for (U.S.) agriculture is characterized by the idea that agriculture is no longer the backward, noncompetitive sector that it once was thought to be (Tweeten forthcoming; Tweeten and Zulauf 1997). Such a sector was thought to require government assistance to deal with the triple jeopardy of weather and disease, concentration in upstream and downstream activities, and a chronic tendency toward overproduction. Instead, the sector (or that part of it that produces the bulk of the output) is now recognized as competitive both with other sectors or the economy (generating roughly comparable return on resources) and on international markets (at least when not facing undue foreign subsidies). Again, government programs are limited to correcting for externalities, particularly in the environmental area, and perhaps to helping farmers manage risk.

If paradigms are the bundle of ideas and analytical concepts with which we approach farm policy, what can we say about policy reform? The conditions under which one would expect to see a significant shift in policy include dissatisfaction with existing policies and instruments and broader ideological and political climate change. Ample evidence exists of such policy reform in many countries, both inside the OECD and outside. This reform has been towards less manipulation of the commodity price level and towards more use of direct payments to farm families. Early signs of this shift in developed countries were the adoption of commodity-based deficiency payments in place of price supports so that consumer demand could respond to market prices. These deficiency payments over time have become increasingly decoupled from present production levels to reduce the incentive to overproduce. The decoupled payments have in some cases been targeted to various environmentally friendly farming practices. Revenue and crop insurance schemes have also been introduced to act as a stabilizing device for rural areas, though these can become incentives to produce when subsidized by governments.

Underlying these changes in farm policy are fundamental trends in the sector itself. Advances in technology and rapid structural change in agriculture increase yields, but domestic market growth is slow in most OECD countries. After the domestic market is satisfied, the protection at the border against competing foreign goods is no longer adequate to maintain farm prices. The traditional options facing policymakers were to control supply, an unpopular solution with farmers and agribusinesses; to store the surpluses against the next period of shortage, which may not come in time; to export with generous subsidies, thus distorting other countries markets; or to lower farm-support prices, with or without paying compensation to those who were disadvantaged. Until the mid-1980s, most countries tried a combination of supply control and surplus disposal, with some attempts to constrain price increases.

The situation began to change in the period after 1985, as countries both adjusted their domestic policies and then later incorporated new trade rules into the Uruguay Round Agreement that reinforced the domestic policy changes. The shift occurred because the old policies were clearly inappropriate in a more open world economy and a less regulated domestic market. Commodity price intervention rewarded farmers according to their output of the product, regardless of the state of the market. Farm output and structures responded to the policy signals rather than consumer demand, and agriculture became dependent upon further government assistance to dispose of surpluses. Moreover, environmental groups began to point out that farming systems geared to high price policies were creating problems for the water supply, for worker safety, for public health, and for the preservation of plant and animal species. Targeting payments to the farmer has the advantage of being more direct, involves less distortion in production incentives, can be tied to environmental standards, and allows the consumer to choose more freely among competing products.

Similarities were apparent in the ways in which countries attempted to change policies. Orden, Paarlberg, and Roe (1999) have framed the reform strategy in U.S. farm policy (and by implication in the European Union as well) as the choice among four options: a "cutout" is a quick termination of price supports without compensation; a "buyout" is a quick termination with compensation; a "squeeze-out" is a slow reduction in price supports without compensation; and a "cash-out" is a slow, compensated cut in prices (Orden, Paarlberg and Roe 1999, 8). In the terminology used in this chapter, a buyout and a cutout are likely to be clear evidence of a reform, though not too many cases exist in the U.S.and the EU where such options have been tried. A squeeze-out is the traditional way in which poorly performing policies get modified, when the forces for change overcome the bureaucratic inertia and the power of vested interest. Calling such changes "reform" may not make much sense. A cash-out has been the instrument of choice in the policy adjustments of the 1990s, both in the United States and in the EU, because it weakens opposition to the squeeze. Whether a cash-out necessarily is evidence of policy reform is debatable, and indeed is at the core of the debate on whether policy trends are reversible. But instruments such as cash-outs clearly are a convenient way of making the transition to a reformed policy if that is indeed what is intended.

But those claiming policy reform cannot be too complacent. Several policies have remained essentially unchanged in the face of significant shifts in the performance of agricultural markets. This is particularly true for policies toward commodities that have enjoyed special status in political circles, the "sacred cows" of farm policy. One could call these policies evidence of "policy lag." It is the tension between policy reform and policy lag that conditions the prospects for further trade liberalization in the WTO and in regional trade agreements.

POLICY CHANGE IN THE OECD COUNTRIES

Much of the discussion of paradigm shifts and policy reform has focused on experience in the United States. But other countries have been grappling with the same problems and many of them have experienced more pronounced policy changes. In fact, the period since 1985 has been one of remarkable change in domestic agricultural policy in most parts of the world. This change was most apparent in the developing and middle-income countries of Latin America, where the overhaul of agricultural policies was part of a package of economic policy changes induced by a combination of external pressures and shifting notions of the role of the state (Williamson 1994). Little doubt exists that both a paradigm shift and a policy reform took place in these countries. Moreover, the two were closely related. Changes in the set of ideas about the role of agriculture in development, as well as those about the linkages between macroeconomic policy and farm policy, were an important precursor to the policy reforms (Krueger et al. 1992). Domestic reforms then allowed countries to bring agriculture into the trade-policy reforms, generally involving the removal of nontariff barriers and the setting of low fixed tariffs against imports. In retrospect, the remarkable fact is that politicians in these countries did not shy away from the inclusion of agricultural markets in the overall reform of economic policy. In most cases, difficult decisions had to be made in the face of opposition from rural constituencies. This agricultural reform implies either a compliant political system that was unable to stop the "technocrats" from adopting the reform measures or a general, if reluctant, acquiescence that such reforms were worth a try.

In the major industrial countries, however, agricultural policy reform came only reluctantly in the period up until 1990 and with a great deal of domestic opposition. The farm policies in these countries had become so entrenched that they seemed almost to be immune from paradigm shifts and external pressures. Among the developed countries, the lead in the policy reform race was taken by New Zealand in 1985. Facing a deepening economic crisis, the new Labor government took a bold approach to economic reform, introducing monetary stringency, deregulation of the economy, and trade liberalization. The reform included a sweeping overhaul of the agricultural programs that had become unwieldy and ineffective.[2] The programs that were cut included the price-support schemes for livestock products, subsidies on capital and on the use of inputs, and tax breaks for landowners. In addition, many of the marketing boards that had been set up as government-sponsored monopolies lost their market power. The wheat and poultry industries were deregulated, and a partial deregulation of the liquid (town) milk sector occurred (Sandrey and Reynolds 1990).

The programs that had been removed were of rather more recent origin than the farm policies of the United States, Europe, or even Japan. The lessons were perhaps more applicable to a middle-income country with favorable agricultural prospects but with a counter-productive mix of agricultural policies giving inappropriate signals to an otherwise competitive farm sector.[3] However, in practical terms, successful agricultural policy reform gave New Zealand a new authority in international discussions of farm policies, as being willing to make dramatic policy changes of the sort being suggested to others.

Some other high-income countries also underwent radical reforms of agricultural policy, following the lead of New Zealand. Australia made a significant step toward reducing market price support for a number of agricultural products in 1988 (OECD 1988, 86). Administered prices for wheat were eliminated and the domestic market deregulated in 1989 (OECD 1991, 115). A process of tariff reduction for sugar was begun in 1991 (OECD 1991, 150), though parastatal marketing of domestic production was allowed to continue. In 1992, the federal support for milk was reduced and an end to export subsidies for dairy products was announced (OECD 1992, 118). Despite some marginal backsliding, market price support in Australia is now minimal—running at about 7 percent of farm output (OECD 1999, 99).

But for every reform in the 1980s, many more examples exist of reform delayed and watered down. The Common Agricultural Policy (CAP) of the European Union was perhaps the most resistant to change. In an account of reform attempts of the 1980s, Moyer and Josling (1990) had to stretch the definition of reform somewhat to include the 1984 and 1988 changes in the CAP. They spent much of their effort explaining why the CAP had survived pressures that might have been thought to be enough to lead to change. U.S. farm policy did change somewhat more noticeably in the direction of "reform" in the 1981 and 1985 farm bills, but again at a pace dictated by domestic political resistance from the agricultural sector. Serious policy lags were clearly predominant on both sides of the Atlantic, even while New Zealand was paving the way toward reform.

The experience of most industrialized countries in the 1980s falls somewhere between the radical shifts of the developing countries and New Zealand and the snails-pace incrementalism of the United States and the European Union. Canada offers a good example of such a country with an ambivalent record over this period. New instruments were introduced that emphasized such elements as crop insurance and direct payments at the expense of open-ended price supports. But supply control still dominated the import-competing sectors, and the export industries still benefited from federal subsidies and state marketing. Japan began to move away from the tight control of internal markets and imports, but only in those sectors in which domestic production was not adequate to meet demand. The basic laws governing agricultural policy still mandated a search for higher degrees of self-sufficiency in spite of new paradigms that defined food security in terms of the availability of food on international markets and the ability of countries to pay.

The situation changed quite appreciably in the 1990s. The 1992 changes in the European Union's Common Agricultural Policy (CAP) and the 1996 farm bill in the United States represented very significant shifts in direction for farm policy and qualify as "reform." Whether one can attribute the reforms to a paradigm shift is less clear. The MacSharry reforms of 1992 were deliberately aimed at preserving the CAP. As Agriculture Commissioner, Ray MacSharry had as his primary objective to put the CAP on a more stable footing, and a compensated price-drop for cereals and oilseeds was a way of removing some of the pressure not only from abroad but also from domestic critics. The connection with the GATT talks has been discussed elsewhere in detail (Coleman and Tangermann 1997) but the causal link between CAP reform and a paradigm shift is more indirect. The nature of the CAP is that it represents a bargain among countries as well as a farm policy for the European Union. Change can come about as a result of changes in the terms of the bargain just as much as a shift in the model of agriculture. Or, to put it differently, each country in the European Union has its own version of the paradigm, and the CAP is a reflection of the bargaining process within the European Union as much as of a coordinated and consistent European view of the appropriate role of government in agriculture.

The U.S. farm bill debate did have elements of "new thinking" in it, but the final bill owed more to a careful calculation by its supporters that it would generate larger returns to farmers than any feasible alternative (Orden, Paarlberg, and Roe 1999). The two agricultural superpowers seemed at last to be following the lead of the smaller exporters and the middle-income countries in adapting their agricultural policies to the new global realities. However, the end of the decade of the 1990s brought indications of backsliding as the United States, along with other countries, tried to protect their farmers from low world prices.

In other parts of North America, agricultural policies began to change somewhat more rapidly. Canada dropped its rail subsidies for western grains in 1995 (OECD 1995, 26) and converted quotas to (rather high) tariffs for sensitive import products such as dairy goods and poultry meat. But tying this development to a conversion to the liberal cause is difficult. The changes were forced on Canada by the exigencies of international trade pacts, NAFTA, and the Uruguay Round Agreement on Agriculture rather than being domestically generated through a revision of orthodoxy. In Mexico, a stronger case can be made that a paradigm shift led to the dramatic reform of agricultural policy that was put in place in 1994, when the PROCAMPO program instituted direct payment in place of high price supports for a number of crops, including maize and beans (OECD 1994, 81). Two years later, Mexico launched another initiative, Alianza para el Campo, which helped farmers purchase capital goods and provided technical assistance (OECD 1996, 59). A constitutional amendment (*ejido* reform) helped to free the land market, and the radical trimming of the role of the main parastatal, CONASUPO, liberalized the domestic market. These changes were not mandated by NAFTA and the WTO, though they made compliance with the terms of these trade agreements very much easier.

In Japan, one of the more reluctant reformers, agricultural policies also began to change significantly in the 1990s. Pressure from the United States had already led to the conversion of quotas to tariffs for citrus and beef imports. The Uruguay Round speeded the process and led to further tariff cuts. The law that mandates the role of the government in agricultural markets was modified in 1996 in the direction of privatization, in particular in the sensitive rice market (OECD 1996, 54). Direct payments were introduced the next year, under the New Rice Policy (OECD 1997, 72), though these instruments have been used more sparingly in Japan than in many OECD countries. As a result, the task of bringing down domestic prices has hardly begun.

Perhaps the most surprising convert to the new paradigm was Sweden, long considered, along with Norway and Finland, to be a bastion of high-cost protected agriculture. In 1991, the government introduced the New Food Policy that included the immediate phasing out of internal price supports, intervention buying, and export subsidies (OECD 1991, 177). However, this liberal policy survived only a couple of years, because by 1993 Sweden had to reintroduce intervention and export subsidies as a part of its entry into the European Union (OECD 1993, 155). Norway, relieved of the obligation to adopt the CAP by its decision not to join the European Union, also began to drift toward direct payments and lower prices (OECD 1993, 152) but with nothing of the abrupt policy shift of its Swedish neighbor. Even

Switzerland, long known as having among the highest levels of agricultural protection in Europe, began in 1992 the process of switching from high price supports to the use of direct payments (OECD 1992, 150). In 1998, this shift was accelerated by a new policy framework (AP 2002) that linked these payments to environmental standards (OECD 1998, 142).

COMPETING PARADIGMS OF OECD AGRICULTURE

Can one conclude from the active period of farm policy reform in the 1990s that all the major OECD countries have evolved a broadly similar view of agriculture and its policy environment? Such a conclusion is almost certainly premature. The extent to which OECD countries have embraced the model of a market-oriented farm sector differs greatly across countries. It is difficult to argue that Japan and Korea, though actively discussing policy reform at home, have fully abandoned either the rhetoric or the policy manifestations of the "old" paradigm of an agriculture that needs support for national objectives that transcend commercial viability. Debate in Norway and Switzerland is still based on the premise that agriculture is basically an uncompetitive sector, hampered by climatic and physical limitations and competing unsuccessfully with an affluent nonfarm sector for labor and capital. The lively debate in the European Union on farm policy reform is certainly not predicated on an efficient sector that can duel with the United States, Canada, and Australia for market shares in emerging regions. It is therefore necessary to examine in more detail what might be other "competing" paradigms of agriculture in the OECD countries. The old paradigm may be on the way out, but that does not necessarily imply that the same new paradigm is being embraced in all OECD countries. In fact, three competing paradigms appear to be vying for the title of successor to the old view of agriculture as a sector in need of support. These are identified in table 13.1, together with the policy objectives and instruments implied by the paradigm. In addition, the implications for trade rules are suggested, as well as the implicit or explicit paradigm for the nature of the world market.

The Dependent Agriculture Paradigm

To highlight the changes inherent in the adoption of each of the three new paradigms, it is useful to briefly characterize the "old" view of agriculture as a dependent sector unable to compete either internally with the nonfarm sector for resources or with other agricultures overseas (table 13.1). The major role of the OECD governments was to find and secure markets for farm products. Over time, this role expanded to managing, through domestic production quotas, the quantities produced. Farmers themselves were in general left to focus on the production of commodities. The implicit social compact was that the government would provide border protection, buy surpluses, and assist with exports if they were in excess of the unsubsidized sales possible on world markets. Such surpluses frequently emerged, as technical change tended to outstrip demand growth. Support for this paradigm came from the mainstream farm organizations and from first-stage processors, whose profits appeared to be linked to the level of domestic production. Even countries that were major exporters of farm products subscribed broadly to the notion that agricultural market supports were necessary for a viable rural sector.

The trade policy implications of such a view of agriculture were patently obvious for years. Trade rules could not be allowed to get in the way of the management of domestic markets. In particular, Articles XI and XVI of the GATT were designed to ensure that agriculture was relieved of some of the disciplines applied to other traded goods.[4] Moreover, the GATT waiver granted to the United States in 1995 and the choice of the ambiguous "variable levy" and "export restitution" instruments by the European Union in 1962 made it clear that countries were not prepared to submit their policies to international discipline. Behind this determination to keep trade rules weak in agriculture was a particular view of the functioning of world markets. Commodity prices were seen as low and unstable, and not a sound basis for domestic policy. This view of a market characterized by chronically weak world prices became self fulfilling, as interventionist policies promoted surpluses that came up against insulated markets.

The Competitive Agriculture Paradigm

The main feature of the competitive agriculture paradigm is the view that the sector can stand on its own two feet (table 13.1). Government policy is defined in this paradigm as performing the twin tasks of giving the competitive farmers a "level playing field" on which to compete while simultaneously putting a safety net under those who cannot play on that field. Eventually the noncompetitive farmers will leave the sector. Supply control is not only unnecessary but also undesirable: Why handcuff a competitive sector and give the market to others? Policy instruments include decoupled payments for when a transition is needed from the previous policy. Emphasis is switched to such issues as risk management and low "safety nets" in lieu of price support. Such policy directions generally have the support of large farmers and their organizations as well as of agricultural traders and processors.

Table 13.1. Four Competing Paradigms of Agricultural Policy in the OECD

	Dependent	*Competitive*	*Multifunctional*	*Globalized*
Nature of Agriculture	Low incomes Not competitive with other sectors Not competitive with other countries	Average incomes Competitive with other sectors Competitive in world markets	Incomes from farming inadequate Producer of under-rewarded public goods	Farmers as part of supply chain Manager of land and livestock resources Consumer-driven sector
Policy Objective	Government needed to find markets Supply control necessary	Move towards free market Relax supply control	Preserve countryside Keep family businesses viable	Establish quality and safety standards Fairness in contractual relationships
Policy Instruments	Border protection Surplus buying State trading Export assistance	Decoupled payments in transition Risk management Low safety-nets	Environmental subsidies Protection against "mono-functional" agriculture	Harmonization of regulations and standards Competition enforcement Protection of intellectual property
Main Supporters	Farm organizations First-stage processors	Larger farmers Agricultural processors and traders	Small-farm groups Farmers in remote areas	Retail stores Specialty farms Food processors
Trade Policy Aims	Avoid restrictive trade rules	Market access Remove export subsidies Constrain domestic support	Moderate pressure on agriculture Allow subsidies under trade rules	Strengthen intellectual property rules Harmonize SPS and TBT rules Ensure competitive conditions
World Market Paradigm	World market unstable Prices depressed and no basis for domestic policy	World market stable and reliable if domestic policies are reformed World prices best guide for domestic policy	World market reflects "mono-functional" agriculture Prices inadequate for supply of public goods	"World market" is often intra-firm sales Instability and uncertainty created by government intervention

This trade-friendly paradigm has been accompanied by (and perhaps reflects) a shift in the composition of the groups that have political influence in agricultural policy (Josling 1999). This is in part connected with the move away from commodity-based support systems. If the government supports the price of a commodity by restricting imports, subsidizing exports, buying up surpluses, or taxing substitute products, then it usually needs to work through the processing sector or the wholesaler to implement the policy. Sugar policies are operated largely through sugarbeet factories and cane refineries. Dairy policy is implemented through the dairies and creameries that take delivery of farmers' milk. Grain policies involve storage and shipping of cereals through merchants and middlemen. Oilseeds policies involve oils and meals from the crushing activities as well as just seeds and beans. When the basis of farm payments is the output of commodities, some coincidence of interests occurs between the processor and the farmer. When the farmer gets paid compensation bonuses on the basis of historical hectarage and regional yield, the processor has less interest in the profitability of the domestic producer and is more willing to import the raw material and look for the cheapest source of supply.

Moreover, the "decoupling" of support from output has in turn had a profound impact on the range of trade policy choices in the sector. It essentially means that a country can recast its trade policy for agricultural goods to be consistent with that for nonagricultural goods. Shifting from price supports to income insurance schemes, for instance, gives the possibility of relaxation of import regulations. One is substituting financial instruments for physical commodity market intervention, generally at lower cost. Similarly, production constraints previously needed to limit government payments on export subsidies would no longer be needed if the compensation payments to farmers were not influenced by market price. The decline of commodity programs liberates trade policy, just as the de-linking of price from income liberates social and environmental policy.

In the absence of externalities in commodity markets, a major role of governments is to run a trade policy that supports the competitive parts of agriculture along with a policy to improve competition at home. This implies that governments agree to restrictions on their ability to intervene in markets in exchange for agreements by others to do the same. Market access needs to be improved by tariff reductions; export subsidies should be removed because they grossly distort world markets; and domestic support payments that escape the disciplines imposed at the border have to meet certain criteria. This is essentially the set of rules that emerged from the Uruguay Round. In other words, the competitive paradigm is enshrined in the Agreement on Agriculture. The implicit world market paradigm that supports this domestic view of a competitive agriculture is of a sector that is essentially stable if domestic policies are reformed.

The Multifunctional Agriculture Paradigm

The primary challenge to the new market-oriented paradigm has come from Europe. In recent years, farm politicians in many European Union members have adumbrated a "paradigm" of their own, that of a "multifunctional" agriculture sometimes referred to as the "European Farming Model" (table 13.1). This argument is often dismissed as merely a mixture of political posturing for position at the WTO and of the repackaging of some bland facts about the significance of farm activities in rural areas. But it is also the case that real differences in perception exist between Europe and (say) the United States about the place of agriculture in the modern economy. Population density and farm size are not unrelated to the political (and economic) relationship between agriculture and other sectors. One does not have to be a closet protectionist to recognize that Holland has a serious problem of groundwater pollution, or to acknowledge that Austrian meadows give pleasure to millions of town dwellers in winter and summer. The question is whether these differences show up as nuances of an essentially similar agricultural model or whether they lead to incompatible policy sets that in turn constrain the reform of the international trade system?

The main characteristic of the "multifunctional" paradigm is that agriculture is viewed as a provider of public goods in addition to, and in many ways more important than, its role as a producer of raw materials for the food industry (table 13.1). The market for commodities inadequately rewards the farmer for such public goods as a pleasant-looking and environmentally friendly countryside, or a stable social infrastructure built on small towns and villages. Appropriate instruments for policy in such an agriculture include environmental payments, either as a reward for providing the public goods or as compensation for the loss of market share to those whose costs are lower as a result of not having to provide those public goods. Support for this view of agriculture comes predominantly from small farmers and those in remote areas. So far, no consensus seems to exist among environmental groups that this is a good way to formulate policy toward the environment: Presumably there is the potential conflict of interest if the total sum that society wishes to devote to environmental stewardship is constrained.

One key question is what are the trade policy implications of a multifunctional view of agriculture? It is often argued that multifunctional agriculture needs protection from the lower-cost supplies from those countries whose agricultural sectors are not asked to bear the burden of countryside preservation, rural development, or societal stability. This is based on a "world market" view of a European agricultural system handicapped by small-sized units and by demands and regulations imposed by a crowded continent, and therefore unable to meet the challenge laid down by the sparsely populated "new world" suppliers. Far from being competitive, border measures are still needed for the viability of such an agriculture. But one can also argue that providing additional public goods through the taxing of private goods at the border is inefficient. In other words, the issue is not whether but how to give the support. And the support could well be given in a way that does not spread the burden to others.

Not all countries in the European Union, or farmers in any one country, fully accept the notion of a multifunctional agriculture. Many European farmers also see themselves as part of a competitive core and have begun to question the restrictions that the CAP has put on them in the name of saving them from "unfair" competition from abroad. This view is particularly prevalent in the United Kingdom, Denmark, and Sweden. It is also possible to argue that the European Commission is using the multifunctional argument for domestic consumption to set up something that can be defended at the bargaining table without having to support the continuation of export subsidies and high tariffs. If multifunctionality means paying producers directly for the amenity value of their farming or their care of animals rather than through distorted prices for output, then the concept is not so inconsistent with the competitive paradigm.

The Global Agriculture Paradigm

As if to escape from the choice between the competitive and the multifunctional agricultures, a third "new" paradigm is emerging with very different implications for trade relations and domestic policy. The main feature of this paradigm is that agriculture is seen as one stage in a global supply chain stretching from chemical and biological input suppliers to retail stores and niche markets (table 13.1). In fact, this view sees agriculture as the "land management" and "animal management" functions in the vertical supply chain. Several sectors of agriculture have gone a long way toward this model, such as pig and poultry farming as well as many fruit and vegetable activities. The emergence of transgenic crops, or rather the opposition to them, has begun to transform the cereal and oilseed sectors as well: The technology encourages the development of the contractual relationships along the chain, and traceability and liability have forced the supply chain to be more transparent.

This "global" agricultural model points policy in a very different direction. The prime role of government is to establish safety and quality standards and to make sure that contractual relations among the elements in the chain are fair and acceptable. Policy instruments are the harmonization of standards and the removal of onerous regulations. Protection of intellectual property and the maintenance of competitive conditions (but perhaps not too competitive!) are the counterpart to the removal of distortions from government price policy. Such a view of agriculture has some resonance within the sector, but its general acceptance will more probably come as a result of the increasing influence of retailers and food processors on the policy process (Josling 1999).

A distinguishing feature of this model is that it is consumer driven. This makes the old style agricultural policy mechanisms less relevant to the needs of the industry. The support of raw commodity prices in the market through the withholding of supplies or the buying up of surpluses effectively breaks the link between producer and consumer. Just as the market is getting more sophisticated and differentiated, the old-style policy sends the message to farmers to "produce low-quality goods for government stocks." Just when supply chains are being set up for the provision of goods to supermarkets, farmers are encouraged under the old policies to take land out of production and live on payments for keeping farmland idle. Clearly the vast expense of farm policy has not helped farmers to meet the challenges of providing for modern consumers. The farmers who have prospered are those who have taken advantage of the changed conditions and begun to service the differentiated market.

Changes in agricultural structures, along with the corresponding changes in the nature of world markets for agricultural products, give this new paradigm an interesting "trade" perspective. The key concept driving such a perspective is of a global market (or a small number of subglobal markets) cooperatively managed and regulated through trade agreements and other forms of accord. National markets cease to be relevant units unless they coincide with other market characteristics. "Countries" do not trade with each other; that is the function of firms. Much "international" trade is in fact transfers of products within a company. Trade policies such as tariffs and export subsidies are undesirable not so much because they distort the resource allocation of a country but because they get in the way of a business transaction among willing partners or branches of the same corporation. Under such a view of the world, market access is a right, subject to well-defined national restrictions, not a benefit to be granted by a government to friends and neighbors. The rules for the global marketplace would include competition laws (including outlawing export subsidies), health and safety regulations, and adequate intellectual property protections to encourage innovation. The scope for domestic action would be largely confined to local and national environmental regulation and rules related to social, health, and educational services that would be provided at the national level.

THE ROLE OF PARADIGMS IN THE WTO AGRICULTURAL TALKS

The Uruguay Round was, whatever domestic politicians may say, the first "global" negotiation on domestic agricultural policy. The completion of the trade negotiations in 1993 marked the first time that effective international constraints had been put on domestic agricultural policies. The Uruguay Round Agreement on Agriculture (URAA) put in place a set of rules that has already gone a considerable way toward shaping the development of such policies. Bound tariffs replaced nontariff import measures, export subsidies were curbed, and domestic programs were codified on the basis of their potential to distort trade.[5] Every significant agricultural policy since that time has been tailored to fit within the new rules.

The attempts at reform of national agricultural policies and the negotiation of modified trade rules in the period since 1985 have been mutually supportive. Though the United States and the European Union adhered steadfastly to the notion that their own domestic policies were not negotiable in trade talks, it was becoming increasingly clear that trade and domestic policies were intimately connected and that progress on one front required movement on the other front as well. Thus the saga of the Uruguay Round negotiations on agriculture formed an important backdrop to the process of domestic policy reform. Though changing trade rules and modifying domestic support are conceptually separate, the politics and the process are closely intertwined.[6]

The role of the competing paradigms identified previously in shaping the current WTO round of agricultural talks is likely to be crucial. The talks themselves are mandated by Article 20 of the Agreement on Agriculture, which sets up the next stage of trade policy reform. But Article 20 also allows for the consideration of "non-trade concerns" in the next step. The focus of much of the debate in Geneva since the start of the new agricultural round in March 2000 (and indeed before that in the lead-up to the Seattle WTO ministerial) has been on the scope that the phrase "non-trade concerns" offers to the proponents of multifunctional agriculture.

The Cairns Group, the most vocal and articulate proponent of the competitive paradigm and the associated view of stable and reliable world agricultural markets, clearly does not want to see multifunctionality enshrined in the WTO. The European Union member states, on the other hand, cannot go home to their electorates without being able to claim some "victory" in this area. Thus much of the real negotiating will be on such seemingly arcane topics as the definition of the green box and the future of the blue box. Should the green box contain some policies that have a positive production impact? At present, this appears impermissible. The Agreement on Agriculture sets up the "fundamental requirement" for support measures to be in the green box that "they have no, or at most minimal, trade-distorting effects or effects on production" (Uruguay Round Agreement on Agriculture, Annex 2:1). However, the "policy-specific criteria" set out in the same Annex states, with respect to environmental programs, that payments are allowed to the extent of the cost of complying with the requirements of the program (Annex 2:12(b)). The apparent conflict between the general and the specific criteria has not been tested in a WTO panel. But it would seem to be vital to the concept of multifunctional agriculture that incidental output effects (presumably, only increases would be of interest) be allowed. For the Cairns Group to swallow this argument, some clear trade-off in terms of market access and export competition would be necessary.

The fate of the blue box may also hang on the outcome of the dialog between the advocates of the competing paradigms. To the Cairns Group, the blue box is a temporary aberration hatched up by the United States and the European Union to avoid the issue as to whether payments linked to supply control were production enhancing. Now that the United States has moved away from supply control in the major arable crops, the only question might seem to be whether the European Union will follow suit. But the European Union and Japan argue that the blue box has been a useful "half-way house" for the transition from coupled to decoupled programs. To close this avenue could make it more difficult for countries such as Japan to reform their policies. And the United States has been reluctant to call for the elimination of the blue box, possibly because to do so would constrain future policy options. In effect, the blue box stands for an established mode of policy, the management of surpluses through supply control, which is out of tune with the competitive paradigm.

Meanwhile the advocates of a "global" view of agriculture will be constantly trying to introduce other elements into the discussion that appear to be outside the realm of the Agreement on Agriculture. They will not so much clash with the other two paradigms as try to move the agenda to consider these "new" issues. One key area would be that of investment policy, not fully covered in the Uruguay Round and unsuccessfully negotiated later in the OECD. Another area would be competition policy, for which at the least a global market should have some global rules. A third aspect of trade policy that is seen as important to this global, consumer-driven agriculture is the establishment of international norms (preferably science based, but certainly consumer credible) for certifying food safety. This, however, threatens to confuse and delay the discussions on market access, export competition, and domestic support. A fourth topic is the possible expansion of the scope for protecting "geographical indications," such as widely-recognized names for particular foods that denote their regional origin. The agricultural talks could well be slowed down by this "agenda creep," to the frustration of those who see the negotiations as being a simple continuation of the process started in the Uruguay Round.

CONCLUSIONS

Agricultural policy instruments in many OECD countries have changed within the last decade, with a convergence toward a set of less trade-distorting programs. These have emphasized targeting and decoupling, replacing the support of commodity markets by payments to farm families on a range of other criteria. In some countries, particularly the smaller exporters such as New Zealand and Australia, and in reforming developing countries such as Mexico, this has reflected a true paradigm shift, involving a sea-change in ideas about the role and scope of policy and the nature of the farm sector. In other countries, including the United States and Canada, the change has been less dramatic and the vestiges of the old policy and the old mental models about agriculture are still apparent. In general, the larger the country, the more resistant are political structures to accept the new paradigm.

In Europe, the old paradigm of a sector needing support is also giving way but is being replaced by one that emphasizes the role of agriculture in providing public goods such as environmental benefits. Other European countries, such as Norway and Switzerland, support this view. This position stands in contrast to that taken by Japan and Korea, where the view of agriculture still reflects more the old paradigm of an agriculture that needs support in order to provide adequate agricultural commodities for domestic consumption—an argument usually cast in terms of food security.

The European Union's multifunctional paradigm and the Japanese concern with food security could soon come to clash with the market-oriented view of agriculture promoted by the Cairns Group and the United States, leading to continued trade conflicts. The WTO talks will provide a ready forum for the articulation of the paradigm clash. The Cairns Group and the United States will be pushing for further reform of the trade rules, in effect trying to eliminate the "exceptions" for agriculture built in to the Agriculture Agreement. The "friends of multifunctionality" will be making sure that enough scope exists in those rules to allow the type of policy that rewards environmental stewardship and promotes agriculture that does not cross the line in terms of public perceptions of animal cruelty. The Japanese will insist that their ability to maintain the agricultural capacity that they consider necessary is preserved, and hope that attention focuses on the exporters for adjustment. Meanwhile, those already in the global supply chain will be pressing for tighter regulations on competition and intellectual property protection, as well as more harmonized standards. The question is whether these divergent demands can be made compatible. The next few years will tell.

NOTES

1. See in particular Orden's Chapter 3 herein and Coleman, Skogstad, and Atkinson (1997). The seminal article on the concept of a paradigm shift in economic policy is Hall (1993).

2. Assistance to agriculture had grown to about 10 percent of the total public expenditure. For a small economy dependent upon livestock export earnings, these subsidies imposed a heavy burden. The political conditions that allowed these reforms are discussed in Williamson (1994).

3. See Valdés (1994) for a discussion of the similarities and differences between the Chile and New Zealand experiences.

4. Article XI allowed quantitative restrictions in cases in which there was domestic supply control, and Article XVI implicitly condoned export subsidies for agricultural products (see Josling, Tangermann, and Warley 1996).

5. See IATRC (1997) for a more complete discussion of the Uruguay Round results and Josling (1998) for the need for further reform of the agricultural trade rules.

6. This interaction has been studied by political scientists, following Putnam (1988), as two-level games (Paarlberg 1997; Moyer 1993) and three-level games (Patterson, 1997) when the national politics of the European Union is involved. Coleman and Tangermann (1997) incorporate the role of secretariats in intermediating these games.

REFERENCES

Coleman, W., G. Skogstad, and M. Atkinson. "Paradigm Shifts and Policy Networks: Cumulative Change in Agriculture." *Journal of Public Policy* 16 (1997): 3.

Coleman, W. and S. Tangermann. "Linked Games, International Mediators, and Agricultural Trade." Paper presented at the IATRC Annual Meeting, San Diego, December 1997.

Hall, P. "Policy Paradigms, Social Learning, and the State." *Comparative Politics*, 25 (April 1993): 3.

International Agricultural Trade Research Consortium. "Implementation of the Uruguay Round Agreement on Agriculture and Issues for the Next Round of Agricultural Negotiations." IATRC Commissioned Paper No. 12, November 1997.

Josling, T., S. Tangermann, and T. K. Warley. *Agriculture in the GATT.* New York: Macmillan, 1996.

Josling, T. *Agricultural Trade Policy: Completing Reform.* Washington, DC: Institute for International Economics, 1998.

Josling, T. "Globalization of the Agri-Food Industry and its Impact on Agricultural Trade Policy." Paper presented to the conference on Agricultural Globalization, Trade, and the Environment, University of California at Berkeley, March 1999.

Krueger, A. O. "The Political Economy of Agricultural Pricing Policy: A Synthesis of the Political Economy in Developing Countries." In A. Krueger, M. Schiff, and A. Valdes, *The Political Economy of Agricultural Pricing Policy.* Baltimore: The Johns Hopkins University Press, 1992.

Moyer, H. W. "EC Decisionmaking, the MacSharry Reforms of the CAP, Maastricht, and the GATT Uruguay Round." Paper delivered to the Third Biennial Conference of the ECSA, Washington, DC, May 1993.

Moyer, H. W. and T. Josling. *Agricultural Policy Reform: Politics and Process in the EC and the US.* Harvester Wheatsheaf, Hemel Hempstead, 1990.

OECD. *Agricultural Policies in OECD Countries,* Paris: OECD, Various Years.

Orden, D. "The Soul of Farm Policy in the 21st Century." Paper presented to the conference on Challenging the Agricultural Economics Paradigm. Columbus, Ohio, September 2000.

Orden, D., R. Paarlberg, and T. Roe. *Policy Reform in American Agriculture: Analysis and Prognosis*. Chicago: University of Chicago Press, 1999.

Paarlberg, R. "Agricultural Policy Reform and the Uruguay Round: Synergistic Linkage in a Two-Level Game." *International Organization* 51 (Summer 1997): 3.

Patterson, L. A. "Agricultural Policy Reform in the European Community: a Three-Level Game Analysis." *International Organization* 51 (Winter 1997): 1.

Putnam, R. "Diplomacy and Domestic Politics: the Logic of Two-Level Games." *International Organization* 42 (Summer 1988): 3.

Sandrey, R. and R. Reynolds (eds.). *Farming without Subsidies: New Zealand's Recent Experiences*. MAF, New Zealand, 1990.

Tweeten, L. "Trade, Uncertainty, and New Farm Programs." Forthcoming in C. Moss et al., eds. *Agricultural Trade, Globalization, and the Environment*. Norwell, MA: Kluwer Academic Publishers, 2001.

Tweeten, L. and C. Zulauf. "Public Policy for Agriculture after Commodity Programs." *Review of Agricultural Economics* 19 (Fall/Winter 1997): 2.

Valdés, A. "Agricultural Reforms in Chile and New Zealand: A Review." *Journal of Agricultural Economics* 45 (May 1994).

Williamson, J. (ed.). *The Political Economy of Policy Reform*. Washington, DC: Institute for International Economics, January 1994.

14

The Changing Economics of Agriculture and the Environment

David E. Ervin and Frank Casey

Private voluntary environmental initiatives appear to be increasing in agriculture, driven by consumer and public demands. They offer the potential to lower the costs of achieving environmental objectives, mostly by granting more flexibility to producers than many government programs. Yet, the private initiatives are not a panacea. Public roles remain essential, but are different from past efforts. Agricultural and resource economists should devote more study to the motivations, advantages, and limitations of private environmental initiatives.

INTRODUCTION

Agriculture, due to its extensive land coverage and often intensive production technologies, can deliver large environmental benefits and costs. Examples are wildlife habitat and nonpoint water pollution. Public education, technical assistance, and compensation programs have been the dominant approaches used to address environmental problems in agriculture. Some regulations, such as for pesticides, also apply. Many of the public programs have been found wanting in efficiency or cost-effectiveness by agricultural and resource economists, who often urge greater reliance on private decision making and market mechanisms.

Private environmental initiative in agriculture indeed appears on the rise. Antle (1999) theorizes that quality will play an increasing role in agriculture. Under the theory, farmers and agribusiness supply products with differentiable quality attributes, such as high-lysine corn. Antle argues, and we concur, that the relevant set of qualities for agriculture includes certain environmental attributes of the production process and products. The main economic driver of the increased demand for environmental quality is rising personal income, reflecting a positive income elasticity of demand. Thus the demand for water quality and other environmental services increases as incomes rise. To date, the effects of the robust demand for environmental quality have been expressed mostly in the political arena via higher standards and new programs. But the influence is increasingly surfacing in the market with firms supplying "green" products. A broadening segment of consumers and investors are rewarding firms that supply competitively priced products that possess environmental qualities. Bottom line: the values of a farm's environmental effects are rising relative to those for food and fiber.

Agricultural economists have been among the leaders in the intellectual development of resource and environmental economics. Much of that effort has been directed at evaluating the costs and benefits of government programs, such as the Clean Water Act and the Conservation Reserve Program (CRP). However, few studies of the forces and effects of privately led agro-environmental management have been conducted (Batie and Ervin 1998). For example, the role of state and local agro-environmental programs with real or threatened compulsory standards in stimulating privately led actions has not been researched (Carpentier and Ervin 1999). There are some exceptions, such as the work of Wu and Babcock (1996, 1999) and Randall (1999a).

In the spirit of this symposium, our chapter is intended to challenge the agricultural economics profession to understand "voluntary" environmental initiatives in agriculture, their advantages and limitations. We first review research on agro-environmental programs to assess opportunities for more private initiative to replace costly or ineffective public approaches. Then we develop a stylized decision model for voluntary agro-environmental management to identify key variables that may be overlooked in standard analyses. We conclude by identifying research topics to understand the motivations for privately led environmental actions, the limitations to such voluntary efforts, and policy implications.

CURRENT VOLUNTARY AGRO-ENVIRONMENTAL PROGRAMS
A plethora of public environmental programs is directed towards working agricultural landscapes, most of which are administered by the U.S. Department of Agriculture (USDA). The evolution of these programs since the mid-1980s reflects a growing desire on the part of the public to improve the environmental performance of agriculture. Some conservation programs have also provided income to producers, mostly in the form of land rental payments. Total federal expenditures on or related to USDA voluntary conservation programs for soil erosion control, improved water quality, wildlife habitat, and other purposes have held fairly steady since 1992 in a range between $3.2 billion and $3.7 billion per year in nominal terms (Zinn 1999).

The largest environmental program administered by the USDA is the Conservation Reserve Program. The CRP was originally designed to retire highly erodible cropland from production through voluntary ten-year contracts that required the use of specified soil conservation practices in return for practice cost sharing and annual rental payments. It has since shifted to include other environmental services as well, such as wildlife habitat. Companion programs to the CRP are the Buffer Initiative and the Conservation Reserve Enhancement Program (CREP). The Buffer Initiative offers producers a cost-share agreement for constructing field riparian buffers to protect water resources. CREP, a recent program based on federal-state agreements to improve environmental resources, has been aimed primarily at water quality and wildlife habitat. Due to the CRP and companion program eligibility and implementation rules, most participation in these programs has been by producers in the Midwest. A major limitation of the CRP and companion programs for enduring environmental problems is their lack of permanent protection.

The Wetlands Reserve Program (WRP) employs three instruments for the recovery and restoration of the nation's wetlands. These instruments include permanent easements, thirty-year easements, and cost-share agreements for rehabilitation. Interestingly, the number of acres allocated by Congress to the WRP under the 1996 farm bill has been nearly filled, with the majority of WRP acres going into permanent easements (Deavers, personal communication, 2000). Although the WRP has enjoyed high participation rates (enrollment rates are now at capacity), the transaction costs for negotiating and monitoring easements have been high.

Two smaller but more targeted programs are the Environmental Quality Incentives Program (EQIP) and the Wildlife Habitat Incentives Program (WHIP). EQIP cost-share funds can be targeted to specific "resource concerns" and "geographic priority areas", but most emphasis has been on water quality. WHIP cost-share projects must be aligned with state-level habitat or species priorities.

Although the various USDA conservation programs have been popular with producers, especially during periods of low commodity prices, their long-term effectiveness for achieving environmental goals is doubtful. Their funding levels have stagnated over the past six years, and have declined from their peak in real terms. Operational constraints raise doubt as to their efficacy, including imperfect targeting of expenditures to maximize the value of environmental benefits, little flexibility in the choice of incentives or management practices, high transaction costs of accessing government conservation programs, the absence of robust monitoring and evaluation to improve environmental performance at the farm or program level, decreased agency capacity to deliver programs, and maintaining research and technology development (R&D) resources for new agro-environmental technologies. Numerous recommendations for increasing the technical effectiveness and economic efficiency of agricultural conservation programs have been made (Casey et al. 1999; Ervin et al. 1998; Heimlich and Claassen 1998; Lewandrowski and Ingram 1999; Lynch et al. 1994; and Ribaudo 1989). An overarching finding is to grant more flexibility to producers in their choice of economic incentives and technical management practices.

Adapting Conservation Programs to Capitalize on Private Initiative
Two essential conditions in the design of more efficient environmental management programs are (1) a clear definition of the biophysical goals to be attained, and (2) a robust program to monitor and evaluate progress towards attaining those goals. Both are required to implement cost-effective adaptive management approaches to environmental protection and to stimulate appropriate private responses. Resource conservation goals can be defined at the farm/ranch or regional (i.e., watershed) level, whatever is appropriate for a given environment problem, and should be consistent with national environmental standards. For water quality, the resource goal may be achieving and maintaining a specified total maximum daily load (TMDL) level.

To increase program efficiency, explicit indicators to monitor and evaluate progress towards achieving environmental goals must be developed. Physical, biological, and economic indicators must be specific, clear, and quantified at the appropriate scale. Furthermore, baseline data on resource conditions and effective monitoring programs to track indicators are required to gauge the efficiency of management practices and economic incentives. Although effective monitoring and evaluation programs can be expensive, the social costs of debating and litigating environmental standards and enforcing compliance with those standards are probably far greater than developing a participatory process of adaptive management based on sound technical and economic data (Randall 1999a).

Conservation Program Structure

Choice of Incentives
To achieve cost-effective environmental protection, producers require flexibility in the choice of economic incentives that best fit their physical and economic circumstances. This task requires producers and program administrators to jointly develop a menu of incentives from which farmers may choose, including full cost reimbursement (green payments), cost-sharing, tax allowances, compliance rewards, deposit/refund schemes, performance bonds, land lease payments, conservation easements for water and land, or tradable development rights. However, the program design must be simple and keep administrative burdens low. Producers would be free to choose the incentive measure that best maximized their household income or overall welfare, consistent with attaining a specified environmental goal.

Institutional Innovations
Several potential innovations for administering resource conservation programs could decrease producer transactions costs and increase cost effectiveness. One concept to explore is a private contract that bundles together diverse programs to restore or conserve environmental amenities on private lands. Such a contract may

be referred to as a Resource Stewardship Agreement (RSA).[1] The RSA would be based on a resource stewardship plan that specifies the environmental resources to be protected, the contract period, stewardship goals, financial instruments and compensation levels, monitoring requirements, penalty clauses, and the responsibilities of all contract parties. The RSA is really a contract to reimburse the providers of certain environmental services for the full costs of supplying those services. The private sector (either for-profit or not-for-profit organizations) could assist with RSA design and administration. For instance, a third party could help administer an RSA by implementing environmental monitoring requirements. Also, crop consultants could represent several producers in designing contracts for a specific geographical area where threshold effects are particularly important.

A second institutional adjustment will be to continue the process of decentralization of environmental planning and program delivery to the state level. Recent examples of decentralization include the design and implementation of the EQIP, WHIP, and CREP programs. However, because federal tax dollars are invested, some federal oversight is still required. For program delivery, cooperative agreements between relevant federal, state, and local government agencies and private entities (including certified crop consultants) can be established to deliver programs. Several states currently use personnel funded by nonprofit groups to implement conservation programs. For example, the Oregon Wetlands Joint Venture Initiative funds a position within the National Resource Conservation Service (NRCS) to process WRP applications.

Few conservation programs currently provide mechanisms for adapting technical recommendations or incentive measures during project implementation. *Ex ante* knowledge of the effects of environmental management practices (especially for water quality and wildlife habitat) is often highly uncertain, a situation that can lead to cost-ineffective resource use if adjustment mechanisms are not in place. Program rules should permit adaptive management responses to changing environmental and economic conditions. A related requirement is to allow producers the flexibility to design, test, and implement (with the assistance of qualified government technical agencies, third party nonprofit groups, and/or certified private consultants) new agro-environmental technologies and management practices that are appropriate to local physical and economic conditions. This provision will foster innovation in management practices, and eventually lead to more cost-effective environmental protection.

New institutions that help producers capture the full social benefits of their environmental management investments require investigation. Alternative market institutions could include the provision of green certification and ecolabeling services, and administrative bodies designed to facilitate habitat or nonpoint source trading. In the case of certification and labeling, there is already substantial experience and continued interest in the private sector in providing these services.

One of the most crucial attributes of the new institutions is to provide producers and consumers with timely, credible information about market opportunities for environmental goods. Potential problems with tradable pollution permits in agriculture, however, are the risk of thin markets and equity issues associated with the potential dominance of large firms controlling a majority of a fixed number of permits (Boggess, Johns, and Meline 1998).

Finally, institutional change to mitigate producer uncertainty concerning environmental regulations is needed to channel resources and investments to their most cost-effective uses. With the assistance of the private sector, public agencies can use various options for providing regulatory certainty in return for verified adoption and effective implementation of agreed-upon resource conservation technologies.[2]

Target Environmental Conservation Funds

Research shows that targeting resource conservation efforts and funds to areas posing the highest environmental risks is a cost-effective use of public conservation funds (Ribaudo 1989; Heimlich et al. 1998; Feather,Hellerstein, and Hansen 1999). The evaluation of individual CRP and EQIP bids on the basis of an Environmental Benefits Index (EBI) has been a step in this direction. However, the EBI is designed to serve as a selection criterion (greatest environmental benefit for a given cost) between competing conservation projects within a state, and not as a means of targeting expenditures to specific resource problems. Targeting expenditures to unique resource concerns, such as water quality and wildlife habitat, has been practiced under the state-federal CREP and EQIP programs, respectively.

Criteria for targeting environmental expenditures have been discussed less thoroughly. Wu et al. (1999) have suggested using threshold effects whereby lands would be selected that maximize the marginal physical product of environmental benefit per unit of investment. One particular complication with this approach is the detailed level of chemical and biological information that is required to measure responses. A second criterion would be to aim project funds to investments providing a high degree of jointness in environmental benefits. For example, investments in erosion control and wetlands restoration generate multiple economic benefits associated with water quality and improved wildlife habitat. Last, for those areas that have serious environmental problems, the number of contiguous landowners submitting joint project applications would serve as another criterion. Having a "critical mass" of participation is essential for such problems as nonpoint source pollution and restoring wildlife habitat.

Technology Research and Development

Ervin and Schmitz (1996) and Swinton and Casey (1999) have provided conceptual arguments for allocating public funds to R and D on environmental technologies and management practices that provide social net benefits. Under those arguments, new R and D should be a continuous process and receive increased public funding to offset the potential for underinvestment due to missing markets. Efforts can be made to partner more effectively with groups of agribusiness firms that are simultaneously investing in environmental technologies. Complementary investments should be made to more effectively link agro-environmental research with technology extension and education services. Again, the private sector can play an important role in communicating and demonstrating technical improvements. The growth in reduced tillage is a relevant example of the heavy influence of private investment in development and extension, not of public R and D.

Conservation Program Administration and Implementation

Agency Capacity Building

Investment to increase the technical and administrative capacity of federal and state agencies to plan and effectively monitor a range of environmental projects is essential. While recognizing that these agencies will take on more of a facilitator role in the future, there remains a need to improve their technical advisory capacity to assist producers, to evaluate returns to public expenditures, and to monitor environmental performance, including ecosystem wide effects.

Streamlining

One the most important sources of inefficiency in the implementation of agro-environmental programs is the high transactions and administrative costs of farmers obtaining information about and accessing the numerous (and often redundant) programs, all of which have different eligibility criteria and implementation rules. Information on this multilayered system of federal, state, and local resource conservation planning and funding programs (USDA, EPA, USFWS, state agencies) could be assembled in ways to allow farmers to easily scan and select the programs best suited to their needs. The RSA concept outlined above is one mechanism to simplify the farmer's decision process and responses.

Enlarging the Private Role

While government conservation programs have provided substantial public benefits, their efficiency or cost-effectiveness often may be improved by relying more on private knowledge and initiative. Several avenues can expand the private business role in the administration of agro-environmental programs, many of which are explored in the next section. However, the definition of the private sector should include not only commercial business, but also formal groups of private citizens and nonprofit, nongovernmental organizations. These entities can assist the business sector and government agencies to implement environmental programs. For example, organized watershed councils, or other regional groups of landowners, can generate resource conservation plans and assist in targeting conservation funds to the most pressing problem spots. Federal agricultural agencies with environmental programs actively seek partnerships and enter into memoranda of understanding with private business and nonprofit organizations to implement resource conservation actions. The most widespread example of partnering in terms of geographic coverage is the National Buffer Council. The Council is made up of some of the largest agribusiness firms and environmental nongovernment organizations (NGOs) in the United States to promote and assist the construction of stream buffers on private agricultural lands. Some environmental NGOs have taken leadership roles in compensating livestock producers from losses resulting from the reintroduction of large predators to their former habitat and from crop losses experienced from the adoption of no-till systems designed to reduce sediment loadings.

ECONOMIC RESEARCH ON VOLUNTARY ENVIRONMENTAL INITIATIVES

Economic research on private "voluntary" environmental initiatives, also termed corporate environmental management (CEM), is young but growing rapidly (Lyon and Maxwell, forthcoming; Reinhardt 2000; and Segerson and Li 1999). A central theme from the nascent literature is that businesses that adopt voluntary environmental initiatives, including agricultural firms, do so to capture gains from flexibility, rather than responding to government programs. However, the possible motives for adopting CEM are broader, ranging from entering new green markets, to gaining "first-mover" advantages over competitors, to persuading regulators to delay inflexible controls, to avoiding trade interruptions, to defusing external criticism. The privately led environmental initiatives need not come solely from the business sector. Indeed, the expanding roles of nongovernmental organizations (NGOs) are one of the least appreciated and studied forces of these changes.

Today's farmer faces a bewildering array of federal, state, and local environmental programs, and a market that increasingly rewards environmental quality. The transaction and administrative costs of responding to the diverse programs, some working at odds, are substantial. One approach by farmers to reduce those costs, retain operating flexibility, and capture market returns from consumers willing to pay for environmental quality attributes is to undertake voluntary actions that meet or perhaps even exceed public environmental standards (Reinhardt 2000). Indeed, as the transaction and administrative costs of diverse public programs increase and the market for green products expands, farmers have increasing incentives to pursue private environmental initiatives as a profit-maximizing strategy, ceteris paribus.

Three types of voluntary initiatives may be pursued:

- *Unilateral initiatives* by individual firms to control pollution or by industry groups to establish industry standards or to self-regulate. An agricultural example is a group of farmers under the Land Stewardship Project voluntarily implementing nonpoint source pollution controls.
- *Bilateral or negotiated agreements* between the government and private firms that usually contain a voluntary environmental target and a timetable for reaching the target. An example from agriculture is the cooperation between the U.S. Pork Producers and the EPA to design a mutually acceptable pollution control strategy for large confined hog operations.
- *Voluntary incentive programs* designed by the government agency to induce participation by individual firms (farms). The traditional approach in agriculture includes educational, technical assistance, financial, or other incentives to elicit participation.

In our judgment, unilateral and bilateral efforts will increasingly spill into agriculture due to the rising public demands for environmental quality, and efforts by farmers and agribusiness to avoid regulations, especially technology-design standards. Moreover, Congress is unlikely to appropriate enough funds for compensatory programs to satisfy the level of public demands for environmental improvement by farming. However, the government remains a key player in voluntary agreements, from setting performance standards, to administering environmental markets, to collaboration on R and D, to other roles. For environmental goods that exhibit high costs of exclusion (such as migratory wildlife), or that are nonrival (such as landscape amenities), the government may provide compensation to assure adequate supplies. However, the old approach of government-led programs, often with input controls, is giving way to more public-private cooperative approaches. These developments have prompted the U.S. Environmental Protection Agency to start a new research program to understand the drivers and efficacy of voluntary environmental initiatives.

Compliance-push and demand-pull forces (Batie and Ervin 1998) influence the private sector's search for voluntary initiatives inside and outside agriculture. The literature on CEM identifies six specific motivations. For any firm, more than one may apply. For example, the owners of Stahlbush Island Farms argue that successful CEM must be a systems approach that links all input, production, and marketing operations (Chambers and Eisgruber 1998).

1. *Improving firm productivity.* Some managers argue that the creation of production and marketing system databases and other tasks necessary to implement a CEM program lead to cost reductions and/or opportunities

for new products (Reinhardt 2000).[3] It is possible that, by evaluating their operations in the light of public or market environmental concerns, firms may find cost savings due to new information and R and D. Boggess and colleagues (1998) estimated productivity gains for a subset of dairy farms that adjusted to higher nutrient pollution control standards for Lake Okechobee. It is natural to ask why the firms did not discover the savings prior to the new standards. In the Lake Okechobee case, cost-sharing was available to offset some equipment costs. In other cases, ignorance, missing markets, or poorly functioning public institutions may hinder learning about such discoveries.

2. *Serve demands by green consumers and investors.* Although still a small portion of the market, retail products and investment funds that emphasize environmental objectives are increasing rapidly by popular press accounts. Successful product differentiation on environmental grounds requires that the firm discover or create (1) a willingness among customers to pay for the quality attribute(s), (2) an ability to communicate credible information about its product,[4] and (3) protection from imitation by competitors (Reinhardt 2000). Environmental actions by firms to gain entry into foreign markets, such as product recycling, fall in this category. Evidence on this motivation in agriculture is sparse. Retailers have begun "natural foods" programs to meet growing consumer demand, such as foods with ingredients produced under systems of integrated pest management (IPM) and organic foods, although part of this growth may reflect perceived reductions in health risks rather than environmental effects (Dimitri and Richman 2000).

3. *Preempt or mitigate future environmental regulations.* Lyon and Maxwell (forthcoming) cite ARCO's introduction of reformulated, cleaner gasoline as an example of this strategy. Such incentives may loom large for agriculture as the public and NGO pressures to remedy water pollution problems, such as the hypoxia problems in the Gulf of Mexico, gather force. However, the transaction costs of building coalitions of diverse farming interests may restrict effective voluntary initiatives in such cases. This motivation also includes actions to assure foreign market access that might otherwise be closed by nontariff trade barriers.

4. *Strategic management of competitors.* Incurring additional costs to improve environmental performance may increase some firms' profits if the actions cause competitors' costs to rise even further than their own (Salop and Scheffman 1983). Firms known as "first movers" may create a strategic cost advantage by forcing their competitors to follow their examples. Part of the reason for Ford's recent decision to push for higher fuel efficiency performance of its SUVs may have been to raise its

competitors' costs. GM's quick response suggests that it may have worked. Likewise, if environmental standards on pesticides and transgenic crops tighten, the incentive for a major agribusiness supplier to capitalize by moving first to "over-comply" increases.

5. *Potential to redefine markets.* To succeed in delivering more value to customers and improving environmental performance, while simultaneously reducing costs, firms need to redefine the markets in which they compete (Reinhardt 2000). Chambers and Eisgruber (1998) describe how Stahlbush Island Farm managers lessened their "environmental footprint" by using diverse crop rotations, growing nitrogen sources, reducing or eliminating pesticides, protecting groundwater, engaging in soil and product residue testing, reusing water, recycling, and composting. The authors claim a key requirement that allows the use of such environmentally progressive techniques is their ability to identify and contract with customers in advance of production. Another requirement is the existence of a market segment willing to pay for the perceived quality attributes. Examples also exist in the wildlife arena whereby some ranchers have differentiated their wool or beef products as made with "predator friendly" production systems (Robles 2000).

6. *Manage risk and uncertainty more effectively.* Reinhardt reasons that this strategy can be effective, particularly if it serves as a source of competitive advantage. A firm is naturally concerned with the risk of financial harm from environmental incidents. The types of financial harms include the cost of cleanup from an environmental accident, legal liability for environmental damage, forgone profits due to the interruption of business practices pursuant to an environmental accident, and losses caused by a damaged reputation in the eyes of government officials, consumers, and the public. If some farms are threatened with large noncompliance penalties and legal suits, preemptive voluntary action may be the least-cost approach.

A summary of the early evidence suggests that the probability undertaking voluntary environmental initiatives increases with (1) larger firms, (2) firms with higher R and D intensities, (3) firms with poorer environmental records, and (4) firms facing increased future government regulation and greater pressure from community, environment, and industry groups (Lyon and Maxwell, forthcoming). In a recent study not reviewed by Lyon and Maxwell, Khanna and Quimio (1999) found that (1) environmental liabilities and penalties for noncompliance with regulatory standards play significant roles in firms' CEM decisions, (2) firms that produce final goods (in closer contact with consumers) and firms in industries with greater competition have more incentive to adopt a comprehensive environmental management system, (3) the incentives are also larger for firms that are more innovative in their production practices and have older assets, and (4) public pressure targeted towards firms with high toxic releases per unit output and higher anticipated costs of compliance created incentives for voluntary environmental management. The major findings related to investor pressure are (1) investors react negatively to higher than expected levels of toxic emissions, (2) firms are rewarded for superior environmental performance, and (3) firms respond to environmentally induced investor pressure by improving their environmental performance. Specific evidence on agriculture is virtually nonexistent.

A STYLIZED FARMER DECISION MODEL FOR VOLUNTARY ENVIRONMENTAL INITIATIVES

A general model of a farmer's decisions about voluntary environmental initiatives helps frame hypotheses about the potential influences and identify needed research. This model differs from typical formulations used to analyze farmer responses to agro-environmental programs in important ways. For example, product quality differentiation (e.g., through vertical integration and contracting) is permitted, and the administrative and transaction costs in environmental management decisions are treated explicitly.[5] Also, the model is cast as a multiperiod decision that includes investments in R and D, equipment, and education.

Consider a typical profit model used to analyze a farmer's environmental management decisions:

$$\text{Max } [(P \times Q) - (PC + CC) + GP] \qquad \text{subject to } EQ \leq EQ^*$$

where, P is a vector of given product prices, Q is a vector of undifferentiated products, PC is a vector of production and marketing costs for the products, CC equals environmental compliance costs for the farm, including private treatment or remediation expenditures and possible public penalties, and GP is government payments for environmental management (e.g., cost-share for practice adoption). EQ is the level of environmental quality achieved by the farm and EQ^* is the public standard. Given a specific technology set, the compliance costs (CC) are a direct function of the level of Q produced. Any attempt to maximize profits $[(P * Q) - (PC + CC)]$ under this formulation leads inevitably to a tradeoff between environmental quality and short-run profit. Public compensation (GP) must be used to offset potential losses. This formulation begs the question of a farmer's economic rationality if the farmer undertakes uncompensated voluntary initiatives.

Now consider a profit function for analyzing a farmer's decision to jointly supply environmental services and food or fiber. The inclusion of the quality attribute (service) requires alterations to the farmer's expected revenues because the quality attribute enters consumer (market) demand functions and to the farmer's cost and market supply functions (Antle 1999). Assume the objective is to maximize the farmer's utility over time horizon T, and that this objective can be approximated (under certain limiting assumptions)[6] as one of maximizing profits.

Max N (t) [(PN * QN) + (PE * ES) − (PCN + ICN + TCN + ACN) + GP]
subject to EQN ≥ E*

where, t is time period t; PN is a vector of prices for a vector QN of differentiated food and fiber products with certain environmental quality attributes; PE is the price received for other environmental services (ES) from the farm, such as wildlife hunting or viewing; PCN is production, processing, and marketing costs, including any efficiency offsets; ICN equals investment costs for system reengineering and human capital (education and training)[7]; TCN equals transaction costs of green production and marketing systems; and ACN is other administrative costs of the green production and marketing system. Note that the environmental (compliance) costs in this formulation become part of the firm's production, marketing, and processing costs, and noncompliance penalties are eliminated because EQN always meets or exceeds the public standard.

The first component (PN * QN) is the market revenue from selling food and fiber products with environmental attributes that consumers are willing to purchase.[8] Farmers have two options to increase their revenue over that possible from selling environmentally undifferentiated (bulk) commodities: (1) create products that have higher environmental benefits or lower environmental costs than comparable products, and (2) create a production process that causes higher environmental benefits or lower environmental costs than competitors.

The second component is the added revenue from selling environmental services, such as wildlife access and carbon sequestration credits. Some farms already do this type of business. For example, some ranchers charge for hunting and fishing access. A number of environmental services may be provided through public programs because of social custom or excessive market operation costs. Conceptually, this component also includes any nonuse values that the operator and family enjoy from farm environmental services, such as landscape amenities. The difficulty in finding monetary values for some of the environmental services hinders the estimation of a profit-maximizing level of provision. Contingent valuation and other nonmarket valuation methods can be employed in situations where credible values can be estimated.

The recent CRP and EQIP compensation rules generally pay farmers for their opportunity costs of providing what are viewed as the highest ranked environmental services. Note such rules are WTO-legal, but do not necessarily reward the farmers for the full social value of the services (Ervin 1999). The acknowledgment that a farm may produce more than food and fiber raises the contentious concept of "multifunctionality." The existence of markets for wildlife hunting on farms and the selling of carbon sequestration credits affirms that farms deliver more benefits than food and fiber. The contention in trade policy is over how the services should be provided when markets are infeasible or more costly to operate than public programs. The conditions under which government interventions can supply agro-environmental services according to WTO rules have been researched (Ervin 1999; Hodge 2000; Runge 1999).

The third component is the production, marketing, and processing (PCN); investment (ICN); transaction (TCN); and administrative (ACN) costs. Generally, PCN and ICN are positive functions of EQ achieved in the short run because added resources are needed to produce the environmental quality attribute. However, there is no firm rule that the sum of the two costs over T will be greater than, equal to, or less than the sum of production and compliance costs under an involuntary compliance strategy. That determination depends on the equilibrium level of EQ chosen under the specification of benefits and costs of voluntary agreements and is therefore indeterminant a priori. It also depends on the existence of "innovation offsets" in which firms lower their long-run unit costs after adopting voluntary initiatives (Porter and van Der Linde 1995; Tebo 1998). Some authors argue that unless noncompetitive or missing markets hinder the search for such production and marketing innovations, such net gains should not be expected (Palmer, Oates, and Portney 1995). The transaction costs under a voluntary environmental initiative, TCN, depend on the search and negotiation costs required under the voluntary strategy. It is conceivable that the transaction costs of successfully implementing a green marketing system could be larger than the transaction costs of the involuntary compliance strategy, because the markets are thin or new with high search costs. The other administrative (paperwork) costs should be lower under the voluntary strategy because compliance is assured.

Reinhardt (2000) theorizes that industry structure and impending changes in that structure, the relative importance of human capital, and the time horizon for evaluation affect the likelihood of firm cost savings under voluntary environmental initiatives. He offers two general observations regarding potential cost savings: (1) the baseline is critical to evaluation—as the price of environmental resources (and noncompliance) rises, it makes good sense to invest in ways to reduce their use; and (2) few short-term private gains are possible, but longer term, the opportunities may be more widespread. Recall also that the green firm may hold a relative cost advantage if it undertakes strategic "first-mover" environmental initiatives to disadvantage its competitors.

The final component is the government payments for reengineering the production, processing, and marketing systems or other expenses. The farming and ranching sector has a long history of receiving cost-share payments to voluntarily install certain practices considered to improve conservation and environmental performance. Studies of past cost-share payment programs often have found that they did not gain the most possible environmental value per dollar of expenditure. The EQIP program rules have improved this allocation by targeting high-priority geographic areas and statewide resources of special concern. State cost-share programs have generally followed the federal model. For example, the Florida program to reduce nutrient runoff into Lake Okechobee contained cost-share monies for new dairy production systems that reduced or reused waste nutrients. This long-standing public program approach likely will not end soon, but it is difficult for operators to predict future payments for making long-term investment decisions about voluntary environmental initiatives. There are two exceptions to the uncertainty embodied in current voluntary federal programs. First, permanent and long-term (thirty-year) easements are options under the Wetlands Reserve Program (WRP). Second, the proposed Resource Stewardship Agreement discussed above could be set up for a period that corresponds to a producer's long-term planning horizon.

CHALLENGES FOR AGRICULTURAL AND RESOURCE ECONOMISTS
Environmental quality and other quality attributes promise to play more influential roles in the future of U.S. agriculture. The agricultural economics profession has mostly analyzed farmers' environmental decisions as passively responding to federal conservation and environmental programs. While the publicly led programs will continue to play an important role, the leading edge of agro-environmental management appears to be shifting to the private sector. Both private for-profit and nonprofit organizations are taking part in this trend. The changing economics of agriculture and the environment requires more research and education in eight topic areas.

1. *Assess the consumer demands for different environmental quality attributes of food and fiber products and production systems.* This analytical task will give insight into the areas where market forces may lead and public R and D and voluntary-incentive programs can play supportive roles. It will also help to understand which product attributes are capable of differentiation.

2. *Estimate the long-run economic returns to production systems that conserve environmental services with rising scarcity values.* We tend to focus on current technologies and present prices in our agro-environmental analyses, but history tells us that neither is a good guide to the future. More theory and empirical work on the effects of rising environmental factor prices is needed.

3. *Analyze and estimate the roles of transaction and other administrative costs in alternative agro-environmental program design and implementation.* This area of economic inquiry has been neglected by our

profession despite Coase's (1960) compelling arguments about the importance of transaction costs in the economics of the real world. Incentives to lower the costs as environmental programs proliferate may be key in stimulating voluntary environmental actions. Potential roles of public, for-profit, and nonprofit organizations should be included.

4. *Analyze the applicability and limitations of privately led initiatives to different agro-environmental situations.* Three conceptual cases exist: (1) voluntary self-motivated action under market forces is most cost-effective in meeting social environmental objectives; (2) voluntary action is most cost-effective, but public standards and some form(s) of assistance are necessary to meet social objectives; and, (3) compulsory programs are necessary to assure the attainment of social environmental objectives. Voluntary initiatives may be limited for many agricultural problems due to nonpoint pollution processes and the diversity and large number of production-natural resource relationships. The lack of traceability for most nonpoint pollution poses a key constraint for public agencies and private firms in implementing voluntary initiatives.

5. *For cases (1) and (2) above, analyze the supportive roles that can be best played by the public sector in providing nonrival and nonexclusive services to further private agro-environmental management, such as ecosystemwide monitoring and R and D for production systems that reduce transboundary environmental wastes or increase environmental services.*

6. *Analyze the potential roles of NGOs in furthering privately led agro-environmental initiatives, such as monitoring, ecolabeling, and certification schemes, and providing technical assistance.* This analysis should include a comparison of the costs and returns of for-profit and nonprofit institutions in providing such services to producers.

7. *Investigate the influence of evolving agricultural industry structure on the feasibility of voluntary environmental strategies.* Under what conditions will increasing horizontal and vertical linkages assist or hinder CEM strategies in agriculture?

8. *Assess the private-sector roles and economic contributions of agricultural producers to supply environmental services.* The growing recognition of the multifunctional nature of agriculture will stimulate policy proposals, and economists must be prepared to analyze their effects.

The shift toward more private environmental initiative, either in markets or government programs, means that farmers and agribusiness have more opportunity to be rewarded for supplying positive environmental attributes that consumers and voters demand, and to be penalized for causing negative effects. A growing list of firms, mostly outside agriculture, have decided that voluntary environmental initiatives are a higher long-run profit strategy than responding to government-dictated programs. If found economical by farmers, this strategy could produce more environmentally and economically sustainable production and marketing systems than many current agro-environmental programs, such as temporary land retirement.

NOTES

1. The RSA concept was first promoted by the Florida Stewardship Foundation as a proactive method to reimburse livestock producers in South Florida for restoring and maintaining habitat for the endangered Florida panther.

2. These may include such options as safe harbor agreements and fast-track permitting.

3. An array of methodologies is available to assist managers in conducting a systematic search for potential benefits from reducing environmental wastes and reusing environmental by-products, such as life-cycle assessments.

4. Where private initiative may fail in certifying environmental attributes, the public sector may establish a certification and labeling system.

5. Transaction costs according to Coase (1960) are the costs of search and negotiation in completing market or nonmarket contracts or agreements. They are part of the administrative costs of running a business, but are separated here for emphasis as they may differ under voluntary environmental agreements.

6. Because the stylized profit function includes terms for nonuse values that the farm operator or family may enjoy, this model could also be cast as a multiple-utility model. Sen (1977), Hirschman (1985), Etzioni (1986), and Casey and Lynne (1999) all suggest that the single-utility, profit maximization model can be restructured and tested to integrate nonuse values associated with environmental actions. Such nonuse values may include what farmers believe is socially desirable, improved landscape attributes, or preservation of threatened or endangered species on ones land.

7. Conceptually, the annual expense from investing in new production equipment, and marketing systems, for example, certification, and training management and labor falls under PC'. However, it is treated separately from PC' for emphasis that the technology set is not fixed but can be reengineered through R and D and other investments.

8. Some might argue that the firm could increase the product price and/or market share by establishing a reputation for excellent environmental stewardship. Reinhardt (2000) mentions several case companies that have pursued such strategies, such as Ben and Jerry's and Patagonia. Such a market "goodwill" factor is difficult to empirically test, of course. If a firm's good will creates an opportunity to enhance price or quantity sold apart from product differentiation, it could be conceptually specified as a factor that is multiplied with PN and/or QN.

REFERENCES

Antle, J. "The New Economics of Agriculture." *American Journal of Agricultural Economics* 81, 5 (1999): 993–1010.

Batie, S., and D. Ervin. "Will Business-Led Environmental Initiatives Grow in Agriculture?" *Choices* (Fourth quarter, 1998): 4–10.

Boggess, W., G. Johns, and C. Meline. "Economic Impacts of Water Quality Programs in the Lake Okechobee Watershed of Florida." In S. Batie, D. Ervin, and M. Schulz, eds., *Business-led Initiatives in Environmental Management: The Next Generation of Policy?*, 165–86. Special Report 92. East Lansing: Department of Agricultural Economics, Michigan State University, 1998.

Carpentier, C. L. and D. Ervin. *Environmental and Health Standards: Compliance Costs for U.S. Agriculture.* Report prepared for the European Commission project "Effects of Environmental and Health Standards on the Trade Competitiveness of European Agriculture vis-à-vis Its Main Competitors." Brussels: Directorate General VI, 1999.

Casey, F., A. Schmitz, S. Swinton, and D. Zilberman, eds. *Flexible Incentives for the Adoption of Environmental Technologies in Agriculture.* Norwell: Kluwer Academic Publishers, 1999.

Casey, F., and G. Lynne. "Adoption of Water Conserving Technologies in Agriculture: The Role of Expected Profits and Public Interest." In F. Casey, A. Schmitz, S. Swinton, and D. Zilberman, eds., *Flexible Incentives for the Adoption of Environmental Technologies in Agriculture.* Norwell: Kluwer Academic Publishers. 1999.

Chambers, K., and L. Eisgruber. "Green Marketing as Green and Competitive." In S. Batie, D. Ervin, and M. Schulz, eds., *Business-led Initiatives in Environmental Management: The Next Generation of Policy?*, 25–36. Special Report 92. East Lansing: Department of Agricultural Economics, Michigan State University, 1998.

Coase, R. H. "The Problem of Social Cost. " *Journal of Law and Economics* 3 (1960): 1–44.

Deavers, L. Personal communication. Resource Conservationist. Washington, DC: Natural Resources Conservation Service, U.S. Department of Agriculture, May 2000.

Dimitri, C., and N. Richman. *Organic Food Markets in Transition.* Henry Wallace Center for Agricultural and Environmental Policy, Policy Studies Report 14. Morrilton, AR: Winrock International, April 2000.

Ervin, D. "Toward GATT-Proofing Environmental Programmes for Agriculture." *Journal of World Trade* 33(April 1999): 63–82.

Ervin, D., and A. Schmitz. "A New Era of Environmental Management in Agriculture?" *American Journal of Agricultural Economics* 78 (December 1996): 1198–206.

Ervin, D., C. Runge, E. Graffy, W. Anthony, S. Batie, P. Faeth, T. Penny, and T. Warman. "Agriculture and the Environment: A New Strategic Vision." *Environment* 40 (July/August 1998): 8–15, 35–40.

Esty, D. "Clean and Competitive: Business-led Environmental Management." In S. Batie, D. Ervin, and M. Schulz, eds., *Business-led Initiatives in Environmental Management: The Next Generation of Policy?*, 37–58. Special Report 92. East Lansing: Department of Agricultural Economics, Michigan State University, 1998.

Etzioni, A. "The Case for a Multiple-Utility Conception." *Economics and Philosophy* 2 (1986): 159–83.

Feather, P., D. Hellerstein, and L. Hansen. *Economic Valuation of Environmental Benefits and the Targeting of Conservation Programs: The Case of CRP.* Agricultural Economic Report Number 778. Washington, DC: U.S. Department of Agriculture. 1999.

Heimlich, R., and R. Claassen. "Agricultural Conservation Policy at a Crossroads." *Agricultural and Resource Economics Review I*, April 1998, 95–107.

Heimlich, R., K. Wiebe, R. Claassen, D. Gadsby, and R. House. *Wetlands and Agriculture: Private Interests and Public Lands.* Agricultural Economic Report No. 765. Washington, DC: U.S. Department of Agriculture, 1998.

Hirschman, A. "Against Parsimony: Three Easy Ways of Complicating Some Categories of Economic Discourse." *Economics and Philosophy* 5 (1985): 1, 7–21.

Hodge, I. "Agri-Environmental Relationships and the Choice of Policy Mechanism." *The World Economy*, February 2000, 257–73.

Khanna, M., and W. Quimio. "Corporate Environmentalism: Regulatory and Market-based Incentives." University of Illinois Program for Environmental and Resource Economics Working Paper No. 19. Champaign-Urbana: University of Illinois, 1999.

Land Stewardship Project. "Measuring Multiple Benefits from Agriculture: Program Overview and Prospectus." White Bear Lake, MN. Mimeo, 1999.

Lewandrowski, J., and K. Ingram. "Policy Considerations for Increasing Compatibilities between Agriculture and Wildlife." *Natural Resources Journal* 39 (Spring 1999): 229–69.

Lynch, S. ed. *Designing Green Support Programs.* Policy Studies Program No. 4. Greenbelt, MD: Henry Wallace Institute for Alternative Agriculture, 1994.

Lyon, T., and J. Maxwell. "Voluntary Approaches to Environmental Regulation: A Survey." In M. Franzini and A. Nicita, eds. *Economic Institutions and Environmental Policy.* Ashgate (forthcoming).

Palmer, K., W. Oates, and P. Portney. "Tightening Environmental Standards: The Benefit-Cost or the No-Cost Paradigm." *Journal of Economic Perspectives* 9 (1995): 119–32.

Porter, M., and C. van der Linde. "Toward a New Conception of the Environment-Competitiveness Relationship." *Journal of Economic Perspectives* (1995): 97–118.

Randall, A. "Providing for the Common Good in an Era of Resurgent Individualism." In F. Casey, A. Schmitz, S. Swinton, and D. Zilberman, eds., *Flexible Incentives for the Adoption of Environmental Technologies in Agriculture*. Norwell: Kluwer Academic Publishers, 1999a.

———. "A New Look at the Old Problem of Externalities." *Choices* (First quarter 1999b): 29–32.

Reinhardt, F. *Down to Earth: Applying Business Principles to Environmental Management*. Boston: Harvard Business School Press, 2000.

Ribaudo, M. "Water Quality Benefits from the Conservation Reserve Program." Report No. AER-606. Washington, DC: ERS, U.S. Department of Agriculture, 1989.

Robles, M. "Incentives for Wildlife Enhancement on Midwestern Farms." Minneapolis, MN: Environment and Agriculture Program, Institute for Agriculture and Trade Policy, 2000.

Runge, F. "A Conceptual Framework for Agricultural Trade and the Environment— Beyond the "Green Box." *Journal of World Trade* 33 (December 1999): 47–68.

Salop, S. C., and D. T. Scheffman. "Raising Rivals' Costs." *American Economic Review* 73(1983): 267–71.

Segerson, K., and N. Li. "Voluntary Approaches to Environmental Protection." In H. Folder and T. Teitenberg, eds. *The International Yearbook of Environmental and Resource Economics, 1999/2000*. Cheltenham: Edward Elgar, 1999.

Sen, A. "Rational Fools: A Critique of the Behavioral Foundations of Economic Theory." *Philosophy and Public Affairs* 6(1977): 317–44.

Swinton, S., and F. Casey. "From Adoption to Innovation of Environmental Technologies." In F. Casey, A. Schmitz, S. Swinton, and D. Zilberman, eds., *Flexible Incentives for the Adoption of Environmental Technologies in Agriculture*. Norwell: Kluwer Academic Publishers, 1999.

Swinton, S., N. Owens, and E. van Ravenswaay. "Agricultural Production Contracts to Reduce Water Pollution." In F. Casey, A. Schmitz, S. Swinton, and D. Zilberman, eds. *Flexible Incentives for the Adoption of Environmental Technologies in Agriculture*. Norwell: Kluwer Academic Publishers, 1999.

Tebo, P. "Business-Led Environmental Management: Good for Business and Good for the Environment." In S. Batie, D. Ervin, and M. Schulz, eds., *Business-led Initiatives in Environmental Management: The Next Generation of Policy?*, 17–24. Special Report 92. East Lansing: Department of Agricultural Economics, Michigan State University, 1998.

Wu J., and B. Babcock. "Contract Design for the Purchase of Environmental Goods from Agriculture." *American Journal of Agricultural Economics* 78 (November 1996): 935–45.

———. "The Relative Efficiency of Voluntary vs Mandatory Environmental Regulations." *Journal of Environmental Economics and Management* 38 (September 1999): 158–75.

Wu, J., R. Adams, and W. Boggess. "Cumulative Effects and Optimal Targeting of Conservation Efforts: Steelhead Trout Habitat Enhancement in Oregon." Draft unpublished manuscript, 1999.

Zinn, J. "Conservation Spending in Agriculture: Trends and Implications." CRS Report for Congress. Washington, DC: Congressional Research Service, Library of Congress, October 6, 1999.

15

Rational Policy Processes for a Pluralistic World

Alan Randall

Economists seek mostly to impose simplicity on a complex world, but it is a losing battle: the diversity of moral intuitions in the real world demands a pluralistic response. Economists' standard micropolicy recommendations—markets and marketlike institutions for resolving "market failures" and benefit-cost tests to impose discipline on public policy—are examined critically in light of existential pluralism, and some features of rational policy processes for a pluralistic world are sketched.

INTRODUCTION

At the conference "Challenging the Agricultural Economics Paradigm," my assignment was to address natural resources issues and, by extension, environmental quality, food safety, genetically modified organisms, and similar concerns that become the focus of "micropolicy." I will address these issues, but not in any systematic way, and often only by implication.[1] My main concern will be with ideas rather than particulars, but these ideas have clear and practical implications for how economists might operate effectively in the policy arenas where these kinds of issues are resolved.[2]

These micropolicy issues have in common the widespread suspicion that uncoordinated individual actions cannot be guaranteed to produce satisfactory outcomes. In response, public policies, programs, and projects are offered in considerable variety and sometimes enacted. Mainstream economists have a standard response to all this: leave it to the market, in those cases where decently well-functioning markets are in place; create new markets, where such markets have reasonable prospects of success; and, where there seems no alternative to a policy response, discipline public policy with a benefit-cost test. These economists typically frame the discussion in terms of economic rationality, a term that in effect, if not always intent, brings to the table some heavy artillery: anyone proposing a different approach risks the charge of irrationality.

Economists are in considerable agreement as to the right way of thinking about these issues and the right policy solutions, but within society at large economists are a tiny minority. People in general exhibit more diversity of moral intuitions and, I am reluctantly coming around to thinking, more creativity in designing institutions.[3] In the real world we find a diversity of moral intuitions—considerations of intrinsic value, moral obligation, and asserted rights of various kinds, as well as preference, come into play; a complex mosaic of institutions—markets and private property, big government, and all manner of intermediate arrangements and processes; and, increasingly, an insistence on direct access to the policy process. If economists are to be effective contributors to micropolicy, they must operate effectively in a pluralistic world, a world of moral pluralism and participatory policy processes.

The task I have set myself in this chapter is not to defend some "muddling through" procedure, on the grounds that reality is not quite rational, and economists had better learn to live with it. Rather, I have undertaken the much more challenging task to make and defend the claim that rationality itself has a different but nevertheless coherent meaning in a pluralistic world,[4] and to provide a sketch of some rational policy processes for that kind of world. Economists' standard micropolicy recommendations—markets and marketlike institutions for resolving "market failures," and benefit-cost tests to impose discipline on public policy—will be examined critically in light of existential pluralism, and some features of rational policy processes for a pluralistic world will be sketched.

MARKETS AND "MARKET FAILURES"
Economists have long been aware that things don't always go well in the public sphere. Public goods don't always get provided, and people have a way of dumping their wastes in the public pond. In agriculture, examples abound: too much soil erosion, nonpoint source pollution, and livestock wastes; too little provision of natural habitats to support biodiversity; and, some would claim, too little assurance that food is safe and GMOs are harmless.

Market Failure, Government Fix
For the first half of the twentieth century, the progressive movement dominated thinking about the role of government in society. While basic policy objectives were the proper province of politics, questions of an instrumental nature were best handled by neutral technical experts. In this way, scientific government would improve the lot of humankind. To this progressive agenda, economists contributed the concept of market failure (Bator 1958), manifested as externality, public goods,

and common property resources. Laissez-faire markets would underprovide public goods, overexploit the commons, and fail to internalize the real costs of externalities, for example, pollution. The progressive solution called for government to step in, typically regulating externalities and access to the commons, and providing public goods with tax dollars (Samuelson 1954). While the market failure concept legitimized a degree of government activism, economists were not agnostic about the instruments progressive government should use: wherever possible, virtual-market instruments should be used: efficient taxes and/or subsidies, rather than regulation, for externalities; privatization of the commons, in many cases; virtual pricing schemes for the provision of public goods; and benefit-cost tests to discipline the whole public undertaking.

In the second half of the century, the market failure paradigm came under scathing attack from libertarians and economists sympathetic to their position, arguing that individual initiative secured by enhanced property rights, not an activist government, provides the appropriate policy response. Virtual markets seemed overly bureaucratic, once real markets supported by real property rights were on the agenda.

The Property Rights Approach

The late twentieth century saw an era of resurgent individualism and concomitant skepticism about public institutions. This shift in thinking was surely boosted by the events of 1989, when the massive experiment in Soviet-style collectivism was exposed as bankrupt, but it began much earlier. Intellectual roots can be found in the work of economists Kenneth Arrow (1951), Scott Gordon (1954), Charles Tiebout (1956), and Ronald Coase (1960), as well as the popular writings of Ayn Rand (1944) during the mid-twentieth century. The Goldwater nomination in 1964 provided an early indication that individualism was starting to catch on with the public.

Coase's (1960) Nobel Prize-winning analysis of externality focused on nonattenuated property rights—that is, fully specified, enforced, and transferable—as a sufficient condition for efficiency. Externality, according to this line of thought, is not sustainable unless accompanied by nonexclusiveness (i.e., the inability to exclude unauthorized users). It follows that privatization would resolve externality problems, eliminating the need for regulation.

This emerging focus on property rights undermined not only the analytics of the market failure paradigm, but also its progressive government activism. As the individualistic resurgence gained pace, it was argued with increasing generality that inadequate property rights are endemic in the public sector itself: government failure is a problem even more pervasive than market failure. It follows that a sustained posture of government activism to rectify market failure is not merely unnecessary, it is undesirable.

THE ISOLATION PARADOX

Nevertheless, an essential reality remains: in many situations, called isolation paradoxes, individual action fails, but (it is possible to find a cost allocation so that) everyone would be better off with coordinated action than with no action at all. Insistence on individual action or none at all leaves everyone isolated and ineffective, but the search for arrangements that make cooperative action beneficial to all concerned may be rewarding.

The intuition that, for an important set of economic problems, coordinated action is essential and may well be stable is hardly new. In 1776, Adam Smith discussed the case of one hundred farmers in the upper end of a valley, beyond the reach of the existing barge canal. While all would benefit from extending the canal, none could bear the cost alone. Yet every single one of them would enjoy benefits larger than one one-hundredth of the cost. Acting alone, each can do nothing, but everyone could enjoy a net benefit from coordinated action. The isolation paradox is the general name given to problems of this kind. An isolation paradox is present whenever individual action fails but there exists a cost allocation (not necessarily an equal sharing of costs, as in Smith's example) such that all parties would be better off with coordinated action than with no action at all. The essential idea is that where an isolation paradox exists, there is in principle the possibility of converting a conflict situation into a sustainable cooperative solution; and we may benefit from exploring that possibility.

Research in game theory, principal-agent modeling, and related fields has demonstrated that, for several important classes of isolation problems, coordinated strategies permit stable, efficient cooperative solutions (Shubik 1981). While this is an important insight, it is not entirely comforting. Coordination is likely to be a costly activity, and complete coordination, especially if it requires consultation among all participants, may be prohibitively costly. Private-goods markets work well because prices convey, in simple signals, sufficient information and incentives to accomplish coordination, and neither centralized management nor direct consultation among all market participants is necessary. The working hypothesis that motivates research on principal-agent models is that signaling devices can be developed for adequate and cost-effective coordination so that cooperative arrangements in large organizations dealing with nonexclusive and public goods are reasonably stable and efficient.

Institutions to Solve the Isolation Paradox

Solutions that break the isolation paradox do not necessarily involve government or (even worse, in today's political environment) big government. Individuals may act together to form and maintain clubs in order to get the job done. Many entities that call themselves clubs, for example, the local health and fitness club, are actually private for-profit enterprises. Today, one can readily imagine a private entity resolving the canal extension problem profitably, an option that did not occur to Adam Smith, just as "city water" is in fact delivered to my home by an investor-owned corporation.

The isolation-paradox concept, then, suggests an openness to solutions that invoke a variety of institutional forms (Hurwicz 1973): private enterprises, voluntary associations, and government from the most local level to the national scale and beyond. Given the centrality of information and coordination, the array of feasible institutions is continually shifting as information, communication, and exclusion technologies develop. For particular problems, the appropriate institutions will be consistent with the dimensions and scale of the problem itself, and with the prevailing technologies and political realities. To protect biodiversity, for example, one can conceive of private for-profit genetic reserves; nature reserves operated by corporations, voluntary associations, or governments; clubs supported by members and donors operating in markets to enhance both private and government conservation efforts; and government operating as facilitator of consensual agreements among stakeholders, as well as legislators, regulators, and resource managers. None of these activities alone will solve the whole biodiversity problem but, in various combinations, they may make real progress. Flexibility is the key, in both institutional forms and the incentives those institutions transmit.

Building on a combination of abstract theory (from game theory, political science, and economics, perhaps among other disciplines) and emerging experience, it is possible to identify some of the characteristics of policies and policy processes that are effective in breaking the isolation paradox.

1. Seek problem-scale solutions. National, one size fits all, solutions to local and regional problems are currently out of fashion, and for some good reasons: sometimes the solutions themselves don't make sense in the local context and, regardless of that, solutions imposed from distant capitals seldom enjoy the local commitment necessary for their success. Indeed, it makes sense to seek solutions scaled to the problem at hand and, to a considerable degree, fashioned by those involved most directly.

   Nevertheless, a framework of national laws and policies remains necessary, to provide parameters within which local solutions can be negotiated. Nationally and internationally mobile industries, for example, have proven more than willing to use the current enthusiasm for state and local institutions to create prisoners' dilemmas for their own benefit.[5] We observe this when states and localities find themselves in destructive competition to attract firms with tax abatements and/or relaxed enforcement of environmental controls.

An effective policy process encourages problem-scale solutions within a framework of national policy. It does not simply set states and localities adrift and wish them well.

2. Establish a long-term process involving all of the legitimate interests. Since the 1970s, public participation has been an important part of the process for resolving resource management issues. Since the 1980s, involvement of all significant stakeholders has been considered essential. What is relatively new is the notion, supported by the theory of repeated games and by practical experience, of committing the participants to a long-term, continuing process. Rather than merely commenting on a solution proposed by professional managers (a typical way of implementing public participation), participants actually work out, over time, solutions to the problems at issue. A long-term continuing process has obvious advantages—it allows time for participants to develop an understanding of each others' interests and objectives, gather and interpret essential information, and develop solutions that will be broadly acceptable—but also an advantage that might not be quite so obvious: after a few rounds, individuals tend to become committed to bringing the process itself to a successful conclusion. If the default outcome is recognized broadly as unsatisfactory, and participants come to see the failure of the process as bad in and of itself, conditions are favorable for a successful process.

3. Establish a shared vision. The process starts by defining goals at the community level and the values that underlie those goals. The objective is to develop and articulate a shared vision: a statement of what it is that the community values and seeks to become. During this process, stakeholders whose most immediate interests would seem to be in conflict frequently discover that their basic values and vision of the future are in fact quite compatible. At this stage, it helps to define the problem set broadly: what does this community seek to become, and how can it get there?

4. Use all of the tools for achieving consensus: deliberation, persuasion, and negotiation. Structured discourse and deliberation can often undermine conflict, and careful consideration of information can erode firmly held priors and open up new possibilities. It would be a mistake—one than an economist might easily make, but still a mistake—to underestimate the value of deliberative processes. Nevertheless, negotiations, real trades, and win-win solutions are often essential to break impasses. Flexible incentives are often an important element of the package, in that they tend strongly to reduce the costs of meeting environmental policy targets, encouraging win-win solutions and easing the pain of compliance in cases

where win-win proves impossible. Depending on particular circumstances, trade in pollution reduction credits, purchases of land or easements, land swaps, mitigation banking, and resources-for-resources compensation can be both efficacious and fair: they help move things toward real solutions that benefit all parties directly concerned. A broad definition of the problem set is helpful at this stage, too, because it increases the scope of potential trades and win-win solutions. As with all negotiations, however, it pays to proceed cautiously. It is not uncommon for parties to proclaim a secure status quo or default position that may in fact be quite shaky, exaggerate the costs and adverse employment impacts of proposed regulatory policies, or understate the costs of proposed public works.

Today's Perspective on "Market Failure"

The economic way of thinking has provided two perspectives on "market failure" during the last century—market failure/government fix, and the property-rights approach—but from today's vantage point both seem unduly restrictive. We are observing a rich stew of institutional innovations based on some of the lessons of game theory and communitarian political theory: innovations that seek problem-scale solutions and replace existing conflict with win-win incentives for sustainable cooperation. Especially at the state, local, and grassroots levels, people are inventing such institutions and making them work, while mainstream economists, bogged down in their market failure paradigm, have (once again) barely noticed what is going on.

BENEFIT-COST TESTS TO IMPOSE DISCIPLINE ON PUBLIC POLICY

Where economists concede a continuing role for government in micropolicy, they usually recommend that government activism be disciplined by benefit-cost tests. Benefit-cost analysis (BCA) already enjoys a considerable role in public affairs. In 1937, the Flood Control Act famously provided that federal funds could be spent on water-resources projects so long as "the benefits to whomsoever they accrue exceed the costs." Ronald Reagan's 1981 Executive Order required BCA for a broad range of regulatory initiatives, and probusiness lobbies have proposed expanding the role of BCA in regulatory policy. Well-entrenched executive agency practice now considers benefits and costs routinely, to evaluate all manner of projects, programs, and policies undertaken by national, state, and local governments.

Over this sixty-year period, textbook benefit-cost analysis has evolved from a relatively crude financial feasibility analysis for capital-intensive public works, to a rigorous application of the economic-theoretic principles of welfare-change measurement based on the compensation test. The test is passed if those who would gain from implementing the proposal could (hypothetically) compensate those who would lose. Actual agency practice has come to incorporate many but not all of the textbook theoretical niceties.

Nevertheless, the current enthusiasm in some circles for an expanded public role for BCA raises some obvious questions: Should a society commit to decide public issues on the basis of benefits and costs? Should a society regard benefits and costs as relevant information for its public decision processes? And, what good reasons can be given to support affirmative answers to either or both of these questions?

ECONOMISTS' JUSTIFICATIONS

When called upon to justify the systematic use of BCA in public decision processes, economists are likely to start talking about the need to impose a marketlike efficiency on the activities of government (e.g., Arrow et al. 1996). After all, efficiency is simply the avoidance of waste, and who could be seriously in favor of waste! However, this justification is not as convincing to citizens at large as it is to economists.

First, a case has to be made that the efficiency of markets is in fact good for society. Perhaps we can do little better than the philosopher Jules Coleman (1987) who has argued that the virtues of market institutions (including, but not limited to their efficiency properties) makes them broadly acceptable for taking care of those kinds of human affairs that are not especially contentious, but that political institutions are required to deal with the really contentious issues of public concern. For example, many human concerns can be resolved through arms-length market transactions secured by property rights, but the definition of those rights—a much more contentious matter—is inherently a function of government. In other words, the justification for market institutions applies to those human concerns that remain after some prior assignment of the contentious issues to the political sphere.

Second, having found some virtue in markets for handling some appropriate set of human concerns, it is then necessary to argue that society ought to require marketlike efficiency in the remaining undertakings that have been assigned for good reasons to government. Mark Sagoff (1981) is most vigorous in rejecting this argument. He asserts that it is a simple category mistake to inquire about the efficiency of a governmental undertaking: government is exactly that institution that human societies invoke when they choose, for their own good reasons, not to be efficient. It is easy to play this argument for cheap laughs ("Of course! What better institution than government, if the goal is to be inefficient!"), but Sagoff's point is not entirely frivolous. Efficiency is a harsh discipline, and one that in practice tends to reinforce the distributional status quo; and it is by no means clear that society ought to impose that discipline on everything that it does.

Third, efficiency economics-style in fact carries a lot of baggage along with its superficially uncontroversial goal of avoiding waste. It judges actions according to the goodness of their consequences; and it judges goodness to the individual in terms of welfare, and goodness to society by adding up welfare changes across individuals, without reference to distributional concerns. BCA uses numerical values based on preference satisfaction weighted by endowments, a practice that makes the preferences of the well-off count for more, thereby reinforcing the distributional status quo (this is true of all the economist's standard value-measures: buyer's best offer, seller's reservation price, and equilibrium price). And, this baggage is itself a source of controversy.

So, we must dig deeper for good reasons to take benefits and costs seriously in public policy. It is useful to start by observing that—despite some legitimate concerns about endowment-weighted values—benefits and costs, when measured rigorously according to the hypothetical-compensation criterion, provide a fairly good account of contribution to preference satisfaction. Then, we can ask why society should take seriously, in its policy decision processes, an account of contribution to preference satisfaction.

Moral Theories

Utilitarianism, the theory that right action is whatever satisfies preferences, is most readily understood as a particular version of the more general moral theory that goodness is a matter of value (Vallentyne 1987): what matters about an action is its consequences, and consequences are valued according to their contribution to preference satisfaction. However, this justification of an efficiency test is unlikely to be entirely convincing to all who take the view that goodness is a matter of value. Many would argue that considerations of value cannot be confined to consequences alone and, furthermore, that there are all manner of consequences that cannot be reduced readily and noncontroversially to welfare but nevertheless are worthy of consideration when evaluating the goodness of an action to an individual or to society. Examples might include consequences for the survival of species, habitats, and rural communities.[6]

Furthermore, the value theory of goodness is but one of the two major branches of Western moral philosophy. The alternative view asserts that goodness is not confined to considerations of value, so there may be good reasons to rule out by constraint some possible actions (Kagan 1998). Some adherents of this view would rule out actions that transgress moral duty, an approach founded in the writings of Immanuel Kant; and others would restrain actions that threaten to violate individual rights. Economists are familiar with the rights-based perspective, which undergirds the contractarian justification of voluntary exchange. However, economists often find Kantian ethics bewildering, with its insistence that moral and prudential reasoning are quite distinct, and that universal moral imperatives can be found from which to deduce rules for action in practical situations. For example, the argument of Kantian origin—that natural entities may have intrinsic value, and humans may have moral duties toward such entities—is quite foreign to many economists but not to many educated laypersons.

To recap, the theory that goodness is a matter of preference satisfaction underpins the economists' justification of BCA, but it is merely one particular and controversial version of one of the two main branches of Western moral philosophy.

Moral Pluralism

A broad acceptance seems to be emerging among Western philosophers that the contest among the competing ethical theories, in their various formulations, is likely to remain inconclusive (Williams 1985). While each has powerful appeal, each is incomplete in some important way, each remains vulnerable to some serious avenue(s) of criticism, and it seems unlikely that any one will defeat the others decisively.

Among those who seek ethical grounding for policy prescriptions, two kinds of pluralism have emerged. The more traditional kind seeks to cultivate an intellectual environment in which people who hold resolutely to different foundational ethics can nevertheless find agreement on particular real-world policy solutions. Agreement might be reached, for example, that real resources should be expended to protect natural environments, among people who would give quite different reasons as to why that should be so. The task of the thoughtful moral agent in the policy arena is, it follows, to find heuristics—rules for action—that can command broad agreement.

Box 15.1

Moral Pluralism

1. Different people invoke different moral frameworks consistently.
   - For me, preference satisfaction is what matters.
   - For you, it's respect for rights.
   - For her, it's obedience to moral duty.

2. One person coherently may invoke different moral frameworks to resolve different issues.
   - For some issues, obedience to moral duty is what matters.
   - For others, it's respect for rights.
   - For everything else, it's OK to be guided by preferences.

The second kind of pluralism imagines thoughtful people calling upon different ethical traditions to answer different kinds of questions in their own lives. To this way of thinking, if the search for the single true, complete, and internally consistent ethical theory is bound to be fruitless, exclusive allegiance to any particular moral theory is hardly a virtue. It becomes coherent to argue that some questions in life are best resolved by reference to moral imperatives, some as matters of respect for rights, and for the remainder it is reasonable to go about maximizing value, perhaps even focusing on consequences and evaluating them in terms of preference satisfaction.

Box 15.2

Moral pluralism implies:

Individuals holding competing moral theories must seek policy heuristics—rules for action—they can agree upon.

For example:
- She believes that humans have a moral duty to preserve "God's creatures."
- You believe creatures have rights.
- I like having a variety of creatures around.

All of us may be able to agree that some public resources should be devoted to protecting biodiversity.

Taking seriously both kinds of pluralism encourages us to think of the policy process as a search for heuristics we can agree upon, and to accept that these heuristics are likely to incorporate insights from various moral theories. The question that motivates this inquiry becomes: would a society of thoughtful moral agents agree to take seriously an account of benefits and costs, within some more-complete set of heuristics?

Public Roles for Benefit and Cost Information

Benefits and costs cannot count for everything. Donald Hubin (1994) asks us to consider benefit-cost moral theory: the theory that right action is whatever maximizes the excess of benefits over costs, as economists understand the terms "benefit" and "costs." It is hard to imagine a single supporter of such a moral theory, among philosophers or the public at large. Instead, we would find unanimity that such a moral theory is inadequate, and an enormous diversity of reasons as to exactly why. Even as we concede that the search for the one true moral theory is bound to fail, benefit-cost moral theory is not a plausible contender.

Benefits and costs must count for something. Preference satisfaction matters morally. It turns out that one cannot imagine a plausible moral theory in which the level of satisfaction of individual preferences counts for nothing at all (Hubin 1994; Randall 1999). Examining a broad array of contending moral theories, preference satisfaction counts for something, in each of them. (Even a thoughtful Kantian would concur that there exists a broad domain of human concerns where happiness may be pursued without violating moral strictures; and, within that domain, more preference satisfaction is better than less.)

So the issue is not whether benefits and costs are morally considerable; clearly they are. The interesting questions are about what sorts of considerations might trump preference satisfaction and in what ways. What else, beyond preference satisfaction, might one want to consider, and in what manner might one want to take account of those things? One approach treats benefit and cost information as simply one kind of decision-relevant information.

Benefit-cost analysis to inform decisions, rather than to decide issues. Since preference satisfaction is a consideration under any plausible moral theory, an account of benefits and costs might be used routinely as a component of some more comprehensive set of evidence, accounts, and moral claims to inform the decision process. The notion that benefits and costs cannot always be decisive in public policy, but should nevertheless play some role, is congenial to many economists (e.g., Arrow et al. 1996, 221). However, it leaves unanswered the question of exactly what role. Are there particular situations and circumstances in which an account of preference satisfaction should be ignored entirely and others in which it should be decisive? How should an account of preference satisfaction be weighted relative to other kinds of information? Can the answers to these questions be principled or must they always be circumstantial?

These unanswered questions are most worrisome if one imagines a technocratic policy process: if we were to entrust decisions to the technocracy, we would surely want to give it more guidance about how to decide. However, if we have a pluralistic policy process in mind, the concern is not so great: these unanswered questions would be resolved in pluralistic discourse.

A benefit-cost decision rule subject to constraints. An alternative way of coming to terms with the idea that preference satisfaction counts for something in any plausible moral theory, but cannot count for everything, is to endorse a benefit-cost decision rule for those issues where no overriding moral concerns are threatened. Benefits and costs could then be decisive within some broad domain, while that domain is itself bounded by constraints reflecting rights that ought to be respected and moral imperatives that ought to be obeyed. This would implement the commonsense notion that preference satisfaction is perfectly fine so long as it doesn't threaten any concerns that are more important.

To free individuals for the pursuit of happiness, constraints securing some well-defined set of human rights seem essential. If the beneficence of reasonably free markets is to be enjoyed, a set of secure property rights is also necessary. People acting together to govern themselves need also to establish a framework of laws, statutes, regulations, and policies to legitimize and also to limit the role of activist government. The constitution was designed with exactly these concerns in mind.

To take this idea beyond the well-known protections for life, liberty, and property, consider a set of policy issues familiar to environmental economists: the protection of habitats, species, and particular ecosystems. A society could adopt the practice of deciding these kinds of issues on the basis of preference satisfaction, but subject to some kind of conservation constraint. A safe minimum standard (SMS) of conservation has been suggested by a variety of authors: harvest, habitat destruction, and the like, must be restricted in order to leave a sufficient stock of the renewable resource to ensure its survival. The SMS constraint makes most sense when cast transparently as a discrete interruption of business-as-usual, imposed to act upon firm—and often nonutilitarian—intuitions that to permit destruction of a unique renewable resource would be foolish and (perhaps) morally wrong.

The general form of such constraints might be: don't do anything disgusting. The basic idea is that a pluralistic society would agree to be bound by a general-form constraint to eschew actions that violate obvious limits on decent public policy. This kind of constraint is in principle broad enough to take seriously the objections to unrestrained pursuit of preference satisfaction that might be made from a wide range of coherent philosophical perspectives. Examples of such constraints might include: don't violate the rights that other people and perhaps other entities might reasonably be believed to hold; be obedient to the duties that arise from universal moral principles, or could reasonably be derived therefrom; and, don't sacrifice important intrinsic values in the service of mere instrumental ends. In each of these cases, the domain within which pursuit of preference satisfaction is permitted would be bounded by nonutilitarian constraints; and these constraints themselves would be determined in pluralistic processes.

The Bottom Line

Moral pluralism will persist and benefit-cost moral theory is not even a serious candidate among moral theories. The pendulum began to swing away from technocracy long before the progressive dream of scientific government had been converted to reality. The issue is not how to perfect BCA, but how to enjoy the services it can provide for us—a reasonably good account of preference satisfaction, itself one valid moral concern among others—without according it more influence than it deserves. We could use BCA to inform decisions rather than to decide issues. Better yet, we could accord substantial influence to BCA, within a domain where preference satisfaction carries a good deal of weight, but bounded by various constraints derived from perhaps different ethical perspectives and adopted for good reason.

CONCLUSION

If we think of society as inherently pluralistic and the policy process as searching for ways that people holding different moral theories (or intuitions) nevertheless can agree on heuristics for policy, it follows that our concept of what is rational will be quite different from the rationality that emerges from standard economics. Economic rationality, a term of fulsome promise, turns out to mean little more than logical consistency in working out the implications of a maintained belief that preference satisfaction is all that matters. As a concept of rationality, it would suffice if the "economic way of thinking" had already won the contest among moral theories, but it is thoroughly insufficient in the context of moral pluralism.

In a pluralistic world, it makes sense to think of rationality as being an attribute of processes. A concept of rationality for a pluralistic world will respect different sincerely held moral intuitions and theories, and seek to identify core values and broadly acceptable rules for action. While committing to objectivity in resolving conflicts among fact-claims, it will respect different perceptions of the facts, until such time as resolution is reasonably complete. The point is that policy issues arise and are dealt with in real time and, concurrently, fact-claims emerge and are contested and evaluated. As all of this is happening, participants in the policy process do well to remember that no particular group has an inherent monopoly on the facts. A rational policy framework will seek avenues of negotiation and agreement, and develop processes for conflict resolution. It will cultivate commitment and allegiance to those processes, because continuity in these evolving processes is essential to provide moral agents operating from different moral commitments and different views of the world with sustainable faith that "the system" will resolve these differences efficaciously and fairly.

NOTES

1. This chapter, while addressed to a different topic, draws on ideas first presented in Randall (1999 and 2000).

2. Some years ago, I received reviewer's comments on a research proposal, replete with references to the "principle investigator." My initial response was irritation—those schooled in the British tradition have a learned intolerance of errors in spelling and grammar—but soon I embraced the term affectionately: that is exactly how I like to think of myself, an investigator of principles; there are many who outperform me in the role of "detail investigator."

3. The idea that, among educated people, economists are different draws some support from the literature (Frank, Gilovich, and Regan 1993). The stylized fact has been established that economists are more likely to report that they would free-ride given the opportunity; and the literature has moved on, to consider alternative explanations: is it years of instruction and mentoring within the discipline

that makes economists that way, or do individuals with this proclivity self-select into specializing in economics?

4. In a similar vein, Michael Farmer and I titled our 1998 paper "The Rationality of a Safe Minimum Standard of Conservation." The SMS is commonly defended as a kind of muddling-through response to ecological crises, while the opposing argument labels it an ad hoc and irrational departure from rational management rules. In contrast, we defend the SMS as a rational policy process, given moral pluralism and limited understanding of the biocomplexity of the real world.

5. A prisoners' dilemma is a situation in which rational individuals acting independently would each chose actions that turn out to make everyone worse off. The original formulation of the problem addressed suspects under interrogation separately by police seeking confessions, but there are many applications in strategy involving business, politics, diplomacy, and war.

6. I know there are economists who readily evaluate such concerns in welfare terms. However, the point is that it is perfectly coherent to resist doing so, if one believes that the relevant values cannot be captured (completely) in welfare terms.

ACKNOWLEDGMENTS

I have had the distinct pleasure of serving as Luther Tweeten's colleague for fourteen years, the last two as his department chair. He is a truly special colleague: brilliant and accomplished, opinionated but rigorous, fearless in debate, and always willing to pitch in for the good of the program. That Luther requested I be included among the keynote speakers at this conference was an unexpected honor, but I will not let it deter me from preparing a chapter for this book that, I imagine, challenges some of Luther's cherished assumptions. I know he expects no less of me.

REFERENCES

Arrow, K. J. *Social Choice and Individual Values.* New York: Wiley, 1951.

Arrow, K. J., M. L. Cropper, G. C. Eads, R. W. Hahn, L. B. Lave, R. G. Noll, P. R. Portney, M. Russell, R. Schmalensee, V. K. Smith, and R. N. Stavins. "Is There a Role for Benefit-Cost Analysis in Environmental, Health, and Safety Regulation?" *Science* 272 (1996): 221–22.

Bator, F. M. "The Anatomy of Market Failure." *Quarterly Journal of Economics* 72(1958): 351–79.

Coase, R. H. "The Problem of Social Cost." *Journal of Law and Economics* 3(1960): 1–44.

Coleman, J. "Competition and Cooperation." *Ethics* 98(1987): 76–90.

Farmer, M. C., and A. Randall. "The Rationality of a Safe Minimum Standard," *Land Economics* 74(1998): 287–302.

Frank, R., T. Gilovich, and D. Regan. "Does the Study of Economics Inhibit Cooperation?" *Journal of Economic Perspectives* 7(1993): 159–71.

Gordon, H. S. "The Economic Theory of a Common Property Resource: The Fishery." *Journal of Political Economy* 62(1954): 124–12.

Hubin, D. C. "The Moral Justification of Benefit/Cost Analysis." *Economics and Philosophy* 10(1994): 169–94.

Hurwicz, L. "The Design of Mechanisms for Resource Allocation." *American Economic Review* 63(1973): 1–30.

Kagan, S. *Normative Ethics*. Boulder, CO: Westview Press, 1998.

Rand, A. *The Fountainhead*. New York: Bobbs-Merrill, 1944.

Randall, A. "Providing for the Common Good in an Era of Resurgent Individualism." In F. Casey, A. Schmitz, S. Swinton, and D. Zilberman, eds., *Flexible Incentives for the Adoption of Environmental Technologies in Agriculture,* 323–38. Norwell, MA: Kluwer, 2000.

———. "Taking Benefits and Costs Seriously." In T. Tietenberg and H. Folmer, eds., *The International Yearbook of Environmental and Resource Economics.* Cheltenham, UK, and Northampton, MA: Edward Elgar, 1999.

Sagoff, M. "Economic Theory and Environmental Law." *Michigan Law Review* 79(1981): 1393–419.

Samuelson, P. A. "The Pure Theory of Public Expenditure." *Review of Economics and Statistics* 36(1954): 387–89.

Shubik, M. "Game Theory Models and Methods in Political Economy." In K. J. Arrow and M. D. Intrilligator, eds., *Handbook of Mathematical Economics.* Amsterdam: North Holland, 1981.

Tiebout, C. M. "A Pure Theory of Local Expenditures". *Journal of Political Economy* 64(1956): 416–24.

Vallentyne, P. "The Teleological/Deontological Distinction." *Journal of Value Inquiry* 21(1987): 21–32.

Williams, B. *Ethics and the Limits of Philosophy*. Cambridge, MA: Harvard University Press, 1985.

Index

AAEA, 52
acquisitions, 127
acreage reduction program (ARP),
Adelman, 240
adjustment capability, 18
adverse selection, 111
AFT (American Farmland Trust),
187, 199
AGR (Adjusted Gross Revenue),
120
Agras, 205
agribusiness concentration, 27
agribusiness market power, 127
agricultural R&D, and trade, 243
Agricultural Risk Protection Act,
109
Alston, 233
Amponsah, 6, 206
AMTA (Agricultural Marketing
Transition Act), 37, 38, 44
Angirasa, 71
Antle, 205, 265, 276
Applebaum, 132
ARMS (Agricultural Resource
Management Study), 71

Arnade, 18
Arrow, 288, 293, 297
ASA (American Sugar Alliance),
160, 162
asset turnover, 73
asymmetric information, 111
Atkinson, 60, 247
Azzam, 132, 137
Babcock, 46, 112, 266
Ballard, 9
Bandyopadhyay, 207
Banker, 71
Barnard, 22
Barrett, 238
Barrows, 195
Barse, 138
Batie, 191, 266, 273
Bator, 287
BCA (Benefit-Cost Analysis), 292
Beale, 25
Beasley, 190
Becker, 148
Bentley, 22
Berck, 240
Bergstrom, 190

Bessler, 176
Black, J., 148
Blend, 192
Bliss, 72
Blue, 2, 14
Boggess, 270
Bonnen, 3, 17
Borrell, 165
Bowers, 189, 198
Boxall, 190
Boyd, 196
Brandow, 147
Breffle, 190
Brester, 132
Bromley, 192, 193, 197
Buffer Initiative, 267
Bullock, B., 128
Bullock, D., 29, 67, 146, 148
Bunce, 185
Burfisher, 9, 10
Calvin, 110
CAP (Common Agricultural Policy),
    250
Carpentier, 266
CART (Conservation, Amenity, and
    Recreation Trusts), 191
Casey, 265, 267, 270, 281
cash flow, 22
CBOT (Chicago Board of Trade),
    122
CCC (Commodity Credit
    Corporation), 35
Chambers, 273
Chapman, 205, 214
Chaudhuri, 206
Chavas, 98
Christian, 161
Claassen, 192, 267
Clean Water Act, 195
Coalition, sugar, 175
coalitions, 160, 166

Coase, 170, 281, 288
Coate, 156
Coble, 105, 112
Cochrane, 3, 47, 49, 65
Coggins, 29, 146
Coleman, 60, 247, 251, 263, 293
Collender, 22
commodity program reform, 56
commodity programs, 1
commodity programs, expenditures,
    1
commodity programs, farm, 1
commodity programs, income
    redistribution, 46
Common, M. , 205
concentration ratio, 135
concentration, market, 128, 129
conduct, market, 127
conjectural variations, 132
conservation compliance, 25, 196
Conservation Security Act, 25
contingent valuation, 191
Cordes, 193
Costanza, 190
cost-benefit analysis, commodity
    programs, 45
CREP (Conservation Reserve
    Enhancement Program), 267
Crop Insurance Reform Act, 109
crop insurance, 92
Cropper, 206
CRP (Conservation Reserve
    Program), 266
CSR (Coalition for Sugar Reform),
    160, 164
Davis, 71
De Bary, 233
de Bruyn, 211
deadweight cost, 7
deadweight loss, 7, 37, 41, 45
Deavers, 267

demand elasticity, 43

Devados, 176

Diamond, J., 233

Dimitri, 274

direct payments, 9

disaster assistance, 115

Dismukes, 120

Dixon, 93

Drury, 17

Duncan, 231

Dupont analysis, 73

Durst, 121

EBI (Environmental Benefits Index), 270

Eckholm, 206

economic efficiency, 4

economic equity, 14

economic rent, 152

Eisgruber, 273

El Obeid, 231

environment, policy, 24

environment, voluntary initiatives, 273

environmental degradation, 204

environmental dilemma, 228

environmental equations, 208

EQIP (Environmental Quality Incentives Program), 25, 267

Ervin, 191, 265, 277

Esseks, 195

Etzioni, 281

Evenson, 233

expenditures, political, 149

FAIR (Federal Agricultural Improvement and Reform Act), 4, 28, 35, 37, 70, 160

family farm, loss, 17

Farm Aid, 37

Farm and Ranch Risk Management Account (FARRM), 24, 121

farm bill,1996, 28, 43, 56

farm fundamentalism, 151

farm operator supply, 26

farm problem, 70, 88

farm production, excessive, 10, 12

farm production, excessive, by source, 10

Farmer, M., 185

Farmer, M., 300

farmland conversion, 184

farmland policy, 184

farmland policy, 186

farmland, beneficiary-pays programs, 192

FCIC (Federal Crop Insurance Corporation), 114

Feather, P., 270

Fernandez, 179

financial performance, 72

financial vulnerability, 26

Finke, 138

fixed asset theory, 3

Flinchbaugh, 161

Flora, 25, 141

Food Security Act of 1985, 93

food security, 26, 231

food security, index, 238

food security, trade policy, 239

food supply, 26

Foster, 185

Frank, 299

Freedom to Farm, 37

Fullerton, 50

Gale, 206

game, noncooperative, 166

GAO (General Accounting Office), 52, 161

Garcia, 98

Gardner, 17, 35, 41, 66, 71, 91, 98, 134, 148

GHG (Greenhouse Gas Emissions), 204

Gilovich, 299
GIPSA (Grain Inspection, Packers, and Stockyards, Administration), 133
Glasser, 93
Glauber, 106, 110, 111, 114
Glickman, 37, 46
global food system, 58
Goodwin, 19, 106, 110, 119, 122
Gopinath, 18
Gordon, 288
government failure, 113
greenhouse gases, 208
Grepperud, 206
Griffiths, 206
GRIP (Group Risk Income Protection), 119
Grossman, 170, 205
GRP (Group Risk Plan), 119
Guither, 29
Gunderson, 30
Haidacher, 134
Hall P., 62, 262
Halstead, 190
Hanley, 189
Hansen, 270
Harberger, 43
Harl, 140
Hart, 170
Harwood, 71, 97, 118
Hazell, 115
Headd, 80
Heady, 21
Heffernan, 128
Heffernan, 139
Heidebrink, 205
Heimlich, 192, 267
Hellerstein, 270
Heneberry, D., 195
Henneberry, 18, 19
Hennessey, 112

Hepp, 71
Hervani, 25, 204
HFCS (High Fructose Corn Syrup), 161
HHI (Herfindahl-Hirschman Index), 135
Hilton, 207
Hirschman, 281
Hodge, 191, 278
Holloway, 134
Holtz-Eakin, 205
Hopkins, 7, 70, 71
Horowitz, 112
Hotelling, 211
household income, farm, 14, 15
household income, nonfarm, 14, 15
household income, source, 8, 15
Hubin, 296
Huffman, 233
Hulshof, 24
Hurt, 98
Hurwicz, 290
imperfect competition, 142
income elasticities, environmental, 231
income elasticity equation, 213
income variability, 79
INFS (Index of National Food Security), 238
Ingram, 267
instability, 23
insurance subsidies, 10, 11
insurance, area revenue, 119
insurance, crop, 100, 109
insurance, regional weather index, 120
insurance, whole farm, 120
insurance, yield-difference, 92
insurance, yield-level, 94
integrated firms, theory of, 170
international competitiveness, 27

Investment Retirement Account, 24
Irwin, 71
isolation paradox, 289
Jagger, 71
Johns, 270
Johnson, D. Gale, viii, 36
Johnson, Glenn, 3, 65
Jordan, 17
Josling, 27, 245, 258, 262
Just, 110, 112
Kagan, 294
Kao, 98
Karagiannis, 112
Kaufman, 215
Khanna, 275
Kliesen, 66
Kline, 185
Knight, 105
Knutson, 161
Koo, 164
Koontz, 139
Kropf, 176
Krueger, 148, 205, 249
Krum, 17
Kuhn, 61
Kunreuther, 111
Kuru, 206
Kurz, 167
Kuznets curves, 204
land, amenity services, 185
land, ecological services, 185
land, state policy, 187
Land-Use Policy and Planning
   Assistance Act, 186
Langemeier, 23
Lee, 233
Lehman, 186
Leibenluft, 138
Levinson, 207
Lewandrowski, 267
Li, 272

Libby, 25, 184, 187,189
Lichtenberg, 112
loan deficiency payments (LDP), 10,
   11, 39, 42, 44
log-rolling, 176
Lopez, 205
loss ratio, 116
Luttrell, 1
Lynch, 267
Lynn, 281
MacSharry reforms, 251
Mahul, 106, 119
Makki, 8, 11, 23, 111
Manshard, 205
market concentration, 128, 129
market failure, 7, 287
market goods, 5, 6
market power, 127, 136
marketing loans, 10, 11
marketing margin, 130, 136
Massey, 93
Matthews, 134
Maxwell, 272
McBride, 47
McIntosh, 51
Meck, 198
Melichar, 71
Meline, 270
Mendez, 206
mergers, 127
Merlo, 192
Miranda, 106, 110, 111, 114
Monke, 121
monopoly, 131
monopsony, 131
moral hazard, 112
moral theories, 294
moral pluralism, 295
Morehart, 7, 70, 71
Morgenstern, 166
Morris, 156

Moseley, 192, 198
Moss, 29, 160
Moyer, 250, 263
MPCI (Multiple Peril Crop Insurance), 100, 105
Musick, 133
Muth, 97, 134
Nakagawa, 17
National Agricultural Land Policy Act, 186
National Agricultural Lands Study, 187
national income, lost, 7
Nature Conservancy, 191
Net Income Stabilization Account, 24
net worth, farm, 15
NGOs (Nongovernment Organizations), 272
Norton, 233
Oates, 278
OECD (Organization for Economic Cooperation and Development), 245, 261
Ogg, 191
oligopoly, 131
oligopsony, 131
Olson, 148
operating profit margin, 73
Orden, 2, 23, 55, 60, 180, 248
Paarlberg, D., 4
Paarlberg, R., 23, 56, 248
Pagoulatos, 132
Palmer, 278
paradigm, competitive agriculture, 254, 255
paradigm, dependent agriculture, 253, 255
paradigm, global agriculture, 255, 258
paradigm, multifunctional

agriculture, 255, 257
paradigm, policy, 1
paradigm, trade, 245
paradigm, new agricultural policy, 1
paradigms, and OECD, 245
Pardey, 233
Patrick, 23
payments, indemnity, 120
PDR (Purchase of Development Rights), 189
Peck, 97
Penn, 161
performance, market, 127
Perlman, 198
Perry, 22
Persaud, 6, 127, 136
Pfaff, 206
Pindyck, 137
plantings, sugarbeet and sugarcane, 177
policy processes, 286
policy, farmland, 186
political economy, 148
politics, farm, 28
Poole, 66
Pope, 98
population growth equation, 212
Porter, 278
Portney, 278
positivistic, 1, 36
precautionary principle, 185
price wedge, 42
Price, M., 43, 50
prisoners' dilemma, 29
production flexibility contract payments (PFC), 83
PSE (Producer Subsidy Equivalent), 178
public goods, 5
Purcell, 128
Quail, 131

Quance, 3
Quiggin, 110
Quimio, 275
Racevskis, 190
Rand, 288
Randall, 2, 185, 224, 266, 268, 286,
    299
rate of return, 7, 70, 79
rate of return, nonfarm businesses,
    79
rates of return
Ravenswaay, 192
Reed, 113, 119
Regan, 299
Reinhardt, 272, 278
Reinsel, 23
reinsurance, 114
reinsurance, area yield, 121
rent seeking, 152
rent-dissipation, 148
rent-seeking, by integrated firms,
    170
Repacholi, 185
resource depletion, 204
returns, large farms, 76
returns, small farms, 76
Reynolds, 250
Ribaudo, 267
Richman, 274
risk, nondiversifiable, 110
risk, systemic, 110, 111
risk, yield, 94
RMA (Risk Management Agency),
    114
ROA (rate of return on assets), 74
Robles, 275
Roe, 23, 56, 248
Roney, 164
Rothschild, 114
RSA (Resource Stewardship
    Agreement), 269

Rubinfeld, 137
Ruddle, 205
Runge, 278
rural community loss, 25
Rutström, 148
Ruttan, 205
Ryan, 22, 147
Sagoff, 293
Salhofer, 67
Salop, 274
Samuelson, 288
Sandrey, 250
Scarpa, 191
SCF (Survey of Consumer Finance),
    74
Scheffman, 274
Schmitz, 29, 160
Schnepf, 99
Schreiner, 22
Schroeter, 132
Schweikhardt, 3, 71
scientific revolutions, 61
Segerson, 272
Selden, 205
Sen, 281
Shafik, 207
Shubik, 289
Skees, 10, 11, 113, 119
Skogstad, 60, 247
Smith, 110, 122, 233
Smith, Adam, 289
SMS (Safe Minimum Standard), 298
Somwaru, 110
Song, D., 205
Spitze, 1, 51
Srinivasan, 206
Stamm, 198
Stanton, 112
Stern, 205
Stewart, 187
Stigler, 148

Stiglitz, 114
Stokey, 206
Stratman, 178
structure, market, 127
sugar policy, 164
sugar program, 160
Sumner, 27, 231, 238
supply elasticity, 43
Suri, 205
Swinton, 270
Tangerman, 238, 251, 262
target price, 38
Tebo, 278
Thurman, 24, 135
Tiebout, 288
TMDL (Total Maximum Daily
   Load), 195, 268
Toman, 190
trade paradigm, 245
treadmill theory, 3
TRQ (Tariff Rate Quotas) , 178
Tucker, 205, 207
Tullock, 148
Turning Point Project, 47
turning point, environmental, 226,
   227
URAA (Uruguay Round Agreement
   on Agriculture) , 248, 259
Valdés, 262
Vallentyne, 294
Van Bergeijk, 207
van Der Linde, 278
van Kooten, 112
Vandeveer, 117
variability, price, 94
variability, yield, 94
Varian, 132
Vecchia, 185
Vercammen, 112
Vercammen, 165
von Nuemann, 166

Ward, C., 129, 131
Warley, 262
Weitzman, 185
Westcott, 9, 10, 11, 43, 50
Whaller, 224
WHIP (Wildlife Habitat Incentives
   Program), 267
Whitlow, 206
Wichelns, 185
Williams, 176, 295
Williamson, 170, 249
Wohlgenant, 134
Wright, 17, 25
WRP (Wetlands Reserve Program),
   267
WTO (World Trade Organization),
   165
Wu, 266, 270
Yandle, 193
yield insurance, 92
yield risk, 92
yield, variance, 119
yield-difference insurance, 92
Young, 9, 10, 11, 117
Zey-Ferrell, 51
zoning, agricultural, 187, 193
Zulauf, 2, 23, 26, 60, 63, 71, 91,
   196, 247

ISBN 0-8138-0899-5

9 780813 808994